Lecture Notes in Computer Science 7008

Commenced Publication in 1973
Founding and Former Series Editors:
Gerhard Goos, Juris Hartmanis, and Jan van Leeuwen

Chunxiao Xing Fabio Crestani
Andreas Rauber (Eds.)

Digital Libraries: For Cultural Heritage, Knowledge Dissemination, and Future Creation

13th International Conference
on Asia-Pacific Digital Libraries, ICADL 2011
Beijing, China, October 24-27, 2011
Proceedings

Springer

Volume Editors

Chunxiao Xing
Tsinghua University
Information Science and Technology Building
Beijing, 100084, P.R. China
E-mail: xingcx@tsinghua.edu.cn

Fabio Crestani
University of Lugano
Faculty of Informatics
6900 Lugano, Switzerland
E-mail: fabio.crestani@unisi.ch

Andreas Rauber
Vienna University of Technology
Institute of Software Technology and Interactive Systems
1040 Vienna, Austria
E-mail: rauber@ifs.tuwien.ac.at

ISSN 0302-9743
e-ISSN 1611-3349
ISBN 978-3-642-24825-2
ISBN 978-3-642-24826-9 (eBook)
DOI 10.1007/978-3-642-24826-9
Springer Heidelberg Dordrecht London New York

Library of Congress Control Number: 2011938816

CR Subject Classification (1998): H.3, I.2, H.4, H.5, C.2, J.1, H.2

LNCS Sublibrary: SL 3 – Information Systems and Application, incl. Internet/Web and HCI

Typesetting: Camera-ready by author, data conversion by Scientific Publishing Services, Chennai, India

Printed on acid-free paper

Springer is part of Springer Science+Business Media (www.springer.com)

Preface

The annual International Conference on Asia-Pacific Digital Libraries (ICADL) series is a significant forum that provides opportunities for researchers, educators, and practitioners to exchange their research results, innovative ideas, service experiences and state-of-the-art developments in the field of digital libraries. Since the first ICADL in 1998, the conference has grown to become one of the premier forums in the digital library community. Based on the success of the first 12 ICADL conferences, the 13th ICADL conference was hosted by Tsinghua University and co-hosted by Peking University and the National Library of China in Beijing, China. In the year of the Tsinghua University Centenary Celebration, ICADL 2011 brought an attractive academic event to the beautiful campus.

ICADL 2011 aimed to further strengthen the academic collaboration and strategic alliance in the Asia-Pacific region for the development of digital libraries in the fields of computer science, library and information science and knowledge management involving applications to environmental sciences, social sciences, humanities and museum studies. The theme of ICADL 2011 was "Digital Libraries—For Cultural Heritage, Knowledge Dissemination, and Future Creation."

The paper topics of ICADL 2011 covered a wide spectrum from various areas, including information visualization, data mining/extraction, cultural heritage preservation, personalized service and user modeling, novel library content and use environments, electronic publishing, preservation systems and algorithms, social networking and information systems, Internet of things, cloud computing and applications, mobile services, interoperability issues, open source tools and systems, security and privacy, multi-language support, metadata and cataloguing, search, retrieval and browsing interfaces to all forms of digital content, e-Science/e-Research data and knowledge management, and cooperative service and community service.

The keynote speakers of ICADL 2011 presented the state-of-the-art developments and challenges in the field of digital libraries. Hsinchun Chen (University of Arizona) talked about "Building a Social Media Digital Library: Collection, Management, and Analytics." Christine L. Borgman (University of California, Los Angeles) gave a speech on "Drowning in the Data Deluge: Digital Library Challenges for Asia." Xiaolin Zhang (National Science Library, Chinese Academy of Sciences) gave a talk on "Developing MetaKnowledge Services: The Next Paradigm for Digital Libraries." Edward Y. Chang (Google Research, China) presented a report on "Mobile Information Management and Retrieval."

ICADL 2011 received 136 submissions from 27 countries on five continents. Each paper was carefully reviewed by the Program Committee members. Finally, 33 full papers, 8 short papers and 9 poster papers were selected. On behalf of the Organizing and Program Committees of ICADL 2011, we would like to express our appreciation to all the authors and attendees for participating in the conference. We also thank the sponsors, Program Committee members, supporting organizations, and volunteers for making the conference a success. Without their efforts, the conference would not have been possible. We look forward to the impact of ICADL 2011 in the promotion of digital libraries in Asia-Pacific and beyond.

October 2011

Chunxiao Xing
Fabio Crestani
Andreas Rauber

Organization

Organizing Committee

General Co-chairs

Lizhu Zhou	Tsinghua University, China
Hsinchun Chen	University of Arizona, USA

Program Co-chairs

Chunxiao Xing	Tsinghua University, China
Fabio Crestani	University of Lugano, Switzerland
Andreas Rauber	Vienna University of Technology, Austria

Tutorial Co-chairs

Aixin Sun	Nanyang Technological University, Singapore
Li Dong	Tsinghua University, China

Publication Co-chairs

Airong Jiang	Tsinghua University, China
Yigang Sun	National Library of China, China

Publicity Co-chairs

Ling Chen	Peking University, China
Schubert Foo	Nanyang Technological University, Singapore
Michael Bailou Huang	Stony Brook University, USA
Ling Feng	Tsinghua University, China

Exhibition Co-chairs

Yi Yang	Tsinghua University, China
Dawei Wei	National Library of China, China

Financial Chair

Chun Zeng	Tsinghua University, China

Workshop Co-chair

Chao Li	Tsinghua University, China
Zhiqiang Zhang	Harbin Engineering University, China

Local Co-chairs

Yong Zhang	Tsinghua University, China
Ming Zhang	Peking University, China

Program Committee

Program Committee Co-chairs

Chunxiao Xing	Tsinghua University, China
Fabio Crestani	University of Lugano, Switzerland
Andreas Rauber	Vienna University of Technology, Austria

Program Committee Members

Akira Maeda	Ritsumeikan University, Japan
András Micsik	Hungarian Academy of Sciences, Hungary
Chao-chen Chen	National Taiwan Normal University, Taiwan, China
Chern Li Liew	Victoria University, Wellington, New Zealand
Christopher C. Yang	Drexel University, USA
Christopher S.G. Khoo	Nanyang Technological University, Singapore
Claudia Niederee	Fraunhofer IPSI, Germany
Dion Goh	Nanyang Technological University, Singapore
Donatella Castelli	Italian National Research Council (IEI-CNR), Italy
Edie Rasmussen	University of British Columbia, Canada
Edward Fox	Virginia Tech, USA
Ee-Peng Lim	Nanyang Technological University, Singapore
Ekawit Nantajeewarawat	Thammasat University, Thailand
Erich Neuhold	University of Vienna, Austria
Frank Shipman	Texas A&M University, USA
Gavin McCarthy	Melbourne University, Australia
Geneva Henry	Rice University, USA
Gobinda Chowdhury	University of Technology, Sydney
Hao-Ren Ke	National Taiwan Normal University, Taiwan, China
Hideyasu Sasaki	Ritsumeikan University, USA
Hsin-Hsi Chen	National Taiwan University, Taiwan, China
Hsueh-hua Chen	National Taiwan University, Taiwan, China
Hussein Suleman	University of Cape Town, South Africa
Ingeborg Solvberg	Norwegian University of Science and Technology, Norway
Jieh Hsiang	National Taiwan University, China
Ji-Hoon Kang	Chungnam National University, Korea
Jin-Cheon Na	Nanyang Technological University, Singapore
Jose Borbinha	IST/INESC-ID - Information Systems Group, Portugal
Jyi-Shane Liu	National Chengchi University, Taiwan, China
Liddy Nevile	La Trobe University, USA
Maristella Agosti	University of Padova, Italy
Michael Nelson	Old Dominion University, USA

Table of Contents

Keynotes

Digital Archives and Preservation

Information Mining/Extraction

Metadata/Catalogue

Distributed Repositories and Cloud Computing

Social Network/Personalized Service

Mobile Services/ Electronic Publishing

Multimedia Digital Libraries

Information Retrieval

Tools and Systems for Digital Library

Posters

Drowning in the Data Deluge: Digital Library Challenges for Asia

Christine L. Borgman

Professor and Presidential Chair in Information Studies, UCLA

Abstract. Scholarly communication no longer consists merely of papers and publications. Research data have become valuable objects to be captured, documented, and shared. Funding agencies are requiring "data management plans" for all new proposals. Libraries, universities, and research institutes are assessing how to manage those data in ways that can be leveraged for future value. But what are "data"? We are drowning in them without being able to define what they are. This talk will explore the shifting landscape of scholarly information, with special attention to how these shifts may influence digital libraries in Asia. Research is disseminated by many formal and informal means, not only by libraries and publishers but also by new media such as preprint repositories and tweets. Access may be faster – if one can separate signal from noise amidst the plethora of communication channels. These changes are the result of the transition from a closed scholarly world to the open Web, the shift in content and context of networked information, the shift in focus from information services for readers to those for authors, and differences between publications and data. If future scholars are to use the scholarly content of yesterday, today, and tomorrow, the digital library community must reclaim information retrieval, rethink partnerships throughout the information life cycle, share responsibility for the information infrastructure, and address policy and incentive issues.

C. Xing, F. Crestani, and A. Rauber (Eds.): ICADL 2011, LNCS 7008, p. 1, 2011.

Building a Social Media Digital Library: Collection, Management, and Analytics

Hsinchun Chen

Director, Artificial Intelligence Lab;
McClelland Professor, University of Arizona

Abstract. In this talk I will present the University of Arizona Artificial Intelligence Lab's recent research in Dark Web, Geopolitical Web, and Business Analytics. Based on funding from the NSF and several other US agencies, the AI Lab has developed techniques for collecting, managing and analyzing large-scale multilingual and multimedia social media contents of relevance to social, geopolitical, and business applications. Our projects aim to study and understand critical social and business phenomena in the cyber world and real world via a computational, data-centric approach. We aim to collect critical social media content generated by various political and business groups, including web sites, forums, chat rooms, blogs, social networking sites, videos, virtual worlds, etc. A social media digital library and portal system has been developed to manage and access these critical multilingual and multimedia contents. We have also developed advanced multilingual data mining, text mining, and web mining techniques to perform link analysis, content analysis, web metrics (technical sophistication) analysis, sentiment analysis, authorship analysis, and video analysis in our research. Selected case studies in geopolitical domains and business intelligence applications will be discussed.

C. Xing, F. Crestani, and A. Rauber (Eds.): ICADL 2011, LNCS 7008, p. 2, 2011.

Developing MetaKnowledge Services: The Next Paradigm for Digital Libraries

Xiaolin Zhang

National Science Library, Chinese Academy of Sciences

Abstract. Science is matching toward a new paradigm of data-intensive knowledge discovery enabled by massive availability of digital data at a time of grand challenges of global scale, interdisciplinary nature, and translational complexity. This combination of events gives rise to great opportunities of meta-knowledge services where the relations, patterns, emerging trends, hidden possibilities, ignored abnormalities, etc., can be revealed and tested.

Several approaches of meta-knowledge services are here today or in near-future. Intelligent monitoring and visualizing of research fields and emerging topics help researchers keep track of development; Literature and patent analysis reveals complicated patterns of research and its competition or cooperation; Output, impact, and portfolio analysis supports official evaluation of research organizations, groups, and individuals; Path exploration and road-mapping are interactively used to build and test research plans; Meta-reading of large amount of data provides students with effective ways to structure knowledge and identify key points.

National Science Library, CAS, as its innovation and future-enabling strategy, has been developing a meta-knowledge-service-centric service structure. On one hand, it arms its analyst teams with sophisticated computational tools of R&D tracking, trends detecting, technology analysis, competition/cooperation analysis, R&D mapping, etc. On the other hand, it re-structures its digital information services into a linked open data based and ontological systems driven discovery platform. These meta-knowledge services require a much different approach from current digital libraries, with the emphasis on the discovery and decision-making utilization of content. A meta-knowledge-driven service cannot be achieved as a simple extension of current digital libraries. Paradigmatic shifts are needed to go beyond the traditional search and retrieval model.

C. Xing, F. Crestani, and A. Rauber (Eds.): ICADL 2011, LNCS 7008, p. 3, 2011.

Mobile Information Management and Retrieval

Edward Y. Chang

Head of Google Research, China

Abstract. The number of "smart" mobile devices such as wireless phones and tablet computers has been rapidly growing. These mobile devices are equipped with a variety of sensors such as camera, gyroscope, accelerometer, compass, NFC, WiFi, GPS, etc. These sensors can be used to capture images and voice, detect motion patterns, and predict locations, to name just a few. This keynote depicts techniques in configuration, calibration, computation, and fusion for improving sensor performance and conserving power consumption. Novel information management and retrieval applications that can benefit a great deal from enhanced sensor technologies are also presented.

Furthermore, the Mobile 2014 research program coordinated by Google Research in China has been funding research projects related to mobile location-based service since 2010. This program has granted several research awards to universities in the US and Asia to conduct work in sensor signal fusion, location-based data service, peer-to-peer protocols, privacy-preserved data mining, and applications assisted by inertial navigation systems. Highlights of this program are enumerated to motivate research into advancing mobile information management and retrieval.

C. Xing, F. Crestani, and A. Rauber (Eds.): ICADL 2011, LNCS 7008, p. 4, 2011.

High Speed Capture, Retrieval and Rendering of Segment-Based Annotations on 3D Museum Objects

Chih-Hao Yu, Tudor Groza, and Jane Hunter

The University of Queensland,
Brisbane St Lucia, Queensland 4072, Australia
(617) 3365 1092
chih.yu@uqconnect.edu.au,
{tudor.groza,j.hunter}@uq.edu.au

Abstract. The aim of the 3D Semantic Annotation (3DSA) system is to deliver a Web-based semantic tagging and annotation service for 3D cultural heritage objects - that enables users to attach semantic tags/annotations to points, surface regions and volumetric segments on 3D digital objects. Specific objectives of the 3DSA system are: support for interactively defined, complex 3D segments; interoperability of the resulting tags/annotations; and fast, efficient capture, retrieval and rendering of annotations on complex 3D fragments. With these objectives in mind, the 3DSA system is based on the Open Annotations Collaboration (OAC) model, which has been extended using X3D fragment identifiers. This paper describes our implementation of the X3D extensions to the OAC data model and demonstrates how this approach significantly improves the speed of capturing, retrieving, downloading and rendering annotations on volumetric segments. The context for this work is the capture of community-generated tags and annotations for cultural heritage artifacts from the University of Queensland Antiquities Museum.

Keywords: Semantic, Annotation, Tagging, 3D, Cultural Heritage, Open Annotation Collaboration, OAC, X3D.

1 Background and Objectives

Advances in 3D data acquisition, processing and visualization are providing museums and cultural institutions with novel methods for preserving cultural heritage and making it more accessible to scholars and the public, via online search interfaces. Increasing numbers of museums are using 3D scanning techniques to overcome the limitations of 2D data representations and to improve access to high quality surrogates of fragile and valuable artifacts via the Internet [1-4]. These 3D surrogates are stored in online repositories that can be conveniently searched and retrieved via Web interfaces. However, as the size of the online collections expands, the ability to search such collections can be increasingly difficult. Museums are finding the cost of providing metadata for their collections prohibitive and are keen to explore ways of exploiting social tagging and annotation services [5]. Just as the Steve.Museum project is investigating social tagging for 2D images, we are

C. Xing, F. Crestani, and A. Rauber (Eds.): ICADL 2011, LNCS 7008, pp. 5–15, 2011.

investigating social tagging tools for 3D digital objects – in the belief that the attachment of high quality annotations and tags to the complete object as well as to specific segments or features, has the potential to significantly improve the precision and relevance of search results.

A recent survey [11] reveals that because 3D annotation is a relatively new topic, only a small number of prototypes with limited functionality currently exist. Moreover, the majority of existing 3D annotation tools are designed for specific disciplines and only support tagging of whole objects, points [6, 7], pre-defined segments (not freely or interactively defined by users) [8] or simple primitive shapes (e.g. boxes, ellipsoids and planes with sketches) [7, 9, 10]. None of these existing approaches enable the user to interactively select and tag volumetric sub-parts of 3D models via a Web interface. Yet the ability to attach semantic tags to interactively-defined sub-parts of 3D museum objects is essential – particularly if we want to apply semantic inferencing rules to automatically infer high-level semantic tags from combinations of low-level segment-based tags. Hence a key objective of the 3DSA system is to provide an easy-to-use annotation service that allows users to attach tags to interactively selected sub-parts on 3D museum objects.

A second key objective of 3DSA is to maximize the sharing, interoperability, re-use and inferencing of the tags/annotations. To achieve this objective, the 3DSA system adopts a Semantic Web/Linked Data approach - the ontology-based tags are represented in RDF, stored separate to the 3D models but linked to them via fragment identifiers and published to the Web via a HTTP URI. In addition, the 3DSA system is based on the Open Annotation Collaboration (OAC) data model, but extended to enable annotations to be attached to 3D fragments that are identified using the Web3D Consortium's X3D standard.

One of the issues that the Steve.Museum project has identified is the poor quality of tags being created by the untrained public. Community-generated tags need to be reviewed by museum professionals [12] prior to publishing, to reduce inconsistency and ambiguity [14]. However, inspecting large quantities of tags on museum collections is a time consuming task and reviewing tags on segments of 3D objects is particularly challenging due to the large size of the complex polygonal meshes that represent the segments. Downloading and rendering the annotations on such segments can be very slow, particularly for users with limited bandwidth. Similarly, uploading and storing the corrected tags/annotations after review can cause further network delays. Hence a final objective of the 3DSA system is to improve the efficiency of storing, retrieving, rendering, reviewing and correcting tags attached to volumetric segments of 3D digital objects.

The remainder of the paper is structured as follows: Section 2 describes Related Work; Section 3 describes the 3DSA system and the case study; Section 4 describes the X3D extensions to the OAC data model; Section 5 describes our evaluation and Section 6 provides a conclusion.

2 Related Work

Most prior work in the field of 3D annotations has focused on the annotation of discipline-specific objects (e.g. architectural and engineering CAD drawings [7], 3D crystallography models [15] and 3D scenes [16]). All of these systems enable users to attach annotations to 3D models and to browse annotations added by others, asynchronously. However, they are limited to the discipline-specific format of the target objects and are not suitable for the museum context.

The latest Adobe Acrobat Reader [6] provides a user interface that allows annotation of 3D CAD models or U3D objects stored in PDF, using proprietary tools. However, Adobe only supports the attachment of annotations to a single point and the annotations are embedded in the PDF document preventing reuse by other non-Adobe formats. ShapeAnnotator [8] enables the attachment of semantic tags (drawn from an ontology) to automatically segmented parts. However, the ShapeAnnotator is not Web-based and it does not enable users to interactively select the sub-parts to be annotated – only pre-identified or automatically generated segments can be tagged. The Arrigo project [9] has developed an interactive 3D museum exhibition that enables visitors to explore 3D models of statues and discover detailed information via annotations attached to specific locations on the 3D models. Annotations are stored in TEI/XML using CIDOC/CRM model and the target objects are represented using the Collada XML format. However, the Arrigo project only supports regions of interest that are spherical. Our requirement is to allow the user to explicitly and interactively define the specific shape or segment that they wish to annotate. MPEG-21[10] provides a URI-based fragment identifier to identify 3D spatial regions. The region specification is integrated inside the URI, using the # tag and media types that are restricted to MPEG-21 formats. The MPEG-21 specification does not provide a concrete example for free-form 3D objects and does not take into account the annotator's viewpoint. Moreover, MPEG-21 is perceived as overly complex, preventing its wide-spread adoption in real-world Web applications.

As far as we are aware, there is currently no open-source Web-based semantic annotation service for 3D museum objects that enables the fast, easy tagging of points, surface regions or 3D segments on a 3D digital object. Moreover, we are unaware of any 3D system that supports annotation interoperability, exchange and reasoning through the adoption of Semantic Web standards - with respect to both the underlying annotation data model and the 3D fragment identifiers.

3 Overview of the 3DSA Project and Case Study

The 3DSA system uses an ontology-directed folksonomy approach to capture community-generated tags on 3D museum artifact – users are provided with suggested and popular tags from an ontology, but still have the option to define their own tags [14]. The annotations are stored in a HTTP-based Sesame RDF repository that is separate but linked via URIs to the Fedora repository that stores the 3D digital objects. The Danno API [21] is used for creating, updating, deleting and querying annotations and replies, and for bulk upload and harvesting of annotations.

3DSA is a HTML5 and WebGL application, that allows users to interactively define points, surface regions or sub-parts on 3D objects (PLY Polygon file format). Users use the drawing tool to draw a 2D polygon around the object parts that they wish to tag. They can choose "select surface" to only include surface polygons or "select through" to include any object polygons inside the projected 2D selection polygon. Users can continue to select more segments/regions and either "include" or "exclude" them from the current selection. When they have completed the selection process, they finish by choosing "tag the current selection" button and entering either a tag and/or free text.

Figure 1 shows some screen shots of our 3DSA annotation tool. The artifact being annotated is a stone head from Palmyra (approx. 200AD) provided by the UQ Antiquities Museum which has been digitized using a Konica Minolta Vivid9i laser scanner. Annotations attached to this object are listed in the window on the right hand side (Figure 1a) and displayed via pins attached to the 3D object in the centre window. To avoid confusion due to overlapping annotations and to improve performance, the complete annotation text and highlighted segments (highlighted in yellow) are only displayed after the user selects an annotation from the list (Figure 1b).

a) Screenshot of 3DSA annotation tool

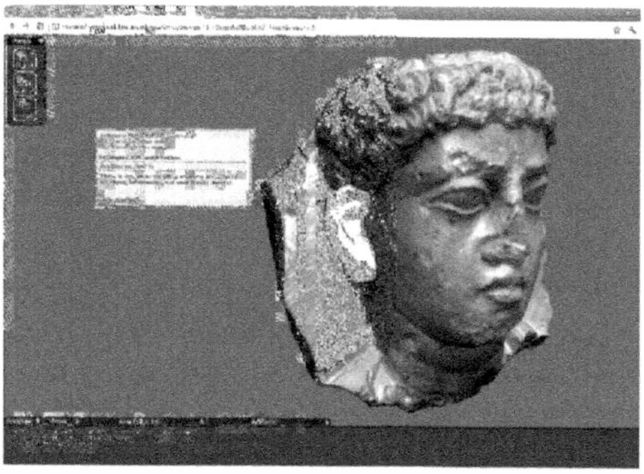

b) Highlighting a segment annotation

Fig. 1.

4 Extending the OAC Data Model Using X3D Fragment IDs

4.1 The X3D Standard

X3D is a royalty-free ISO standard that provides a XML-based file format for storing, exchanging and representing 3D graphics [17]. The X3D standard is maintained by the Web3D Consortium, which has cooperative agreements with W3C. Its open design, modular architecture and extensible light-weight approach based on XML – makes it an ideal candidate for storing and identifying 3D segments in the Web environment and, more specifically, within the 3DSA system.

4.2 The Open Annotations Collaboration Data Model

The Open Annotations Collaboration (OAC) was established to facilitate the emergence of a Web and resource-centric interoperable annotation environment that allows leveraging annotations across the boundaries of annotation clients, annotation servers, and content collections [18]. To this end, an annotation interoperability specification consisting of an annotation data model has been developed. The OAC model offers a common but extensible model that can easily be extended and refined to support the interoperability requirements of 3D cultural heritage collections [18]. The

OAC specification recommends an approach for annotating regions within 2D images using SVG (see Figure 2) – but does not provide any recommendations on the best approach for capturing annotations on 3D fragments. Both SVG and X3D are XML-based file formats – so it is relatively easy to modify the example in Figure 2 to support annotations on 3D fragments using X3D. Fig. 3 demonstrates how the OAC model can be extended to annotate a 3D fragment of a file of XML MIME type *model/x3d+xml*. The X3D segment is given a unique identifier (URI) at the time of creation, and HTTP GET can then be used to retrieve the segment data and associated annotation.

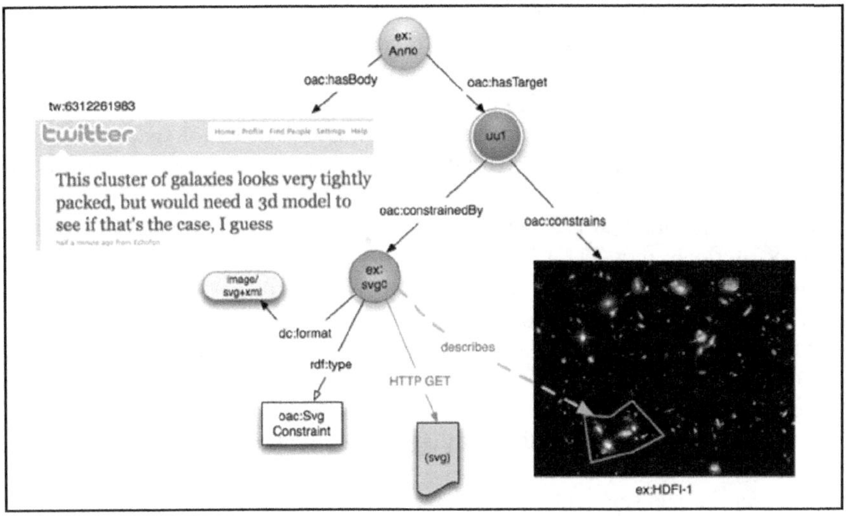

Fig. 2. Attaching an annotation to an SVG region of an image using OAC [19]

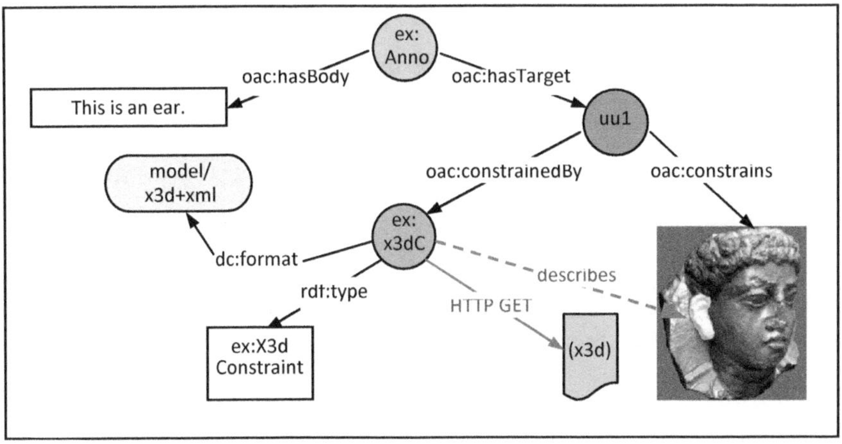

Fig. 3. Extending the OAC model to support annotations attached to X3D segments

4.3 Implementation of the X3D Extensions to the OAC Data Model

Our X3D extension to the OAC data model comprises two parts. The first part (Part1) specifies the polygons that comprise the annotated segment and that should be retrieved/highlighted by the annotation client. Part 1 also specifies the user's viewpoint at the time of annotation. The second part (Part2) includes detailed geometric data (e.g. vertices and face indexes) for the segment, encoded in X3D format – the second part is not retrieved unless such data is not already available in memory. It is anticipated that in most cases, the geometric data is loaded into memory, when the original 3D object is first displayed. As a result, retrieval of the second part of the extension will be unnecessary unless there is no equivalent geometric data cached in memory.

In addition, by using the X3D *colorIndex* attribute to represent the shape segment instead of the detailed geometric data, we can save significant space and time. The "*colorIndex*" is a standard X3D attribute which we employ to identify which polygons are selected (e.g. "0" = "black" = "not selected", "1" = "yellow" = "selected"). This method uses an array of singular values of "0" and "1", rather than an array of floats (e.g. [12.1238, 21.1231, 312.4345...etc]). The number of colour indexes must equal the total polygon count for the object but this approach also allows us to easily attach a single tag to multiple disconnected segments.

Our X3D extension is demonstrated below; the fields with italicized comments are the fields that contain application-specific data. Part 1 uses the X3D "*Viewpoint*" node to record the position of the camera and the rotation of the scene and the X3D "*Transform*" node is to record the position and the rotation of the 3D object.

Part1 – user defined segment – polygons with *colorIndex* set to 1:

```
<X3D profile='Interchange'><Scene>
  <Viewpoint position='0 0 11.5'/> (Camera position)
  <ExternProtoDeclare name='artifact234'
  url='http://…/ear_of_artifact234.x3d' (Second Part's URL)
  DEF='ear_of_artifact234'/>
   <field name='colorIndex' type='MFInt32'accessType='initializeOnly'/>
  </ExternProtoDeclare>
  <Transform rotation='12.12 0.34 0'> (3D object's rotation)
   <ProtoInstance containerField='geometry' name='artifact234'>
    <fieldValue name='colorIndex' value='1 1 0 0 1 1......'> (Colour indexes)
   </ProtoInstance>
  </Transform>
</Scene></X3D>
```

Part2 - associated object's geometry data encoded in X3D:

```
<X3D profile='Interchange'><Scene><ProtoDeclare name='artifact234'>
 <ProtoInterface>
  <field name='colorIndex' type='MFInt32' accessType='initializeOnly'/>
 <ProtoInterface>
 <ProtoBody><Shape>
  <IndexedFaceSet colorPerVertex='false' solid='false'
  coordIndex='0 1 2 -1 0 3 2 -1......'/> (Face indexes)
   <IS><connect nodeField='colorIndex' protoField='colorIndex'/></IS>
   <Color color='0 0 0 1 1 1 0 1'/> (Highlight colour - yellow)
```

```
    <Coordinate point='0.123 1.23 2.34 0.123 1.24 2.56.....'/> (Vertices)
   </IndexedFaceSet>
  </Shape></ProtoBody>
</ProtoDeclare></Scene></X3D>
```

The X3D standard requires the "*colorIndex*" data in Part 1 to be associated with the geometric data in Part 2. In the case above, the "*ProtoDeclare*" and "*ProtoInterface*" nodes in Part 2 together with the "*ExternProtoDeclare*" in Part 1, specify that the "*colorIndex*" for the object specified in Part 2 is over-ridden by the colorIndex values in Part1. Part 2 of the extension is implemented by migrating the geometric data of the 3D object into its annotation instance. It is arguable that such a migration will enlarge the annotation file size. However the X3D geometric scene is required to be loaded only once and all of other annotations that are attached to the same 3D object reference the same X3D scene.

When a user selects a 3D segment annotation from the list (RHS of Figure 1a) - the annotated segment is highlighted by extracting the colorIndex data from Part1 and changing the colour of those polygons for which the colorIndex is set to 1, to the highlight colour. Part2 of the extension is only loaded if there is no identical geometric data in the memory. Part2 of the extension is also used to ensure that the complete geometric details for the annotated segment are accessible and displayed correctly, even if the original annotated 3D artifact is no longer available or has been transformed to a different resolution or format (as illustrated in Figure 4).

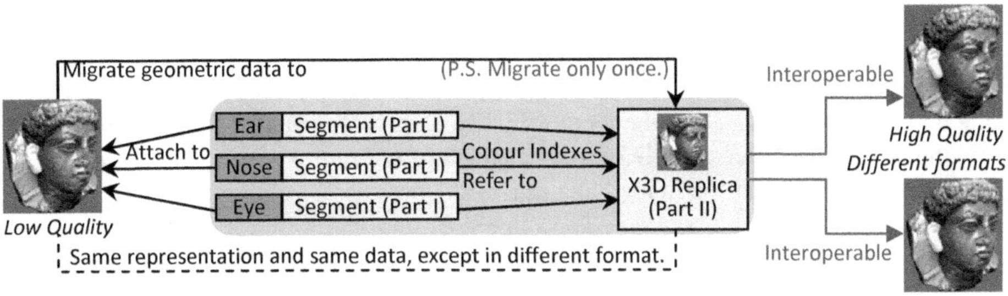

Fig. 4. Visualisation of the extension, showing how it relates to different 3D representations

5 Evaluation

One of our objectives was to support interoperability of the segment annotations across clients/browsers. The simplest method for evaluating this goal (whilst simultaneously testing the system's X3D compliance) is to attempt to display the annotated segments using multiple X3D viewers – both offline and online. Fig. 5 demonstrates how annotated segments captured using the 3DSA system, are able to be displayed in the BS Contact, Flux Player and Octaga players. This also suggests that annotation segments captured using the 3DSA system, can also interoperate with other annotation clients that are based on X3D, despite the fact that the 3DSA tool is a WebGL application which uses PLY format for displaying 3D.

Fig. 5. 3DSA's annotation segments can be displayed in existing offline/online X3D viewers – BS Contact, Flux Player and Octaga

A second objective was to improve the performance associated with up/downloading and displaying 3D segment annotations. Performance is measured by comparing the download time and filesize for annotation segments attached to a given object - for both the *unindexed vertices* approach and our compressed *colorIndex* approach (described in Section 4). The 3D object used in the evaluation consists of 65,000 polygons, the download times and file sizes are measured using Firebug v1.7.1 and the network connection is Cable/ADSL2.

Table 1 compares the unindexed vertices method with our colour index method. The results show that it is generally much more efficient to represent segment annotations using an array of colour indices. Three vertices have to be stored for each polygon – "[x1, y1, z1], [x2, y2, z2], [x3, y3, z3]". Each x,y,z value consists of a decimal number (e.g. "10.12345" = 8 characters). So each polygon requires 24 characters. On the other hand, the colour index method uses a single "1" or "0" character to represent "selected" or "not selected". These values are associated with the geometric data loaded in the memory. Hence, the colour index method results in a much smaller data store, smaller network transfers and improvements in the speed and performance of annotation upload and retrieval.

Table 1. Evaluation results – comparing storage of vertices to colour indexes

Type	File-size (Vertices)	Response Time (Vertices)	File-size (Colour Index)	Response Time (Colour Index)
Small	100.3kb	0.203sec	131kb	0.250sec
Small-medium	788.9kb	1.17sec	131kb	0.250sec
Medium	1589.7kb	2.40sec	131kb	0.219sec
Large	6357kb	10.11sec	131kb	0.234sec

However, the *colorIndex* method stores a colorIndex value for every single polygon in the 3D object – while the *unindexed vertices* method only stores vertices for those polygons that form the segment. Hence the *unindexed vertices* method is more compact for small segments. As the segment size increases, the file size and response time increase linearly for the *unindexed vertices* method. For the *colour index* method, all segments require the same amount of data to be stored, so the storage requirements and network traffic remains small and consistent for all segment sizes.

According to our evaluation results, an average 2.1MB of data store and 3.23 secs of network delay have been saved per tag retrieved using our *colour index* method.

Assuming 5 tags are attached to each artifact and a total of 677,687 artefacts (the number of objects in Taiwan's National Palace Museum [20]) our method can save up to 5.4 TB of server storage. This takes into account the storage of X3D replicas but does not include backups. Over 3042 hours of idling time are saved using our *colour index* method. Assuming 8 man-hours/day, the colour index methods reduces the time to retrieve all of the tags associated with a collection by 380 man-days.

Table 2. Results after HTTP compression

Type	Original (Vertices)	Compressed (Vertices)	Original (Colour Index)	Compressed (Colour Index)
Small	100.3kb	15.1kb (6.6× smaller)	131kb	1.2kb - (109.2× smaller)
Small-med	788.9kb	129.6kb (6.1× smaller)	131kb	1.1kb - (119.1× smaller)
Medium	1589.7kb	288.6kb (5.5× smaller)	131kb	3.1kb - (42.3× smaller)
Large	6357kb	1200kb (5.3× smaller)	131kb	3.2kb - (40.9× smaller)

HTTP compression (based on deflate/gzip) is a common capability that is built into the majority of web servers and clients, to minimise network transfer. Combined with our *colorIndex* method, HTTP compression further reduces the size and retrieval time of 3D segment annotations. Table 2 shows the results after HTTP compression. Our colour index method has significantly higher compression rates compared with unindexed vertices because the data stream only contains "0" and "1" and is highly repetitive - duplicate string elimination leads to higher compression rates. The *unindexed vertices* approach has a significantly lower repetition rate and takes less advantage of HTTP compression, making the data size and network delay differences between the two approaches even greater.

However, interoperability challenges remain with regard to automatic mapping of 3D segment annotations across different resolutions of the one 3D object (e.g., low, medium and high resolution). This problem occurs because the 3D segment is defined/bound by the polygonal structure of the source 3D object. One solution is to generate a transparent low quality 3D object overlay on top of the high quality 3D object and to base all selected and highlighted segments on the single low resolution polygonal structure. Although this approach, combined with our OAC+X3D data model will enable annotation interoperability across different resolution formats of the same 3D model – the quality and precision of segment selection are sacrificed to support interoperability. The future challenge is to find a way to precisely define the segment of interest on a 3D artifact that is not bound by the polygonal structure and that will persist across the 3D object independent of resolution and format.

6 Conclusion

The primary aim of the 3DSA system described here is to optimize Web 2.0 social tagging services for 3D museum artefacts to enhance the discoverability of 3D collections. More specific objectives were: to enable users to interactively select precise, complex 3D segments and attach tags/annotations to them; to maximize the sharing, re-use and interoperability the attached tags/annotations by using a common data model and Semantic Web/Linked Data approaches such as RDF and OWL; to investigate the suitability of the OAC data model, extended using X3D to identify 3D fragments; to improve the efficiency and performance associated with capturing, retrieving and rendering annotations associated with 3D segments. We have shown here that our innovative approach to implementing the X3D extensions to OAC can significantly improve the speed and efficiency of uploading and retrieving annotations. We have also demonstrated that our approach enables interoperability across annotation clients. However, we are aware of the difficulties of making 3D fragment annotations persistent - across different resolutions of a single 3D object. The future challenge is to discover ways to define 3D surface regions and segments that are independent of the resolution and file format. To conclude, the 3DSA system and the OAC+X3D extensions described in this paper deliver a high performance 3D annotation service that will benefit both cultural heritage institutions and the general social tagging community. It enables faster, more efficient capture of rich semantic metadata for describing and discovering 3D collections, whilst simultaneously paving the way for 3D annotation interoperability in the future.

References

1. Koller, D., Frischer, B., Humphreys, G.: Research challenges for digital archives of 3D cultural heritage models. Journal on Computing and Cultural Heritage 2(3), 1–17 (2009)
2. Ikeuchi, K.: 3D digital preservation of cultural heritages (2002), `http://www.cvl.iis.u-tokyo.ac.jp/gallery_e/`
3. Hunter, J., et al.: Using the semantic grid to build bridges between museums and indigenous communities. In: Proceedings of the GGF11-Semantic Grid Applications Workshop, Honolulu, Hawaii, pp. 46–61 (2004)
4. Rowe, J., Razdan, A.: A Prototype Digital Library for 3D Collections: Tools To Capture, Model, Analyze, and Query Complex 3D Data. In: Museums and the Web 2003: Proceedings, Charlotte, North Carolina, USA. Archives & Museum Informatics, Toronto (2003)
5. Chun, S., et al.: Steve.museum: An Ongoing Experiment in Social Tagging, Folksonomy, and Museum, Albuquerque, New Mexico. Archives & Museum Informatics, Toronto (2006)
6. Adobe: Using Adobe Acrobat X Standard., Adobe Systems Incorporated, San Jose, California, pp. 304–305 (2011)
7. Jung, T., Gross, M.D., Do, E.: Annotating and sketching on 3D web models. In: Proceedings of the 7th International Conference on Intelligent User Interfaces, San Francisco, California. ACM, New York (2002)
8. Attene, M., et al.: Characterization of 3D shape parts for semantic annotation. Computer-Aided Design 41(10), 756–763 (2009)
9. Havemann, S., et al.: The arrigo showcase reloaded—towards a sustainable link between 3D and semantics. Journal on Computing and Cultural Heritage 2(1) (2009)
10. ISO/IEC (2006) Information technology—MPEG-21, part 17: fragment identification of MPEG resources. ISO/IEC 21000-17:2006
11. Spagnuolo, M., Falcidieno, B.: 3D Media and the Semantic Web. IEEE Intelligent Systems 24(2), 90–96 (2009)
12. Trant, J.: Tagging, Folksonomies and Art Museums: Early Experiments and Ongoing Research. Journal of Digital Information 10(1) (2009)

13. Trant, J., Wyman, B.: Investigating social tagging and folksonomy in art museums with steve.museum. In: World Wide Web, Edinburgh (2006)
14. Hunter, J., Gerber, A.: Harvesting community annotations on 3D models of museum artefacts to enhance knowledge, discovery and re-use. Journal of Cultural Heritage 11(1), 81–90 (2009)
15. Hunter, J., Henderson, M., Khan, I.: Collaborative annotation of 3D crystallographic models. Journal of Chemical Information And Modeling 47(6), 2475–2484 (2007)
16. Kadobayashi, R., et al.: 3D Model Annotation from Multiple Viewpoints for Croquet. In: The Fourth International Conference on Creating, Connecting and Collaborating through Computing, C5 2006, Berkeley (2006)
17. Daly, L., Brutzman, D.: X3D: extensible 3D graphics standard. In: ACM Siggraph Asia 2008 Courses, Singapore, pp. 1–6. ACM, New York (2008)
18. Hunter, J., et al.: The Open Annotation Collaboration: A Data Model to Support Sharing and Interoperability of Scholarly Annotations. In: Digital Humanities 2010. King's College, London (2010)
19. Sanderson, R., Sompel, H.V.d.: Open Annotation Alpha3 Example: Hubble Deep Field Image (2010), `http://www.openannotation.org/spec/alpha3/examples/hubble.html`
20. Hu, L.J.: Pleasures of the Imperial Treasure - Gaining historical insight at the National Palace Museum of Taiwan. My Favourite Museum (2010)
21. Crawley, S., Chernich, R.: Danno/Dannotate Overview (2010), `http://metadata.net/sites/danno/`

Digitization and Value-Add Application of Bamboo Weaving Artifacts

Kuo-An Wang[1,*,**] Ya-Chin Liao[3,**], Wei-Wei Chu[1],
John Yi-Wu Chiang[4], Yung-Fu Chen[2,5,***], and Po-Chou Chan[1,***]

[1] Department of Management Information Systems &
[2] Department of Healthcare Administration,
Central Taiwan University of Science and Technology, 40601 Taichung
{gawang,bjjem,wwchu,yfchen}@ctust.edu.tw
[3] Department of Commercial Design,
National Taichung Institute of Technology, 40402 Taichung
liaoyachin@yahoo.com.tw
[4] Department of Computer Science & Engineering,
National Sun Yat-sen University, 80424 Kaohsiung
chiang@cse.nsysu.edu.tw
[5] Department of Health Services Administration,
China Medical University, 40402 Taichung, Taiwan

Abstract. Chinese people used bamboo to make bamboo weaving utensils for hunting, farming, fishing, and even transportation before. However, bamboo weaving utensils are no longer needed in daily life nowadays. The craft of bamboo weaving utensils is gradually losing people's attention that few craftsmen can still work on it. In this study, a folklore hobbyist, a craftsman, a horticulturist, and an interior decorator were invited to digitize bamboo weaving artifacts and crafts, as well as to develop value-add applications of the artifacts. Among the 1200 collected bamboo weaving artifacts, 150 artifacts accompanied with 20 weaving patterns have been digitized and stored with image and video formats, respectively. The value-add refers to the adoption of the bamboo weaving artifacts as flower vases for orchid planting and flower arranging with artworks designed by the horticulturist, which were then adopted by the interior decorator to decorate restaurants to elevate the environmental quality. The digitized contents were also used as part of the e-learning materials in a community college. The questionnaire surveys show that the digitized material is useful for learning bamboo weaving craft and flower arrangement skills for students. It was found that combination of bamboo weaving artifact and flower arrangement highly promote quality of service of restaurants.

Keywords: Bamboo weaving, Flower arrangement, Service quality, Value-add application.

---

* The author is also with Dept. of CSIE, National Chi Nan Univ.

** Co-first author.

*** Corresponding authors.

C. Xing, F. Crestani, and A. Rauber (Eds.): ICADL 2011, LNCS 7008, pp. 16–25, 2011.

1 Introduction

Recently, in addition to historic and artistic values, the selection criteria of culture heritage have been modified by including cultural value [1]. Folklore and traditional culture were recommended by UNESCO as one of the tangible heritage. A tangible monument bears the value of memory recognizing the aspects that are pertinent to human deeds and thoughts [1]. Folklore reflects on the ancestral missions that have shaped a people and the inherited values reflecting on their daily lives and passing to the future generations [2]. It may refer to unsubstantiated consciousness, such as beliefs, legends and customs, currently existing among the mind of a people [4] as well as substantiated artifacts, crafts, skills, and rituals widely governing the living style of the people [3]. Recently, digital contents of cultural heritages have been widely developed around the world. However, most of these contents emphasized on static artifacts rather than the crafts in making, skills in operating, or rituals in using them [5]. It is believed that the artifacts may be lost, deteriorated, or damaged no matter how well the preservation and exhibition environments are controlled. Since the deteriorated or damaged artifacts are not easy to recover and the manufacturing procedures, functions, and usages of the artifacts might no longer be preserved generation after generations if not being inherited, it is very important to preserve the intangible heritage, such as crafts, skills, and ceremonies (or rituals). In addition to folklore artifacts [3], folklore activities [5] have also been digitized during the past 5 years by our team. The crafts in making as well as the skills in using these artifacts may involve delicate finger, hand, and body operations, which makes the digitization work and preservation task very valuable with regards to culture heritage.

In a precious report, we classified folklore craft, skill, and ritual into 7, 5, and 6 categories, respectively [5]. Among them, weaving crafts were mankind's commonly owed cultural assets since ancient times. Various natural environments breed different plants with fiber material for carving and weaving, which in turn nurtured regional cultures. Because of different material resource and skill development, diversity in weaving skills with local characteristics is generally found. In the early time, bamboo utensils could be seen everywhere in China, Taiwan, and other Asian countries People used bamboo utensils in hunting, farming, and even transportation because the bamboo material, was easy to get and people could make bamboo utensils by hand. However, as time goes by, bamboo utensils are no longer needed in daily life. Instead, many different types of instruments have been produced massively after the machines were invented for automated manufacturing. Therefore, the craft of making bamboo utensils is gradually losing, and very few craftsmen can adopt and carry on with it.

To evoke the public's attention about traditional bamboo weaving craft. In this study, we aimed at digitizing the collected bamboo artifacts and the crafts in making them with the assistance of experienced bamboo weaving craftsmen. The first objective is to present the digitized artifacts to provide people the opportunity to appreciate the beauty and delicacy of bamboo weaving artifacts. The second objective is to preserve the skill inherited by the experienced craftsmen. Finally, we combined bamboo weaving artifacts and follower arrangement as an example of value-add applications for interior decoration of restaurants and homes.

1.1 Collection of Traditional Bamboo Weaving Utensils

The quantity of bamboo in China and Taiwan is extremely abundant. The early ancestors who were proficient in traditional Chinese hand weaving skill discovered that it is easy to get bamboo as weaving materials. They adopted bamboo weaving craft to make farm tools, fishing tools, and other daily necessities. Hence, bamboo utensils played an important role in the daily life of ancient time.

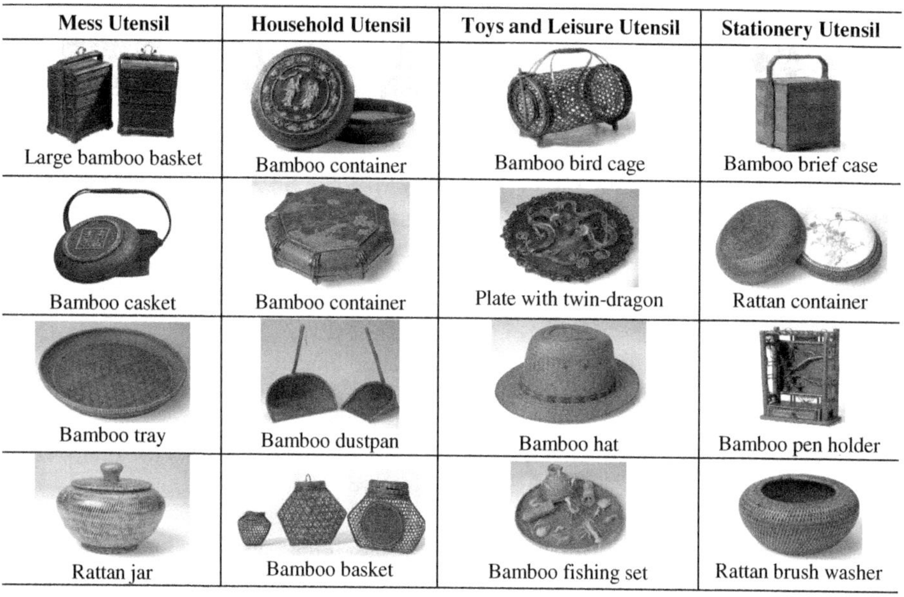

Fig. 1. Illustrations of 4 categories of traditional bamboo weaving utensils

Taiwanese bamboo craft inherited from professional craftsmen immigrated from Fujian Province of Mainland China, which was then experienced a huge change at Japanese occupation period. The Japanese Governor thought it had a high economic value and a bright future, so the officials gave support to Taiwanese craftsmen. Owing to the help of Japanese Governor, the bamboo weaving skills had made a lot of progress, especially from basis to exactness. People began to design different shapes of bamboo fruit plates, flower baskets and some fine works for decoration. And these high-quality works were sold almost in Taiwanese tourist sites that some were even shipped to Japan. Most of the 1200 bamboo weaving artifacts mentioned in this study were collected from southern Fujian province, and the rest manufactured by Taiwanese craftsmen. Figure 1 illustrates examples of traditional bamboo weaving utensils classified into 4 categories, i.e. mess, household, toys and leisure, and stationary utensils.

1.2 Cultural Heritage of Bamboo Weaving Crafts

A craftsman must manage bamboo selection, scraping, degreasing, and bamboo shaping before starting to weave any bamboos. In addition, the skills of weaving the bottom, drawing in the mouth, building the base, making the handle are all steps craftsman have to be familiar with. An example of step-by-step demonstration of bamboo weaving is described in Table 1. Each step was recorded and stored as video clip in QuickTime format.

Table 1. An example of step-by-step illustration of bamboo weaving

Video Clip	Step and Description				
 **CWBK_01**	**1. Scraping**: Place the blade edge on one of the joints of the bamboo and scrape it away moving the blade in a clockwise direction. Then ensuring that the top end of the bamboo is secure against something, scrape away the green bamboo veneer. Remember to scrape from top to bottom to prevent damaging the bamboo veneer.				
	**Step2. Splitting into strips CWBK_02**		**Step3: Splitting into thinner strips CWBK_03**		**Step4. Trimming width of strips CWBK_04**
	**Step5. Trimming thickness of strips CWBK_05**		**Step 6. Round mouth weaving CWBK_06**		**Step7. Weaving the bottom CWBK_07**
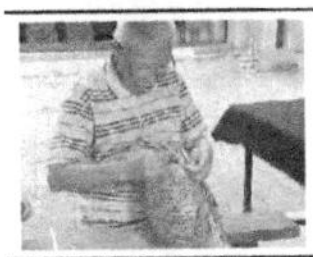	**Step8. Drawing in the mouth CWBK_08**		**Step 9. Making the base CWBK_09**		**Step10. Making the handle CWBK_10**

As shown in Table 2, there are 20 primary weaving patterns widely applied for making bamboo weaving utensils. Owing to many delicate methods, the traditional bamboo weaving craftsman had to take a long-term practice before he could weave bamboo utensils with high quality. The skills in weaving these patterns were recorded and used as e-learning materials.

2 Digital Archive for Bamboo Weaving

Figure 2 demonstrates the system interface of the archive system for preservation and demonstration of bamboo weaving artifacts and crafts. As shown in this figure, functions including demonstration of digitized bamboo weaving artifacts and crafts, bamboo weaving knowledge, weaving patterns, and value-add applications are illustrated. Notice that the utensil shown in Fig. 2(c) can be weaved with two primary weaving patterns, i.e. BWP_11 and BWP_18, shown in Table 2.

Step-by-step procedure to weave a bamboo utensil was demonstrated by the folklore specialists and recorded by a professional photographer. Additionally, primary weaving patterns (Table 2), which are the basic skills in learning bamboo weaving, were also demonstrated and recorded. A bamboo utensil may be weaved by applying several primary weaving patterns following the weaving procedure. Video clips of individual steps were linked with other related information using metadata compatible with the Dublin core standard. Metadata designed based on the Taiwanese folklore artifacts [5] were extended to link weaving patterns and weaving procedures by using the "Relation" element which contains two quantifiers, i.e. "Has Part" and "Is Part Of". The element is used to interlink between the main (parent) metadata record and its children metadata consisting of individual operating steps [3] or different weaving patterns.

Table 2. Primary bamboo weaving patterns for weaving bamboo utensils

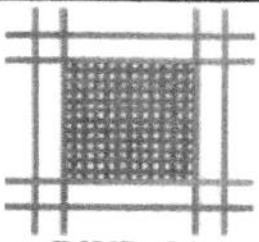 BWP_01. Square pattern	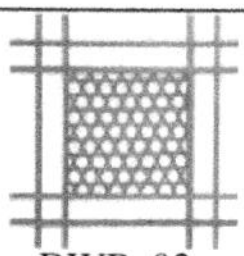 BWP_02. Hexagonal pattern	 BWP_03. Triple-ply herringbone pattern	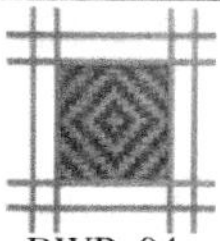 BWP_04. Double-ply herringbone chevron pattern	 BWP_05. Triple-ply herringbone chevron pattern
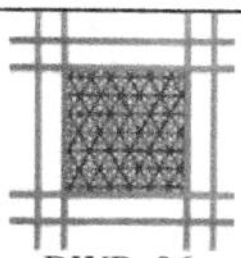 BWP_06 Ramie hexagonal pattern	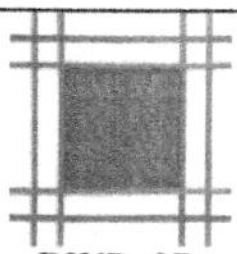 BWP_07. Triple-ply pattern	 BWP_08 Hexagonal straight pattern	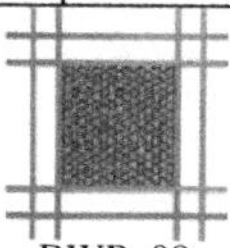 BWP_09 Mixed hexagonal and triangular pattern	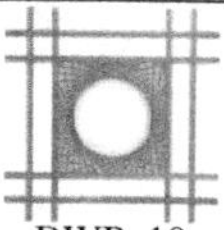 BWP_10 Wheel shape pattern
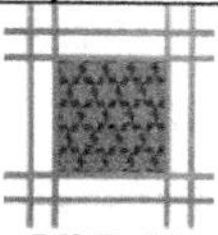 BWP_11 Windmill pattern (left and right)	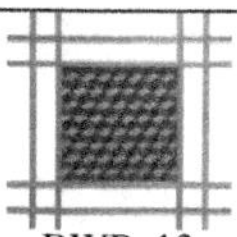 BWP_12 Windmill pattern (3-D)	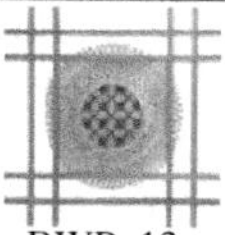 BWP_13. Concentric wheels pattern	 BWP_14 Triple-ply meander pattern	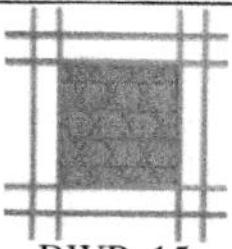 BWP_15 Chrysanthemums pattern
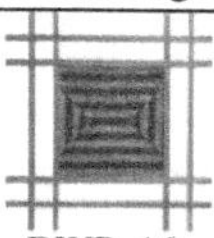 BWP_16. Triple-ply rectangular pattern	 BWP_17 Star pattern	 BWP_18 Fan pattern	 BWP_19 Union jack pattern	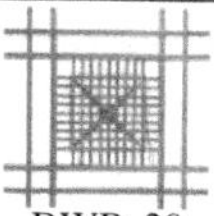 BWP_20 Rectangular backdrop pattern

For example, as shown in Fig. 3, the main (parent) metadata which records a bamboo weaving artifact links the related weaving patterns using the "Has Part" quantifier in the "Relation" element for delineating the bamboo utensil consisting of two weaving patterns (BWP_11 & BWP_18). Furthermore, each step of the bamboo weaving example demonstrated in Table 1 embeds a corresponding video clip that two Quantifiers; i.e., "Has Part" and "Is Part Of" of the "Relation" element adopted for describing the sequential relation between the parent and children metadata records. The "Reference source" Quantifier is applied for expressing its relationship with other artifacts or folklore activities. The "Has Part" Qualifier is used for the parent metadata record to relate its children steps, while the "Is Part Of" for the child steps to trace back to their parent. With this mechanism, all the child steps can be tightly connected to their parent so that the ASP webpage design program can support flexible interaction between the users and the browsers for easy navigation. The Quantifier "Sub-Collection Type" was added to the Element "Type" in the metadata proposed in [5].

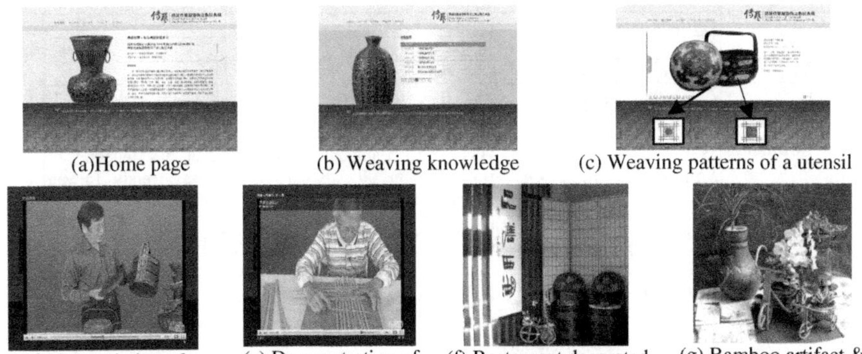

(a)Home page (b) Weaving knowledge (c) Weaving patterns of a utensil

(d) Appreciation of bamboo weaving artifact (e) Demonstration of bamboo weaving craft (f) Restaurant decorated with bamboo artifact (g) Bamboo artifact & flower arrangement

Fig. 2. Graphic user interface of the digital archive system for preservation and demonstration of bamboo weaving artifacts and crafts

3 Folklore Education and Value-Add Application

In order to promote the usage of digitized materials and the value-add applications of the artifacts, the digitized bamboo weaving artifacts and crafts were adopted as e-learning materials for folklore education in a community college, as well as for promoting value-add application specific to interior decoration of restaurants.

Digitized materials for folklore education: It is widely believed that as more senses are involved in the learning process, the greater the impression is retained and the longer the information stays with the learners [6,7]. It was also reported that an enriched environment with multi-sensory would create a thicker cortex within the brain, more dendrite branching, more growth of spinal nervous, and larger cell bodies enabling the neurons to communicate more efficiently [8]. Hence, the image- and sound-rich video contents are suitable for operation-intensive craft education, especially for the aged students. In this study, digitized bamboo weaving artifact and crafts were adopted as part of the e-learning materials for folklore education in a community college.

Service-quality elevation of restaurant and home: The conceptualization of service quality perceptions is still the most debated topic to date. Among many service quality theories, the service quality model including 3 dimensions: interaction quality, physical environment quality, and outcome quality proposed by Brady & Cronin have been widely used for evaluating service quality in various sites [9].

In this study, we mainly focused on physical environment quality to evaluate the value-add application with regards to applying bamboo weaving artifacts combined with flower arrangement in interior decoration of restaurants and homes. A horticulturist was asked to design the flower arrangement artworks by utilizing bamboo weaving artifacts for follower arrangement design. The designer adopted the bamboo weaving artifacts as flower vases for orchid planting and flower arranging with patterns designed by the horticulturist. Then the artworks were used by the interior decorator to decorate restaurants to elevate their quality of service.

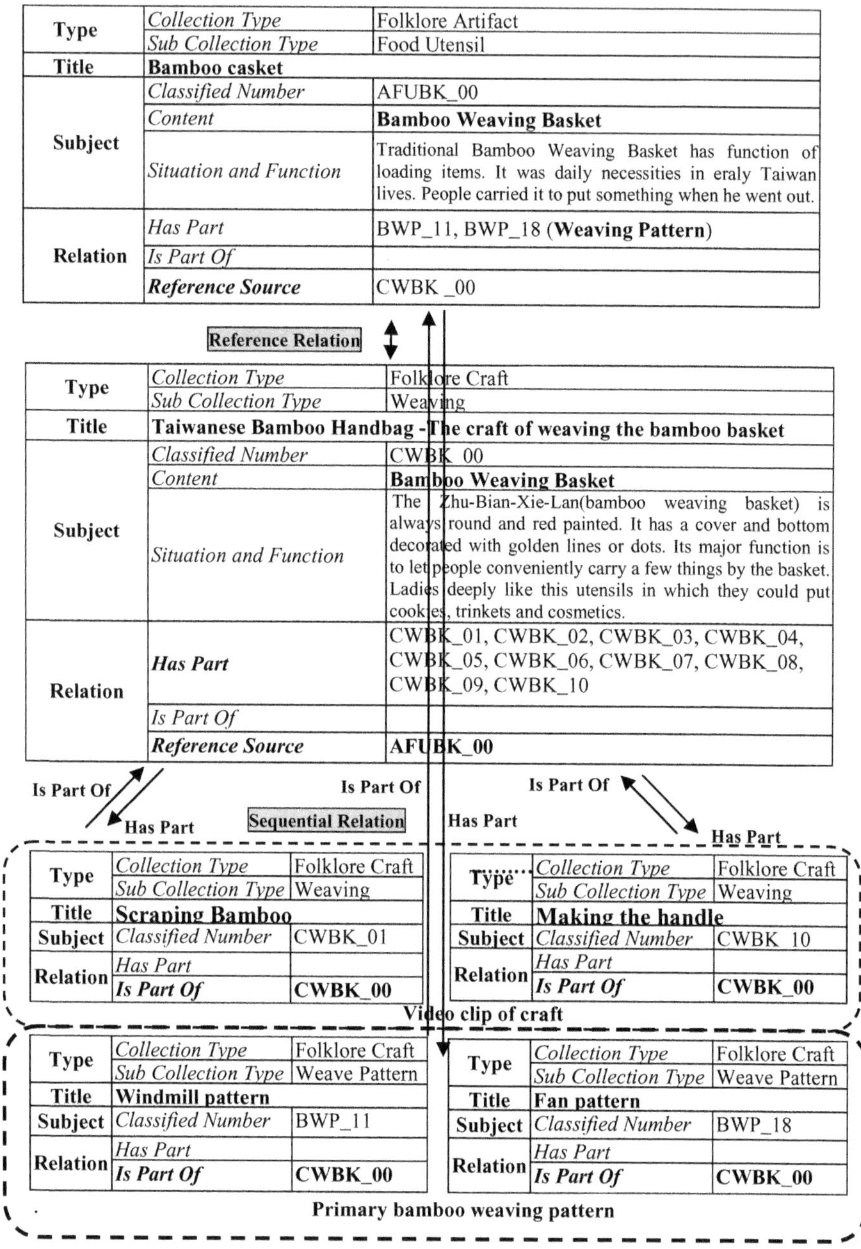

Fig. 3. Metadata structure for implementation of sequential and reference relations

Table 3 shows several examples of the flower arrangement artworks designed by the horticulturist. Figs. 2(f) and 2(g) demonstrate the flower artworks used for interior decoration of a restaurant.

Table 3. Examples of combining bamboo weaving artifacts with flower arrangement

(a)	(b)	(c)	(d)	(e)	(f)

4 Evaluation

A total of 84 students taking the course "Folklore Craft Training" were requested to use the digitized contents as supplementary materials. A questionnaire (Table 4) based on the modified technology acceptance model (TAM) [10,11] was used to evaluate perceived usefulness (PU), perceived ease of use (PEU), and behavior intention (BI) of the website and its accompanied digitized bamboo weaving artifacts and crafts. In Table 4, the scores of PU, PEU, and BI tested with one-sample *t*-test are all significantly greater than 3 ($p<0.05$) indicating usefulness of the digitized bamboo weaving contents, ease of use of the website , and positive attitude toward using the contents in folklore craft training. Brady & Cronin [9] divided service quality into 3

Table 4. Descriptive statistics of modified TAM

Technical Acceptance Model (N=84)	Mean(SD)
A. Perceived Ease of Usefulness	3.85(0.73)
1. I found it is easy to operate Bamboo Weaving Content (BWC) Repository	3.85(0.70)
2. I found BWC Repository is easy to do what I want it to do	3.80(0.73)
3. I found the user interface of BWC Repository is clear and understandable	3.72(0.77)
4.I found the interaction with Bamboo Weaving Content Repository is flexible	4.04(0.82)
B. Perceived Usefulness	3.83(0.76)
5. I agree BWC can facilitate self-learning and accomplish tasks more quickly	3.96(0.72)
6. I agree Repository can decrease learning time and increase productivity	3.78(0.81)
7. I agree BWC Repository can elevate learning wiliness and effectiveness	3.81(0.74)
8. I agree BWC Repository can provide information for different age groups	3.69(0.62)
9. I agree BWC Repository can promote folklore activities	3.77(0.71)
10. I agree BWC is useful for making folklore course materials	3.88(0.82)
11. I agree BWC is useful for learning the current course	3.86(0.84)
12. I agree BWC is useful for understanding Taiwanese folklore	3.86(0.79)
C. Behavior Intention	3.87(0.74)
13. I intend to use BWC as frequently as I need	3.85(0.82)
14. I will continue to use BWC whenever possible in suitable circumstance	3.86(0.79)
15. I expect to use BWC in other related activities and courses in the future	3.91(0.76)

dimensions, i.e. interaction quality, physical environment quality, and outcome quality. The physical environment quality was further classified into 3 sub-dimensions including ambient conditions, design, and social factors. In this study, physical environment quality was adopted to evaluate influence of the restaurant environment decorated by adopting bamboo weaving artifacts as vases for designing flower arrangement artworks, which were then used for interior decoration of restaurants. Again, as depicted in Table 5, the questionnaire surveys of ambient conditions, design, and social factors are all significantly greater than 3 (p<0.05) indicating that integration of bamboo weaving artifacts and flower arrangement highly improve customer satisfaction with regards to physical environment quality.

Table 5. Descriptive statistics of physical environment service of a restaurant

Service Environment Quality (N=40)	Mean(SD)
Ambient conditions	3.73(0.85)
1.At this restaurant you can rely on there being a good atmosphere.	3.85(0.80)
2.The restaurant's ambience is what I'm looking for	3.65(0.77)
3.The restaurant understands that its atmosphere is important to me.	3.68(0.92)
Design	3.78(0.94)
1.The service provider's layout never fails to impress me	3.75(0.82)
2.The restaurant's layout serves my purpose	3.70(1.02)
3.The restaurant understands that the design of its facility is important to me	3.88(0.91)
Social factors	3.65(0.81)
1.I found that restaurant's other customers consistently leave me with good impression of its service.	3.50(0.78)
2.Customers do not affect restaurant's ability to provide me with good service.	3.88(0.94)
3.The restaurant understands that other patrons affect my perception of its service.	3.58(0.78)

5 Discussion and Conclusion

It is believed that folklore and traditional culture are endangered, marginalized, or misunderstood fields that UNESCO recommended as one of the tangible heritage [1]. Folklore artists are encouraged and honored to study and inherit the skills building upon earlier generation [10].

In this study, a folklore hobbyist who collected the bamboo weaving artifacts, a folklore artist who is the expert of bamboo weaving, a horticulturist who used the artifacts to design flower arrangement artworks, and an interior decorator who endeavored to adopt the artworks for restaurant decoration collaborated in providing their expertise to develop the useful digital contents and valuable value-add application. They also serve as folklore educators to prepare introductory materials and answer questions regarding their private collections, crafts, and expertise. Similar function was also recently integrated in MOSAICA project in which dual objectives of preservation and presentation of diverse cultural heritage have been achieved [3]. It is believed that a platform containing abundant and diverse digital folklore contents and folk artists' knowledge can stimulate students' interests and motivations in learning.

Information quality and system integration are two important factors which highly influences perceived usefulness and post adoption of an information system [12].

In our study, the originality, category, and function of each artifact were studied, examined, and recorded by well-known Taiwanese folklore specialists, which greatly ensures quality of the digital contents [3]. Furthermore, integration of the folklore activities and folklore artifacts were achieved through "Relation" element of the metadata [5].

In conclusion, the paper presents the digitized contents of bamboo weaving artifacts and crafts, which not only are useful for education of folklore crafts but also valuable for value-add application. Surveys based on the TAM and environmental quality suggest that the developed digital bamboo weaving contents are useful in e-learning and value-add application by adopting the artifacts as vase to design flower arrangement artworks for interior decoration.

Acknowledgments. This work was funded in part by National Science Council of Taiwan under grants NSC96-2422-H-039-002, NSC97-2631-H-166-001, NSC99-2631-H-166-001, & NSC100-2410-H-166-007-MY3.

References

1. Vecco, M.: A definition of cultural heritage: From the tangible to the intangible. Journal of Cultural Heritage 11, 321–324 (2010)
2. Randall, M.: Unsubstantiated belief: What we assume as truth, and how we use those assumptions. Journal of American Folklore 117, 288–295 (2004)
3. Chan, P.C., Chen, Y.F., Huang, K.H., Lin, H.H.: Digital Content Development of Taiwanese Folklore Artifacts. In: Fox, E.A., Neuhold, E.J., Premsmit, P., Wuwongse, V. (eds.) ICADL 2005. LNCS, vol. 3815, pp. 90–99. Springer, Heidelberg (2005)
4. Bronner, S.J.: The Meanings of Tradition: An Introduction. West Folklore 59, 87–104 (2000)
5. Chen, Y.F., Chan, P.C., Huang, K.H., Lin, H.H.: A Digital Library for Preservation of Folklore Crafts, Skills, and Rituals and Its Role in Folklore Education. In: Sugimoto, S., Hunter, J., Rauber, A., Morishima, A. (eds.) ICADL 2006. LNCS, vol. 4312, pp. 32–41. Springer, Heidelberg (2006)
6. Flaherty, G.: The learning curve: Why textbook teaching doesn't work for all kids. Teaching Today 67, 32–33 (1992)
7. Jensen, E.: Teaching with the brain in mind, The Association for Supervision and Curriculum Development, Alexandria, VA (1998)
8. Chan, P.C., Liao, Y.C., Wang, K.A., Lin, H.H., Chen, Y.F.: Digital Content Development of Folklore Artifacts and Activities for Folklore Education. In: Li, F., Zhao, J., Shih, T.K., Lau, R., Li, Q., McLeod, D. (eds.) ICWL 2008. LNCS, vol. 5145, pp. 332–343. Springer, Heidelberg (2008)
9. Brady, M.K., Cronin, J.J.: Some New Thoughts on Conceptualizing Perceived Service Quality: A Hierarchical Approach. Journal of Marketing 65, 34–49 (2001)
10. Liao, Y.C., Wang, K.A., Chan, P.C., Chen, Y.F.: Digital Folklore Contents on Education of Childhood Folklore and Corporate Identification System Design. In: Chowdhury, G., Koo, C., Hunter, J. (eds.) ICADL 2010. LNCS, vol. 6102, pp. 15–24. Springer, Heidelberg (2010)
11. Davis, F.D.: Perceived usefulness, perceived ease of usefulness, and user acceptance of information technology. MIS Quarterly 13, 319–340 (1989)
12. Saeed, K., Abdinnour-Helm, S.: Examining the effects of information system characteristics and perceived usefulness on post adoption usage of information systems. Information and Management 45, 376–386 (2008)

Digital Archive "Dao Fa Hui Yuan" for Daoism Research

XiaoXiao Feng, Koichi Matsumoto, and Shigeo Sugimoto

Graduate School of Library, Information and Media Studies,
University of Tsukuba,
Kasuga1-2, Tsukuba, Ibaraki, Japan
{fengxx,matsuko,sugimoto}@slis.tsukuba.ac.jp

Abstract. Dao Fa Hui Yuan (道法会元), is an important resource for the study of Daoism. It is a compilation of a number of *Fu*s used in Daoism. A *Fu* is expressed as a complex graphical symbol composed of one or more constituent parts. This paper presents the Dao Fa Hui Yuan digital archive named Digital DFHY which is designed to help researchers explore relationships among sacred symbols of yet unknown meaning and associated natural language text. Some of the descriptions of *Fus* in DFHY include only the name and the graphic expression of a *Fu*, and others include descriptions of the constituent parts of a *Fu* in addition to its name and graphic expression. This paper describes the Digital DFHY, and how it can be used with a network analysis tool to identify relationships among the *Fus* and their constituent parts.

Keywords: Digital archive for humanities, Daoism *Fu*, Dao Fa Hui Yuan, Analysis of *Fu*s, Network analysis.

1 Introduction

Daoism is a religion originated in China. It refers to a philosophical and religious tradition that has influenced the people of China and surrounding areas in East Asia for more than 2.5 millennia. The word *Dao*, is often used for "path" or "way", but with a myriad of nuances in ancient mythology and Chinese philosophy. More than 10 schools existed during the Song and Yuan Dynasty. How to identify and distinguish the schools has become a fundamental subject for the researchers of Daoism [1].

Dao Fa Hui Yuan (DFHY) is one of the most important Daoism manuscripts. This manuscript has a long history and had a profound influence in Daoism. It contains programs for Daoism ceremonies, dogmas and sacred symbols of Daoism prevalent from 10th century to 13rd century. The version surviving until today was published during the Yuan or Ming Dynasty [2]. DFHY is a very important resource for a Daoism priest to inherit the creed from his predecessor and for a researcher as a core collection of *Fu*s [3]. (*Fu* is pronounced like foo.) DFHY is composed of 268 volumes and 2651 pages. More than 6000 *Fu*s represented as graphic symbols are included in DFHY. A *Fu* is expressed primarily as a graphic symbol. In DFHY the many *Fu*s are presented in a few different styles – one named *Juxing-Fu* shows a graphic symbol and associated texts, and another named *Sanxing-Fu* has descriptions of the constituent components of

C. Xing, F. Crestani, and A. Rauber (Eds.): ICADL 2011, LNCS 7008, pp. 26–34, 2011.

the graphic symbol. However, as all of them are written in ancient Chinese even trained researchers need a lot of time to understand those texts. On the other hand, because each school of Daoism has tried to hide their own religion from other schools and people, their charms and spells have been expressed in sacred codes and mysterious symbols. This secrecy of the schools made it harder to understand their religion. Thus, *Fus* in DFHY are complicated but important subjects of Daoism research.

With the development of digital technologies, computers are indispensable to searching, preserving, managing and analyzing digital content for humanities research. Thus, Digital DFHY is demanded by Daoism researchers using information technologies. The primary goals of this research are to preserve and manage the contents of the DFHY in digital form and to support Daoism researchers by providing not only the digitized content of DFHY but also elaborated tools for research.

In this paper, we focus on the organization of Digital DFHY and its management. We also show a content analysis of DFHY as a case study. The rest of the paper is organized as follows. Section 2 describes the related work and the context of this study. Section 3 describes the steps in creating the Digital DFHY. Section 4 discusses the analysis method of the contents of DFHY, e.g., the texts and the graphics, and shows some use cases of Digital DFHY. Section 5 concludes this paper.

2 Requirements Analysis for Digitization of Dao Fa Hui Yuan

2.1 Related Works and Research Context

Many digital archives have been built in different subject domains. Digital archives provide users with convenient tools not only to find and browse the contents but also to help users analyze them. Digital archives are not just a digitized collection of resources but an integration of the digitized resources and services which add value to the contents. Large scale digital collections such as the American Memory of the US Library of Congress, *Europeana* in Europe, the National Digital Archive Program in Taiwan and the Digital Library from the Meiji Era of the National Diet Library of Japan are generally designed for researchers, educators and learners. These are primarily digitized collections of primary materials - books, manuscripts, maps, physical objects, pictures, movies, sound recordings, etc. On the other hand, there are other types of digitized collections specially designed in accordance with the primary materials – the Humi project at Keio University designed for digitization of Gutenberg's Bible [4] and the Cuneiform Digital Library project hosted at UCLA [5].

For those classic cultural resources like DFHY, which contain not only texts but also figures associated with the texts, a key aspect in building digital archives is how to identify semantic relationships among the texts and figures. The Archimedes Palimpsest Digital Project [6], for example, converted the manuscript Archimedes Palimpsest into a digital form and organized the contents of Palimpsest as a dataset to provide authoritative data, and derived information such as transcriptions, and also to offer a standardized product for future users. In general, the important aspects for digitization projects for humanities research are how to deal with the context of the manuscript, how to preserve the digitized contents for a long period, and how to use the information for content analysis. Thus, content analysis using digitized materials is a common goal.

2.2 Requirements Analysis for Digital Dao Fa Hui Yuan

Digital archives are not just a collection of digitalized documents but a newly organized resource collection for users of today and in the future. We need to take the future development into account in developing a digital archive, e.g. management and analysis of the contents. Because *Dao Fa Hui Yuan* is a well-used Daoism document, which is a compilation of *Fu*s, the goal of this study is not only to convert DFHY into a digital form but also to design the data set of Digital DFHY so it will be usable for researchers as a primary resource of Daoism.

Generally, every digital archive has its own features designed in accordance with the source resource(s), audience and main goal. For example, in the case of digitization of books whose main goal is to provide a new reading environment as digital books, crucial factors are user-friendliness for users and reasonable costs to realize the environment. For another example, very high quality mark-up texts of a source resource would be required in the case of digitization of books for humanities scholars who heavily rely on text analysis. In the case Digital DFHY, the primary purpose is to build a database that helps researchers study *Fu*s.

For simple understanding of *Fu*, we assume here that a single *Fu* is a piece of paper on which a graphic symbol and associated texts are drawn. Each *Fu* has a graphic symbol, which expresses a religious statement. Each graphic symbol is composed of more than one constituent part. There are two types of *Fu*s – with and without descriptions of its constituent components. We call the former *Sanxing-Fu* (SF) and the latter *Juxing-Fu* (JF), respectively. Fig.1 includes JF and SF, respectively. Meanings of some *Fu*s are still unclear because not all of the *Fu*s are accompanied by a description of their constituent parts.

The primary users of Digital DFHY are people who have deep knowledge about Daoism but are not familiar with information technologies. In DFHY, a single *Fu* could be explained in one or more pages and a single page could include one or more *Fu*s. This feature made our digitization process quite complicated. In the early stage of this study, the database contents had to be organized by collaboration between Daoism researchers and information technology researchers in order to appropriately define the database schemas. Since the early attempts of Digital DFHY [6][7], we have re-designed the schemas several times in accordance with requirements identified in the collaboration.

3 Building Databases for Digital DFHY

Prior to this study, in the early 2000s, the second co-author of this paper, Koichi Matsumoto led a project to digitize DFHY. In his project, he digitized about 30% of the DFHY. The digitization process has been done in several separate phases and is still continuing. The first digitization was done more than 10 years ago. At that time, the research focus was on how to organize and build the database. After the experiences in the first several years of the development, we started to apply the database to a specific application. The database of DFHY presented in this paper is the latest development of this digitization process.

3.1 Digitalizing DFHY

To build Digital DFHY, we digitized all of the 2651 pages in DFHY. An example page is shown in Fig.1 which shows a *Fu* titled as *Fei-Ren Zhuo-Long Fu*. Following the title, there are the *Sanxing-Fu* and *Juxing-Fu* of *Fei-Ren Zhuo-Long Fu*. Every *Fu* is saved as a JPEG file page by page.

We wanted to use automatic text recognition software to convert scripts in DFHY into text files but it was not feasible. We manually converted the original scripts into text files that were reviewed under the guidance of Daoism researchers.

Each graphic symbol was extracted and saved as a JPEG file. Each graphic symbol was associated with the texts attached to it and stored in a database. Every graphic component was also extracted manually and stored in a database with its associated information. Each of the graphic symbols and components is given a unique identifier during this digitization process. This task is crucial factor because some components look very similar but have different meanings.

For instance, in the case of the example *Fu*s shown in Fig.1, the page number is marked on the top of the page. (Here we remind the reader that in DFHY pages are read from right to the left.) Two *Fu*s are shown in this page. When *Fu*s occupy more than one page, we located them on the page where the *Fu*'s title was. We gave each page a sequence number for digitization. The page image shown in this figure is the 20th page of volume 56. This image file is named 05620.jpg. The graphic symbol of the *Fu* is the #1. *Fu* appearing in this page, so it was saved in a file named after the identifier, e.g. 0562001.jpg. The naming scheme means this *Fu* is expressed as the graphic symbol No.1 on page 20 of volume no. 56.

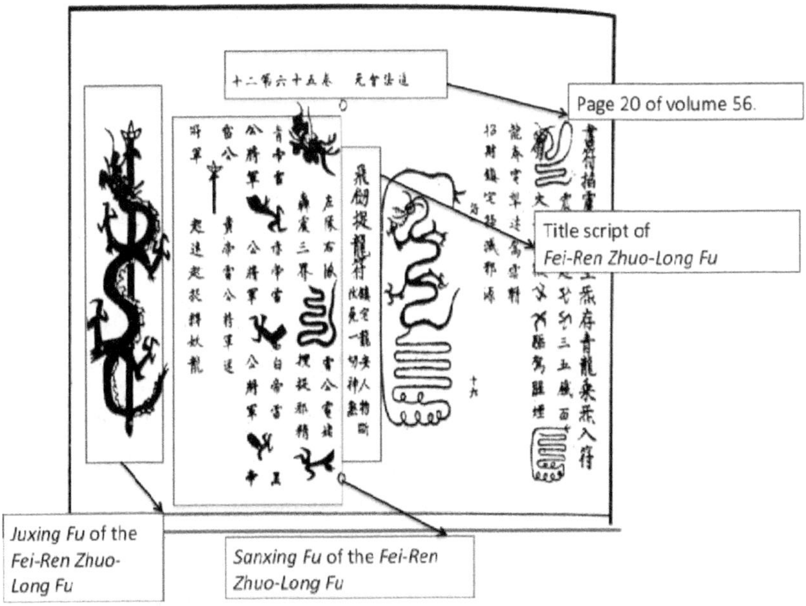

Fig. 1. Page 20, volume 56 of DFHY

3.2 Framework of Digital DFHY

Figure 2 shows the framework of Digital DFHY. In this framework there are two major components. The left part of figure 2 shows the core database of Digital DFHY – the database of images and texts of the *Fu*s and constituent parts. Images are saved as JPEG files. Texts are taken from titles of *Fu*s, scripts attached to *Fu*s, and other texts included in the pages. This set of data is used to search *Fu*s and the constituent parts. This database is used in several different applications as shown in the right hand side of figure 2. For example Hayagawa classified the shapes of parts using this dataset [7], Uda classified the shapes of parts [8], and the authors are using the data set for relationship analysis of *Fu*s and their constituent parts. The outputs of these studies are included in Digital DFHY as a database or a tool to use the database. Thus, we are augmenting Digital DFHY by adding new functions and databases from the application research. From another viewpoint, this augmentation is a collaborative and reciprocal process by Daoism and IT researchers - Daoism researchers help IT researchers create the new functions and databases from the core databases, e.g., *Fu* ontology, and, in turn, the new functions and databases are used to enrich the core databases.

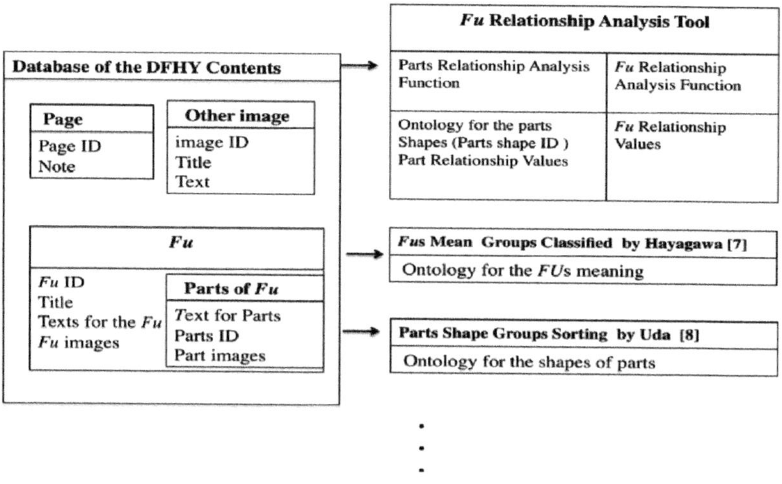

Fig. 2. The framework of the Digital DFHY

3.3 Experiences in Digital DFHY

Using the digitized contents of DFHY, several projects have been carried out. One of the early projects was the gathering and sorting of the *Fu*s using the digitalized pages and their simple metadata including titles of the *Fu*s. The database helped researchers to find the contents much more effectively than before - they were enabled to put all of the *Fu*s names and the position of them together to find out the distribution of the titles, and to analyze the characters statistically [7]. In another project, the meanings of the *Fu*s noted in the metadata were used to classify the *Fu*s by their meanings. The result of the research was reflected to the Digital DFHY [8]. Analysis of the frequency of the word usage and the word meanings was also performed as part of the Digital DFHY project [9]. These experiences were used to augment the function of the core database of DFHY as shown in figure 2.

4 Use Case: Analysis of *Fu*s and Their Components

We have selected a set of *Fu*s stored in the database for an experimental use of the digital contents. Because of the long history of Daoism and secrecy of knowledge maintained separately by each Daoism school, not all of the meanings of the *Fu*s included in DFHY are known. Inconsistency is natural with this kind of resource but it is still worth trying to analyze the relationships among the *Fu*s and their components using a consistent procedure. This section shows an analysis of the relationships among *Fu*s and their components, which cannot be carried out without the database of DFHY and computers.

4.1 Some Previous Analysis of *Fu*s

In the earlier studies conducted by the second co-author of this paper, they assumed that the constituent parts of *Fu*s would determine the semantic relationship between a pair of *Fu*s [8]. In other words, *Fu*s sharing the same parts would share meanings, and those sharing fewer or no parts would share less or no meaning. They counted the pairs of *Fu*s that share parts. For that study, how to sort parts was found to be a crucial problem. It was found that each part has its different importance according to the Daoism researchers [1]. To sort parts by shape more precisely, we tried to make a new sorting method to reflect the meaning of every part and a position of a part in a *Fu* [7].

4.2 Analysis of Relationships *Fu*s and Parts

From our previous studies we got the result that *Fu*s of similar meanings share the same constituent parts. We did a graph-based analysis to analyze relationships between *Fu*s and their parts. We assumed that the number of identical parts in a pair of *Fu*s stands for the strength of the relationship. SF is useful to represent the relationships of *Fu*s and their parts because it is a simple task to parse a *Fu* into their components and SF includes an explanation of every constituent part. However, not all *Fu*s are presented in the form of SF in DFHY.

In this study, we arranged a set of the *Fu*s into a graph where *Fu*s and parts are represented as a graph node. *Fu*s which share an identical constituent part are connected via the part. Thus, we created a graph composed of *Fu*s and their constituent parts, i.e. a network of *Fu*s and parts of DFHY.

In order to analyze the *Fu* network, we defined the algorithm below, where Distance between two parts i, j (d_{ij}): the number of the *Fu*s on a path between two parts on the graph.

The shortest path length between two parts i, j (t_{ij}): the minimum distance between two parts. i.e. min(d_{ij})

Number of shortest paths between two parts i, j (w_{ij}): the number of paths whose lengths are equal to t_{ij}.

Relationship between parts r (P_i, P_j) is determined by the next expression. here, a is a heuristic value and we have supposed it as 1.2. (Subscripts i, j are omitted.)

$$(r) \; : \; r = a^{w-1} / t \tag{1}$$

In this way, the relationship values are in inverse proportion to the distance of each pair of the parts. The relationship value between two *Fu*s (F_A and F_B) is defined by the expression below:

$$r(F_A, F_B) = \frac{\sum_{\substack{1 \le i \le m \\ 1 \le j \le n}} r(p_{Ai}, p_{Bj})}{m \times n} \tag{2}$$

In this way, we can build the relationship network of *Fu*s, and the relationship values of any pair of *Fu*s can be calculated.

4.3 Evaluation of the Analysis Results

To show the data of the digital contents and relationships of *Fu*s, we have implemented the analysis system on Digital DFHY. This system helps researchers search the contents in DFHY, and browse the analysis results based on the relationship values. As of June 2011, Digital DFHY has 499 *Fu*s. We applied this analysis tool to Digital DFHY.

Users can use this system to search any page and image they want to read. Fig.3 shows an example for the search results of *Fus* that have more than one identical part contained in *Hei-Sha Da-Jiang Fu*.

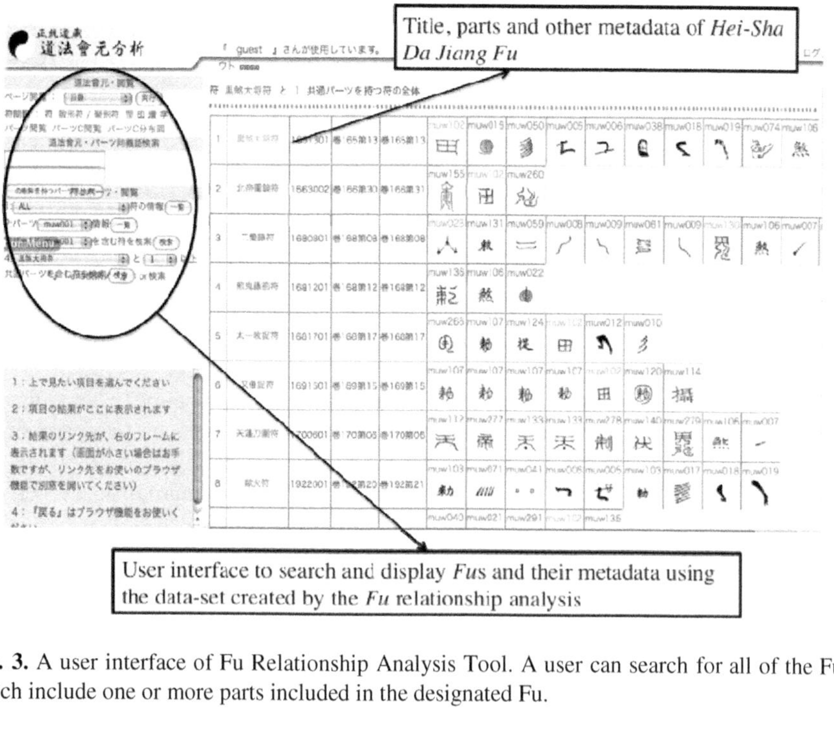

Fig. 3. A user interface of Fu Relationship Analysis Tool. A user can search for all of the Fus which include one or more parts included in the designated Fu.

In Digital DFHY, we can also check the analysis results of relationships between *Fu*s. Table 1 shows the 5 pairs of *Fu*s that have the highest relationship values among all pairs used in this experiment. Most of the pairs of *Fu*s share the same parts but the fourth pair does not have the same parts. However, the parts in these two *Fu*s are related with other parts, which present together frequently, so the value of the relationship is higher. *Jian-gui Zhuanlu-Fu* means kill all of the ghosts by fire and *Huang-Zhu Ding-Yuan Fu* means the god will kill all of the ghosts. We found that both of the *Fu*s of the 4th pair have the meaning "Kill all of the ghosts."

Table 1. Examples of *Fu*s analysis results

Rank	*Fu* pairs	Shared parts	Relationship value
1	魁字符 (*Kui-Zi Fu*) 本身符 (*Ben-Shen Fu*)	1	10
2	魁字符 (*Kui-Zi Fu*) 九醜大魔攝魂附體符 (*Jiu-Chou Da-Mo shehun Fu-Ti Fu*)	3	3.25
3	本身符 (*Ben-Shen Fu*) 九醜大魔攝魂附體符 (*Jiu-Chou Da-Mo shehun Fu-Ti Fu*)	1	2.84
4	煎鬼籙箚符 (*Jian-gui Zhuan-Lu Fu*) 皇主定元符 (*Huang-Zhu Ding-Yuan Fu*)	0	2.33
5	大赤皇定元真符 (*Huang-Zhu Ding-Yuan Fu*) 丹天太皇真符 (*Dan-Tian Tai-Huang zhen Fu*)	2	2.2

The relationship values help researchers gain an overall view of the distribution of *Fu*s. If a pair of *Fu*s has a high relationship value, they might have the similar meaning, the same function. By using these analysis results, researchers can check out not only about a particular Fu but also other *Fu*s related to the Fu. Researchers can locate related *Fu*s and find the description of the *Fu*s. The example in table 1 shows that the relationship is determined not only by the number of shared parts. As the relationship values can be browsed with images stored in the database, this tool helps users understand relationships among many *Fu*s and their components.

5 Conclusion

This paper presents the Digital DFHY, a digital archive created from Dao Fa Hui Yuan – a very important resource for Daoism researchers. Digital DFHY is designed as a tool to support Daoism researchers. A *Fu* is expressed as a complex graphical entity composed of more than one constituent part. The goal of this study is to build an environment based on Digital DFHY to help the Daoism research community analyze the structure of all *Fu*s and relationships among *Fu*s.

There are several problems left for our future work. We have created a tool to calculate and show the relationships between any parts. In this tool, we assumed a uniform weight for all constituent parts. However, the parts are sometimes given different meanings depending on their position in a *Fu* [10]. In other words, we need to establish a new tool that can reflect rules to compose *Fu*s, say a *Fu* grammar. Finding such a grammar is a research challenge for the Daoism research community. We hope collaboration with Daoism researchers will help us solve this problem. We believe that further development of Daoism and classic materials is crucial for the Humanities researchers.

Acknowledgement. We thank Prof. Norihiko Uda and Ms. Misa Hayakawa for their support and contributions. This study is supported in part by JSPS Grant-in-Aid-for Scientific Research (A) #22240023 and (C) #40154489.

References

1. Li, Y.: Shen Xiao Lei Fa: The History and Thought of Daoism Shen Xiao Lei Fa, pp. 338–470. Sichuan People's Publishing House, Sichuan (2003) (in Chinese : 神霄雷法: 道教神霄雷法沿革与思想)
2. Nikaido, Y.: The Change of the Marshal God in the Daoism and Folk Beliefs, p. 256. Osaka, The Institure of Oriental and Occidental Studies, Kansai University, Osaka (2006) (in Japanese: 民間信仰における元帥神の変容)
3. Matsumoto, K.: Dao Fa Hui Yuan—A Great compile of magic, Reading the Classic Daoism Book, pp. 236–248. Taishukan Publishing, Tokyo (2001) (in Japanese「道法会元」一道教呪術の集大成、道教の経典を読む)
4. Introduction of the Humi project, `http://www.humi.keio.ac.jp/en/index.html`
5. Introduction of the host of the Humi project, CDLI Cuneiform Digital Library Intiative, `http://cdli.ucla.edu/`
6. Introduction of digital archive for the Archimedes Palimpsest project, `http://archimedespalimpsest.org/`
7. Hayakawa, M., Matsumoto, K., Uda, N.: Classifing Terms Used in Taoist Charms Appearing in "Dao-fa Hui-yuan". Journal of Information and Media Studies 8(1), 57–69 (2010)
8. Uda, N.: Fumi, Tamezawa., Matsumoto, K., Yoshihiro, Nikaido.: Charm Analysis Support System for "Dao-fa Hui-yuan" Research. In: Proceedings of JSIK (May 24, 2004)
9. Hayashi, H.: The Digitizing of History Documents for Supporting the Research, pp. 25–27. The Graduation Thesis of Tsukuba University (2001) (in Japanese: 研究支援を目的とした歴史資料の電子化)
10. Feng, X., Matsumoto, K., Sugimoto, S.: Computerization Analysis of Charms Structures: Using "Dao-fao Hui-yuan" as an objects, pp. 9–16, JPST SIG Technical Report 2008-CH-79 (2) (2008)

Libraries in a Digital Frontier: Preserving Chinese Canadian Cultural Heritage

Allan Cho and Yu Li

University of British Columbia Library

Abstract. As a three-year community-based research project at the University of British Columbia, *Chinese Canadian Stories: Uncommon Histories from a Common Past* is government grant-funded project by the Community Historical Recognition Program (CHRP) that brings together the expertise and resources of a wide range of UBC Library units and off-campus partners: from the digitization of archival material of UBC Library's Rare Books & Special Collections; to the digital storage infrastructure of UBC's Digital Initiatives; to the community outreach and digital technology of the Irving K. Barber Learning Centre; to the Chinese language online resources and community historical preservation expertise of the Asian Library. Through partnerships with community and civic institutions nationwide, this UBC-library led project focuses on three initiatives: a one-stop web portal, a series of community workshops, and digital interactive cultural game using cutting edge technologies. This paper is a progress report of the project.

Keywords: Chinese Canadian. history. libraries. archives. digitization.

Value of *Chinese Canadian Stories*

To understand the project's value and impact on the Chinese community in Canada, we must first appreciate the complex history of its people. As Canada is a country of immigrants since its founding, migrants from China have long resided in the regions of Canada for more than 200 years. As such, Chinese in Canada have shared a unique history with Canada's other "founding" peoples [1]. Despite immigration restrictions during the "exclusion period" in the years between 1885 and 1947 when Chinese migration was restricted by government regulation, the kind of histories which have not been recorded or recognized by traditional Canadian history textbooks and educational curricula have provided researchers with a goldmine of historical research and materials yearning to be uncovered. According to 2006 statistics, 43% of Vancouver's population is ethnic Chinese, with heterogeneous histories ranging from fourth and fifth generation Chinese [2]. Although Chinese have lived in Canada even before it was founded, comprising a large number of its people, their histories have not been well recorded or documented.

However, through the power of easily accessible and engaging new digital media technologies, *Chinese Canadian Stories* [3] provides a platform for families and historians to give voice to the unknown stories and often overlooked histories of immigrants and their families that often never make it beyond the ivory walls of academic discourse. While academic libraries have traditionally concentrated on

C. Xing, F. Crestani, and A. Rauber (Eds.): ICADL 2011, LNCS 7008, pp. 35–38, 2011.

building collections, providing research support to students and faculty, and offering information literacy instruction, they have always been integrated into the broader aspirations of the university [4]. As the academic library can be natural focal point for this interaction and exchange between academia and community, *Chinese Canadian Stories* helps to build UBC Library as a gathering place for community outreach and community-based research.

Web Portal

At the heart of this project is the web experience. Until now, there has never been a one-stop web portal dedicated to collecting, archiving, and accessing information about Chinese Canadian history. The vision is for any user from the public – whether a student, researcher, or just the curious -- wanting to learn about the "Chinese experience" in Canada to easily find this information online. As a one-stop portal, the Chinese Canadian Stories provides a gateway to Chinese Canadian history materials belonging to libraries, museums, and cultural heritage collections throughout Canada. The website is fully indexed by Google. With a description and URL containing the words "University of British Columbia" and "ubc.ca," the site immediately "leaps out" as academic and most useful in terms of a classroom assignment.

The underlying architecture of the web portal is leveraging UBC Library's institutional repository, cIRcle [5], as the storage mechanism for the digital materials. In using cIRcle, which already has an existing work flow that organizes and preserves UBC's online content by making it accessible to the community partners who deposit their digital objects, the project doesn't require a reinvention the wheel. As an "Ingest Tool" based on the DSpace software, which accepts the oral histories, photographs and documents of the community into a long-term preservation initiative, the tool allows community members to share and preserve their digitized archives with UBC, while allowing them keep their original physical and digital copies themselves.

Moving beyond cIRcle is the public interface of the web portal, requiring an "editorial process" in which librarians with subject expertise will select the most appropriate content to be transferred to CONTENTdm, the digital collection management solution that offers scalable tools for archiving collections of any size.

To break the language barrier and make the portal's digital content accessible to a broader audience, the project develops a bilingual user interface and metadata. Student employees with both English and Chinese language skills are hired to do the translation and metadata creation work under the supervision of librarians.

Except for creating the bilingual content, an even bigger challenge of building the portal collections is to find the best subject headings for the materials. For most of British Columbia and Canada's history, racial discrimination and exclusion has resulted in missing gaps in our historical archives and thus there is lack of relevant subject headings in the established subject authorities for the project librarians to choose from. Therefore, one of the most powerful opportunities this project brings is a whole new set of thesaurus terms that the LC subject headings and other established authorities do not accurately capture. Through this initiative, UBC Library is making a difference in preservation of the Chinese Canadian history and in correcting the past's erasures in Canada's national memory.

Community Workshops

UBC librarians and researchers have worked with the Gee How Oak Tin Benevolent Association and Ng Suey Sun Associations, two of the early Vancouver Chinatown clan associations, in creating outreach workshops on collecting both physical materials and oral histories. The Gee How Oak Tin is a particularly historic association, as members can be traced back to the region of Taishan in Mainland China, which comprise much of the Chinese migrant population in Canada's early years. The UBC workshops facilitated not only the gathering of family histories among the elders of the association, it also allowed for the digital recording of the Taishanese dialect, a rich cultural heritage with its own grammar and syntax that is slowly disappearing in China, as older generations pass on while Mandarin has displaced this "peasant dialect" to become the lingua franca.

Yet, the main challenge with making these materials available is that there are so few local people who can read the old-style Chinese writing. While the characters are no different than those which have been used for thousands of years, the classical writing style of the time required fewer characters than are used today to express the same idea, making interpretation an incredible challenge. Not only are many of the handwritten characters to decipher, the letters also used "invented" phonetically-sounding Taishanese characters that are unintelligible to contemporary Chinese.

These ancient letters from the Taishanese are extremely significant to researchers, not only for the preservation of language and history, but also because the Taishanese comprised the majority of Canada's Chinese population in the early twentieth century, thus giving us a clear picture of the ethnic composition of early Chinese Canadians. While the ultimate goal is to collect the digital stories for the web collection, the project contextualizes these materials by providing the stories (and their metadata) behind these seemingly unintelligible scripts of the forgotten past.

Highly Interactive Cultural Game

To better engage young people in meaningful learning experiences, the project creates a video game called "Gold Mountain Guest" for secondary school students which will be integrated into the BC school curriculum.

The idea of the game is by asking the user to go step by step through their voyage from China to Canada in the shoes of this virtual migrant from the past, *Chinese Canadian Stories* realizes the fullest details of the migration experience from leaving the village to the travel to a port such as Hong Kong, to the boarding of the ship and passage across the Pacific, to arrival in Western Canada to the registration and payment of the Head Tax. For each step of the virtual tour, the user's voyage of discovery of Chinese Canadian history introduces him to the concrete details of what life was like in the late 19th and early 20th century for Chinese migrants and for Canadians of all backgrounds—from the building of the CPR to the anti-Chinese legislation that Chinese Canadians overcame.

Conclusion

For most of British Columbia and Canada's history, racial discrimination and exclusion has resulted in missing gaps in our historical archives and thus in our abilities to remember the true diversity and multicultural engagements of our collective past. Through this project, it can be said that UBC Library is making a difference in correcting the past's erasures in Canada's national memory, working with academics, libraries and archives as well as the diverse communities of Canada whose stories are still underrepresented in historical records, innovatively creating a different approach to library and archiving that brings together communities, researchers, and students at UBC Library.

The Chinese Canadian Stories is actively creating new ideas, new concepts, and new fields of historical study, it's increasingly challenging to classify and catalog the Chinese Canadian historical materials.

References

1. Yu, H.: Global Migrants and the New Pacific Canada. International Journal, 1011–1026 (Autumn 2009)
2. Ahadi, D., Murray, C.A.: Urban Mediascapes and Multicultural Flows: Assessing Vancouver's Communication Infrastructure. Canadian Journal of Communication 594, 34 (2009)
3. Chinese Canadian Stories: Uncommon Histories from a Common Past, `http://www.chinesecanadian.ubc.ca`
4. Singh, S.: Librarians as Information Providers and Facilitators: The Irving K. Barber Learning Centre as a Model for the Expansion of the Role of Academic Libraries in University-Community Engagement, One Road In, Many Roads Out; Education as a Catalyst for Regeneration, Conference, pp. 1–8. University of Limerick (2010)
5. University of British Columbia Library Institutional Repository, `http://www.circle.ubc.ca`

Automated Preservation: The Case of Digital Raw Photographs

Stephan Bauer and Christoph Becker

Vienna University of Technology, Vienna, Austria
http://www.ifs.tuwien.ac.at/~becker

Abstract. In digital preservation, a common approach for preservation actions is the migration to standardized formats. Full validation of the results of such conversion processes is required to ensure authenticity and trust. This process of *quality assurance* is a key obstacle to achieving scalability for large volumes of content. In this article, we address the quality assurance process for the preservation of born-digital photographs and validate conversions of raw image formats into standard formats such as Adobe Digital Negative. To achieve this, we rely on a systematic planning framework. We classify requirements that have to be evaluated according to their measurement needs. We extend an existing measurement framework using a combination of tools, image similarity algorithms, and purpose-built plugins. By combining metadata extraction, image rendering and comparison, and perceptual-level quality assurance, we evaluate the feasibility of automating the core part of quality assurance that is often the most costly part of preservation processes.

1 Introduction

In our society so strongly relying on digital data and communication for all purposes, digital preservation has become an important challenge: Only a functioning and appropriate technical environment can create a meaningful interpretation of digital objects. In the case of static content, where data structures specify the full semantics of the intellectual meaning of data streams, a common solution is to convert data into different representations to enable rendering in the environment of a certain user group. For preserving images, this corresponds to conversion to a standardised image format with long expected life time and widespread, solid tool support. These *migration* processes need to be fully validated to ensure that the intellectual meaning of the content is fully and authentically preserved. The migration itself can be relatively easily automated and parallelised to achieve scalability on large volumes of content. The key bottleneck lies in verifying the migration results.

Today, professional photographers take literally tens of thousands of pictures per month[1]. They normally rely on a so-called *raw* format that stores the raw

[1] http://www.nationalgeographic.com/educator-resources/presidents-photographer/window-into-white-house/

C. Xing, F. Crestani, and A. Rauber (Eds.): ICADL 2011, LNCS 7008, pp. 39–49, 2011.

data recorded by the camera's imaging sensor. The fundamental difference between the raw data and the interpreted data in a JPEG file is that the raw data itself needs interpretation in order to create a meaningful and presentable image, much like a negative film requires developing and printing to produce a photograph. Born-digital photographs captured in raw formats thus present a particular challenge, since the semantics of data captured by the sensors in digital cameras require complex interpretation processes to deliver a meaningful and authentic rendering. Without a trustworthy evaluation of different possible migration paths, no decision and conversion process can be trusted to deliver authentic and usable photographs. There is a vast amount of objects involved in large collections of raw camera data stored in proprietary formats containing specific camera profiles and the inherent uncertainties of migration. Hence, it is clear that such a conversion into standardised digital still image formats can only be successful with reliable and meaningful methods for automated Quality Assurance (QA). QA in this context refers to the process of measuring the significant properties of content and delivering measures of similarity according to specified criteria of interest in a structured form.

This article leverages a preservation planning framework to address this key issue in providing automation and scalability for the preservation of born-digital photographs. We elicit requirements that have to be evaluated and classify them according to their measurement needs. By extending an existing measurement framework using a combination of tools and purpose-built plugins for metadata extraction, image rendering and comparison, and perceptual-level quality assurance, we demonstrate automated QA for preserving digital photographs. We discuss experiment results and outline challenges lying ahead.

The remainder of this paper is organised as follows: Section 2 provides an overview of related work in image preservation and comparison. Section 3 discusses challenges and requirements posed by born-digital photographs. Section 4 demonstrates how to leverage an existing evaluation framework and extend it to achieve automated QA using existing software and new tools. The results of a case study are presented in Section 5. Finally, Section 6 draws conclusions and gives a short outlook on future work.

2 Related Work

In the digital libraries field, the discourse about preservation of images has largely focused on migration strategies converting scanned image content to standardised still formats such as TIFF-6 or JPEG 2000 [8]. A recent contribution discusses issues with JPEG 2000 as a preservation format [10]. Considering born-digital photographs, it must be noted that the term *raw format* does not describe one specific format. Each major vendor has multiple different types, most of which are partially proprietary and not standardized. To increase standardization, Adobe created an open raw format called Digital Negative (DNG).

The format is an extension of the TIFF-6 format and compatible with the TIFF-EP ISO standard [1] submitted for ISO standardisation[2]. The DNG format supports lossless compression as well as uncompressed storage of image data. The suitability of raw formats and DNG as archival still image format for digitised cultural heritage content was discussed in [4]. The authors provide a detailed discussion of the advantages of DNG, but do not propose or perform experiments to evaluate whether migration processes to DNG are preserving the significant properties of the images in question.

In the practitioner's community of digital photography, a certain awareness for the importance of digital preservation is emerging. The initiative *dpbestflow*[3] suggests best practice for storage and migration of digital photos. The recommended strategy is migration to DNG. However, no systematic analysis of the significant properties to be preserved is presented. Some limited experiments have evaluated whether conversion to DNG preserves the image content correctly[4]. However, the small scale of this experiment and the lack of a systematic evaluation approach result in insufficient evidence for drawing conclusions.

The systematic evaluation of preservation actions is the core strength of the preservation planning approach implemented by the planning tool Plato[5] [2]. It relies on controlled experimentation on sample content to enable evidence-based assessment of candidate preservation actions against scenario-specific requirements in a repeatable, well-documented workflow. The challenges currently in the focus of DP research are dominated by an urgent need to make operative preservation actions scalable, so that they can be applied to large volumes of content without the need for human intervention [3]. This requires a combination of tools, procedures and quality assurance mechanisms.

The DP community has systematically analysed the issues and factors influencing digital preservation operations. However, the degree of automation that can be achieved when executing preservation actions such as migration or emulation still strongly depend on the type of content and is in fact low for most object types and virtually non-existent for emulation [3].

Automated QA for image preservation requires methods to measure the similarity of images. Most of the common image similarity metrics focus on error related concepts such as Absolute Error (AE) or Mean Squared Error (MSE). These metrics have a solid mathematic background and can be used for discovering errors in conversion processes. However, they do not properly describe the human perception [11]. A recent promising approach is the structural similarity metric (SSIM) which corresponds more closely to the human visual system [12]. SSIM was originally designed for grayscale images. However, [9] compared the performance of

[2] See `http://www.dpreview.com/news/0805/08051503adobeDNG.asp` and `http://www.iso.org/iso/standards_development/technical_committees/other_bodies/iso_technical_committee.htm?commid=48420`

[3] `http://www.dpbestflow.org/`

[4] `http://www.libraw.org/articles/2-ways-to-nowhere.html`

[5] `http://www.ifs.tuwien.ac.at/dp/plato`

SSIM on different channels in comparison to subjective assessment. They showed its applicability to color channels and propose to apply SSIM on the HSV (Hue, Saturation, Value) color model.

3 Preservation of Born-Digital Raw Images

Multiple interpretation steps are required in order to get a meaningful rendering of the raw data recorded by a camera through an appropriate rendering software [5]: (1) *Demosaicing* interpolates missing color information. (2) *White balance adjustment* defines the ratio between red, green, and blue that shall be rendered as neutral white. (3) *Colorimetric interpolation* maps the arbitrary color values to a certain color space. (4) *Tone mapping* delinearises the data and performs a mapping to the desired bit depth. (5) Finally, *Image enhancements* may include noise reduction, adjustment of color balance and saturation, antialiasing, and sharpening. This process is commonly referred to as *developing* the raw file, analogous to the chemical process of developing exposed film in a dark room. It can be influenced by multiple parameters such as exposure compensation, contrast, and sharpness. There is no *a priori* correct representation of a raw file, and the results of interpretation heavily depend on the parameters and the tool used. The photographer strives to create a subjectively optimal interpretation by manipulating the parameters, thus creating a parameter set. The final result is often a set of derivative files in an image format such as JPEG or TIFF. Given the raw file, the interpretation information, and software capable of interpreting raw files, it must be possible to reproduce an interpretation of the recorded scenery that is exactly identical to the photographer's view.

Our case study is based on a collection of approximately 30.000 photos from about five different cameras. The sample test set contains raw files recorded by different cameras from different vendors using different settings, ranging from a file from a 2002 camera to brand-new raw files. Based on the scenario outlined above, we consider possible preservation strategies. Keeping the original raw files will be sufficient for short periods of time only, since raw capable software is often not supporting older raw file formats. For example, the raw processing engine in *Mac OS* does not support raw files created by *Canon Powershot S45* in version 10.6.7 anymore. However, storing the data in an interpreted format will not preserve the authentic form of the digital image. Normalization of raw files to DNG is potentially an ideal solution combining authenticity with contextual interpretation information, but requires in-depth validation.

Based on the taxonomy described in [3], we can classify the relevant decision criteria that need to be evaluated. Most of the critical criteria belong to the category **outcome object**, which describes desired properties of the outcome of a preservation action. Here, we can distinguish between significant properties of the context and of the content, as discussed in depth below. The characteristics of the chosen **object format** include format adoption, openness of specification, standardisation, stability, compression, the relative file size tendency, embedding of additional metadata and preview images, and whether or not the content is already interpreted. For scalability reasons, **action runtime** criteria describing the performance and resource usage of conversion processes have to be considered. **Static action** characteristics include the ability to create error reports and performing file format verification and support for XMP sidecar files.[6]

The measurement of these categories has been discussed elsewhere [3], where outcome object properties were identified as the critical obstacle to scalability: They require in-depth analysis of content and generally constitute the dominating factor of trustworthy preservation decisions. The majority of these criteria refer to *significant properties*, i.e. *'the characteristics of digital objects that must be preserved over time in order to ensure the continued accessibility, usability, and meaning of the objects, and their capacity to be accepted as evidence of what they purport to record'* [6]. We will thus focus on the measures required to judge the degree of success in preserving these significant properties by performing fully automated QA instead of the manual validation prevailing to date.

4 Automated Quality Assurance

When transforming raw files into a different format, we need to ensure that this new representation is authentic, i.e. that it represents the same photograph with all its significant properties. Fundamentally, we need to distinguish between QA of content and of context related criteria. Table 1 lists the most important complementary metrics used to verify that content and context are fully retained.

Since the original and the migrated raw file are encoded in entirely different representations models, no formal equivalence relation can be defined. For example, during the course of DNG conversion, specific camera-dependent color matrices are embedded into the DNG file. When creating an interpretation of the DNG file, these are used to perform a mapping from the arbitrary camera color space to the well defined CIE XYZ space [1]. However, we may assume that *two raw files can be called equal if (and only if) they can be transformed into the same interpreted form.* In other words, the original and the migrated raw file can be called equal if they are rendered the same way in a reference rendering environment. In the absence of ground truth for verifying conversion and renderings, we need to rely on a standarized and widely used rendering environment.

The open source command line tool *dcraw*[7] supports full raw interpretation to different uninterpolated as well as fully interpreted derivative formats. For each conversion, dcraw can thus create two pairs of uninterpolated and interpreted image representations in 16-bit high quality TIFF. Subsequent comparison of these using a range of complementary metrics enables automated evaluation. The error-related criteria on the top rows of Table 1 verify that the pixel data remain unchanged. They apply to the technical representation of the uninterpolated image. But as discussed above, equality of uninterpolated data is not sufficient:

[6] These standalone files in Extensible Metadata Platform (XMP) format can include metadata which is added by raw capable software, mostly used for interpretation information and descriptive metadata.

[7] `http://www.cybercom.net/~dcoffin/dcraw/`

Table 1. Key content and context properties to be measured

Property	Description
relative AE	describes the percentage of pixels that differ when comparing the uninterpolated images
relative MSE	calculates the mean squared error between two uninterpolated images divided by the bit depth
SSIM Value	the raw data is transformed into fully interpreted 16-bit TIFF, and the brightness of the HSV representation is compared using SSIM
SSIM Saturation	SSIM is applied to the S-channel of HSV for comparison
SSIM Hue	SSIM is applied to the H-channel of HSV for comparison
Exif (exposure)	ExposureTime, FNumber, ExposureProgram, ExposureMode, ISOSpeedRating, SubjectDistance, Flash, FocalLength, ...
Exif (technical)	ImageWidth, ImageHeight, CFAPattern, Make, Model, ...
Exif (location)	GPSLatitude, GPSLongitudeRef, GPSAltitude, ...
Exif (generated)	Metadata added during migration to enable correct rendering: ColorMatrix, AsShotNeutral, BaselineExposure, ...
IPTC	Creator, ContactInfo, SceneCode, Location, City, Province-State, Country, Headline, Description, Keywords, ...
Dublin Core	Title, Subject, Description, Source, Coverage, Creator, Rights, Date, Format, Identifier, Audience, ...
Private Tags	e.g. ColorTemperature, WhiteBalance, InternalSerialNumber, AFPointsSelected, ...
XMP	XMP contains Exif and IPTC metadata as well as custom information as used by Adobe Camera RAW and others

Interpretation parameters, such as the color transformation to apply, constitute an integral part of the semantic structure of the photographs. We thus need to compare fully interpreted images as well. This comparison should be aligned to the human perception and thus relies on the SSIM metric, which is applied to the three HSV channels Hue, Saturation and Value.

An in-depth analysis of the applicability of the commonly used image comparison tool *ImageMagick compare*[8] to the presented case reveals a number of shortcomings. Color systems different from RGB are not readily supported, and parsing the results is error prone as there is no standardised output format. The tool is very sensitive when handling TIFFs. TIFF files with embedded metadata cause numerous warnings and can often not be compared successfully. Stripping metadata in advance may be an option, but increases the overhead and introduces potential error sources. Moreover, color profiles are not correctly handled, as can be demonstrated with test images that use a degenerated color space[9]. Comparing two different images that have the same visual appearance due to their ICC profile are wrongly classified as very different. Finally, the comparison does not support perception-based metrics such as SSIM. On the other hand, Java provides features to perform correct color management and Java Image IO API plugins that handle TIFFs. We have thus implemented all content related

[8] `http://www.imagemagick.org`; (version 6.6.1-5).
[9] `http://www.gballard.net/psd/go_live_page_profile/`

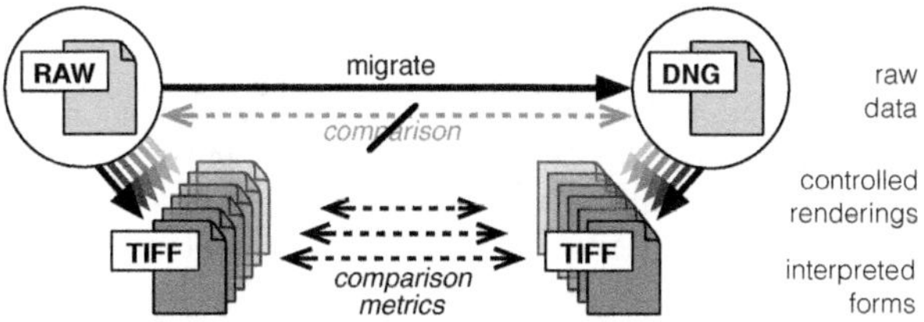

Fig. 1. Perceptual-level QA relies on controlled renderings for comparing objects

comparison metrics using the Java Advanced Imaging API[10]. The presented solutions are fully integrated in the upcoming release of the planning tool Plato.

The context related properties of a raw file contain different types of metadata information, such as information about the picture taking conditions (Exif[11] as well as private tags, i.e. vendor-specific metadata fields), technical descriptions of the equipment in use, and additional contextual information (IPTC[12], especially Dublin Core). Table 1 lists examples for relevant metadata criteria.

These properties are verified by extracting each field from the source and target objects and comparing the features, thus extending the extraction functionality described in [3]. Specifically, the QA methods included in Plato define evaluator plugins that can measure certain properties. These are defined by a URI-like identifier that contains the measurement domain, a unique property pathname and an optional metric [3]. The identifiers are linked to the corresponding criteria in Plato in order to automatically perform evaluation and report the results. In order to verify the preservation of metadata in a migration process, we have integrated the widely used *ExifTool*[13] for metadata extraction. Each property of interest is assigned an identifier, which is mapped to the corresponding metadata and processed by the evaluation plugin. Metadata can be grouped to cover related fields without the need for explicit descriptions of each specific field. For instance, for verifying if *makernotes* are retained, the comparison tool in its current configuration automatically checks whether all private tags are retained.

5 Experiment Results and Discussion

Space constraints our discussion on two common migration options from RAW to DNG, *Adobe DNG Converter 6.1* (ADC) and *digiKam 0.6.0-svn*. Table 2 shows exemplary results of automated evaluation performed on random samples of different source formats and contrasts these results with manual validation in Adobe Camera Raw (ACR) and dcraw. Problematic migration paths are shaded.

[10] `http://java.sun.com/javase/technologies/desktop/media/jai/`

[11] The 'Exchangeable image file format' describes structure and tags for use in images.

[12] The 'Information Interchange Model' by the 'International Press Telecommunications Council' defines structure and metatags for various file types.

[13] `http://www.sno.phy.queensu.ca/~phil/exiftool/`

Table 2. Conversion to DNG: Adobe DNG Converter (ADC) and digiKam (dKam)

	Canon Powershot S45 *(CRW)*		Canon EOS 5D Mk2 *(small CR2)*		Canon EOS 40D *(CR2)*		Nikon D80 *(NEF)*	
Criteria	ADC	dKam	ADC	dKam	ADC	dKam	ADC	dKam
dimensions (dcraw)	ident.	ident.	ident.	-	ident.	ident.	smaller	ident.
dimensions (ACR)	ident.	larger	ident.	-	ident.	larger	ident.	larger
relative AE	0%	0%	99.9%[a]	-	0%	0%	0%[b]	0%
relative MSE	0	0	0.00791[a]	-	0	0	0[b]	0
SSIM Value	1.0	0.84	0.85	-	1.0	1.0	0.92[b]	1.0
SSIM Saturation	1.0	0.67	0.85	-	0.91	1.0	0.64[b]	1.0
SSIM Hue	1.0	0.76	0.9	-	0.94	1.0	0.94[b]	1.0
manual (dcraw)[c]	ident.	diff.	simil.	-	simil.	ident.	simil.	ident.
manual (ACR)	ident.	diff.	ident.	-	ident.	diff.	ident.	diff.

[a] *dcraw* delivers a color image instead of a grayscale image
[b] comparing the cropped area of the DNG
[c] manual evaluation distinguishes between different, similar and identical

Generally speaking, the Adobe DNG Converter achieves much more consistent solid results in comparison to the free open source tool *digiKam* (dKam). A comparison of the SSIM distances shows that in several cases, differences are introduced that merit closer attention. In the case of the NEF file, the SSIM saturation metric goes as low as 0.64. However, manual evaluation does indicate that results are very satisfying, since the human perception is much less sensitive to changes in saturation than, for instance, to shifts in color space. While the exact correlation of manual evaluation results and thresholds for SSIM on the different channels will depend on context and risk aversion, the SSIM can provide a good indicator of problematic areas that require in-depth validation.

Migrating Canon S45 files with dKam results in files that are equal to the original on the uninterpreted level, but show strong differences when color interpolation is included. The high SSIM value indicates that the grayscaled image itself is retained, but the low SSIM saturation and hue values indicate strong differences in color. Even when manually correcting white balance in Adobe Camera RAW (ACR), reconstruction of the original color appearance is not possible. In-depth analysis reveals that the underlying reason is the incorrect color matrix in the Exif section, which does not properly represent the camera's profile.

The small CR2 file produced by the EOS 5D, which is downscaled to one fourth of the sensor's original size in the camera, is especially difficult to handle. It cannot be converted at all by dKam. For the file created by ADC, dcraw is not able to correctly extract the raw data without color interpolation. Although no official information exists, it is assumed that Canon's sRAW stores intensity values similar to a YCbCr model, whereas the classical RAW data is based on a RGB model [7]. A comparison of the color interpolated images indicates minor differences, but manual evaluation using ACR shows no noticeable differences.

For the EOS 40D photographs, comparing the uninterpolated data shows no differences between CR2 and DNG. For the interpreted rendering, there are minor differences in the DNG created by ADC. The large color-related SSIM score indicates that the difference is not perceivable, which is confirmed by both manual checks. The DNG produced by dKam gets perfect scores based on the renderings created by dcraw, but manual validation against the rendering of ACR reveals strong differences. A possible explanation is that the DNG file created by the ADC contains two Exif color matrices, whereas the DNG created by dKam contains only the second of these. Given two matrices, a raw converter may perform color mapping by interpolating between these matrices and using additional calibration matrices. Additionally, the white balance information is represented differently by ADC and dKam in this case. As a result, rendering the DNG created by ADC in dcraw produces an incorrect white balance adjustment, as does rendering the DNG created by dKam in ACR.

An unexpected level of difference occurs in the case of NEF files: The resulting images are not even of the same dimensions. The reason lies in the fact that color interpolation is very difficult at the sensor borders, as there are not enough neighbors to interpolate easily. In-camera software often discards these border pixels, while raw processing software may use more processing-intensive algorithms and take those pixels into account. This may result in differently sized images, as seen for the NEF file in our case study. For the DNG created by ADC, the un-interpolated image exported by dcraw is slightly smaller than the export of the original NEF. Only by ignoring the extra rows contained in the original is a direct comparison possible. The uninterpreted image created by ADC is then equal to the original. The color interpolated files still differ, since the DNG created by ADC is slightly brighter. Manual validation using the interpretation rendered by ACR, however, shows that the DNG created by ADC is identical to the original NEF, whereas the DNG created by dKam is slightly brighter. The most reasonable explanation is that ACR correctly respects the exposure metadata, whereas dcraw ignores it.

From the experiment results, a number of observations can be drawn. The correctness and reliability of the intermediate representations used for QA is a crucial element in the evaluation chain. Yet, several shortcomings of existing tools are evident. Nevertheless, we can use each of these tools to discover potential failures in migration. Using the SSIM metrics, we can further distinguish between minor and major differences and classify the difference in terms of structure related (low SSIM Value) and color related (low SSIM Saturation, low SSIM Hue). Comparison metrics can be used more successfully to *falsify* conversions than to *verify* them: False negatives are more likely than false positives. Furthermore, different renderings are required to uncover the variety of potential errors. In this sense, it is advisable to rely on a combination of metrics and interpretation processes to uncover hidden gaps in conversion and evaluation.

6 Conclusion and Outlook

This article has described an automated preservation experiment on born-digital raw photographs that combined controlled experimentation, integration of existing metadata extraction and image comparison tools, and implementations of state of the art image comparison algorithms. In the case of photographs, metrics and algorithms exist to validate conversion processes, but a direct comparison of the artifacts resulting from preservation actions to the original content is not possible. Hence, we rely on perceptual-level QA to validate conversions. By combining controlled rendering paths and comparison metrics, it is in principle feasible to fully automate preservation operations on non-trivial content. This can be done without violating fundamental requirements such as a full validation of the significant properties of content. However, this validation approach relies on intermediate processing steps with varying reliability. If the interpretation step itself contains conversion errors, this may result in false negative reports of differences between identical files or, even worse, false positive case where a converter creates two identical images (e.g. all black) out of different images. The former case may be identified by cross-checking against different renderings, the latter by checking image properties such as histograms for unusual patterns. Although confidence in the results can be improved by comparing the results of multiple conversion tools, absolute certainty would depend on ground truth. As in many comparable cases in digital preservation, this utter lack of well understood test corpora with known ground truth is a significant impediment against full validation of interpretation and processing algorithms [3]. Such corpora would be a massive enabler for improvement of digital preservation processes.

Current work is focused on developing a modular service deployment of the employed QA mechanisms to prepare large-scale experimentation on the entire data set using a workflow engine. This will be combined with a detailed assessment of the correlation between SSIM and manual validation.

Acknowledgements. The authors wish to thank Michael Kraxner, Gi-Won Chung and Sara Pourmanouchehri for carrying out some of the work that led to this article. Part of this work was supported by the 7th Framework Program, IST, through the SCAPE project, Contract 270137.

References

1. Digital negative (DNG) specification. version 1.3.0.0. Technical report, Adobe Systems Incorporated (June 2009)
2. Becker, C., Kulovits, H., Guttenbrunner, M., Strodl, S., Rauber, A., Hofman, H.: Systematic planning for digital preservation: Evaluating potential strategies and building preservation plans. IJDL 10(4) (December 2009)
3. Becker, C., Rauber, A.: Decision criteria in digital preservation: What to measure and how. JASIST 62(6) (2011)
4. Bennett, M.J., Wheeler, F.B.: Raw as archival still image format: A consideration. In: Haag, D. (ed.) Archiving 2010, pp. 185–193. Society for Imaging Science and Technology, The Netherlands (2010)
5. Fraser, B., Schewe, J.: Real World Camera Raw with Adobe Photoshops CS4. Peachpit Press, Berkeley (2009)
6. Grace, S., Knight, G., Montague, L.: InSPECT Final Report. InSPECT (December 2009), `http://www.significantproperties.org.uk/inspect-finalreport.pdf`
7. Kerr, D.A.: The canon sraw and mraw output formats. The Pumpkin (1.1) (March 2010)

8. Kulovits, H., Rauber, A., Brantl, M., Schoger, A., Beinert, T., Kugler, A.: From TIFF to JPEG2000? Preservation planning at the Bavarian State Library using a collection of digitized 16th century printings. D-Lib Magazine 15(11/12) (2009)
9. Okarma, K.: Colour image quality assessment using structural similarity index and singular value decomposition. In: Bolc, L., Kulikowski, J.L., Wojciechowski, K. (eds.) ICCVG 2008. LNCS, vol. 5337, pp. 55–65. Springer, Heidelberg (2009)
10. van der Knijff, J.: JPEG 2000 for long-term preservation: JP2 as a preservation format. D-Lib Magazine 17(5/6) (2011)
11. Wang, Z., Bovik, A.C.: Mean squared error: love it or leave it? - a new look at signal fidelity measures. IEEE Signal Processing Mag. 26(1), 98–117 (2009)
12. Wang, Z., Bovik, A.C., Sheikh, H.R., Simoncelli, E.P.: Image quality assessment: From error visibility to structural similarity. IEEE TIP 13(4), 600–612 (2004)

Image Tagging by Exploiting Feature Correlation

Xiaoming Zhang and Zhoujun Li

School of computer, Beihang University, Beijing, 100191, China
yolixs@cse.buaa.edu.cn, lizj@buaa.edu.cn

Abstract. Image tagging is a task that automatically assigns the query image with semantic keywords called tags. Since tags and image visual content are represented in different feature space, how to merge the multiple features by their correlation to tag the query image is an important problem. However, most of existing approaches merge the features by using a relatively simple mechanism rather than fully exploiting the correlations between different features. In this paper, we propose a new approach to fusing different features and their correlation simultaneously for image tagging. Specifically, we employ a Feature Correlation Graph to capture the correlations between different features in an integrated manner, which take features as nodes and their correlations as edges. Then, a revised probabilistic model based on Markov Random Field is used to describe the graph for evaluating tag's relevance to query image. Experiments on large real-life corpuses collected from Flickr indicate the superiority of our proposed approach.

Keywords: Image Tagging, Feature Correlation, Image Retrieval.

1 Introduction

With the rapid advance in the technology of digital cameras and the Internet, there is an explosive growth in the amount of available image data in our daily lives, which requires an effective retrieval and browsing technology for large volume of images [1]. With the prevalence of photo sharing websites such as Flickr and Piscasa, which host numerous of digital images with user provided keywords called tags, tag-based image retrieval and browsing have become potentially popular. As a result, there is a growing research interest in exploiting increasingly available social tags to facilitate high quality image tagging algorithms. Compared with tagging text documents, tagging multimedia objects generally faces more problems, e.g., the so-called "semantic gap" between the lower level visual features and the higher level semantic features. Since social images are usually associated with other information such as tags and description and so on, there are always some correlations existing between features with the same or different types, i.e., semantic features and visual features, which can then provide useful clues to dealing with the challenge described above.

A few recent works have been proposed to utilize feature correlations. One paradigm mainly exploits the inter-relations between different tags or objects [2]. The other paradigm uses a semantic context model to integrate the semantic correlations and correlation between semantic concepts and images by treating an image as a

C. Xing, F. Crestani, and A. Rauber (Eds.): ICADL 2011, LNCS 7008, pp. 50–59, 2011.

whole [3]. However, this paradigm has a limit perform in social media environment where the objects are with wide variety, i.e., two semantically similar images may have different background or two visually similar images may be semantically diverse.

Unlike previous approaches, we fully merge multiple features through their correlations in an integrated manner for tag relevance evaluation. Specifically, we first introduce a new structure Feature Correlation Graph (*FCG*) to fuse different types of features with their correlations. Each feature is represented as a node in the graph, and two feature nodes are linked by an edge if there has correlation between them. Each tag or query image is represented by a *FCG*. To evaluate the relevance of a candidate tag t_o to the query image I_q, we employ a probabilistic model, i.e., Markov Random Field [4], to calculate the joint distribution probability of such a graph that represents compatibility of tag t_o given the features of image I_q. A large value of joint distribution probability means high relevance of the tag to the query image. Then, the tags with the greatest relevance values are assigned to the query image. The experiment results on real-life dataset how the effectiveness of our approaches for image tagging.

2 Related Works

A significant amount of work has been devoted to the problem of image tagging. Some works use model-based learning method to tag the query images. Wu *et al.* [5] formulate image tagging as multi-label classification which is treated as a regression model with a regularized penalty. Shen *et al.* [2] formulate image tagging as object classification, and a multi-task structured SVM algorithm is developed to leverage both the inter-object correlations and the loosely-tagged images for achieving more effective training of a large number of inter-related object classifiers. Timothée *et al.* [6] present a graph-based random walk scheme founded on the GCap method to perform automatic image tagging. To exploit the concept correlation, some pioneering works have been done recently. Xiang *et al.* [3] propose a discriminative Conditional Random Field (CRF) model for semantic context modeling in image tagging, which is built over semantic concepts and captures the interactions between semantic concepts. Lei *et al.* [7] propose a multi-modality recommendation model based on both tag and visual correlation of images.

Some other works use the example-based or knowledge-based approach to tag the query images. The DCMRM [8] is a dual cross-media relevance model for automatic image tagging, which estimates the joint probability by the expectation over words in a pre-defined lexicon. Li *et al.* [9] propose a neighbour voting tagging method, in which the tags which appear in neighbours frequently are selected to tag the query image. Siersdorfer *et al.* [10] propose an automatic video tagging method based on video duplicate and overlap detection, and the assigned tags are derived from the videos which are overlap with the test video. Liu *et al.* [11] first estimate the initial relevance scores for tags based on probability density estimation, and then perform a random walk over a tag similarity graph to refine the relevance scores.

3 Problem Formulation

For each image, there are two types of feature vector to represent it i.e., semantic feature vector and visual content feature vector. The semantic feature vector is a binary vector whose element indicates the presence or absence of the corresponding tag in the tag vocabulary $T=\{t_1,t_2,\ldots, t_{|T|}\}$. For the visual content, there are different types of low-level features can be used, such as color, texture and edge. While these features are not effective for evaluation their correlation with semantic features, we adopt middle level features "visual word" to represent visual content, which are generated by clustering image blocks and can perform better than raw features as reported in [12]. Then the element of visual content feature vector indicates the weight of the corresponding "visual word" in the visual word vocabulary $V=(v_1, v_2,\ldots, v_{|V|})$. Thus, within a set of social images, there exist various types of correlations between features, i.e., two different tags are semantically similar or have frequent co-occurrence, two visual words are visually similar, a tag and a visual word co-occur frequently. Since tags and visual features belong to different representation space, how to fully utilize the feature correlation to evaluate the relevance of a tag to the query image which is only represented by visual features is an import problem.

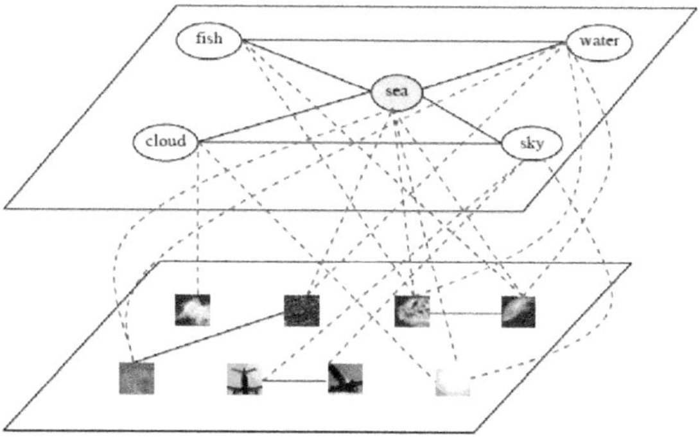

Fig. 1. An example of FCG built for tag "sea"

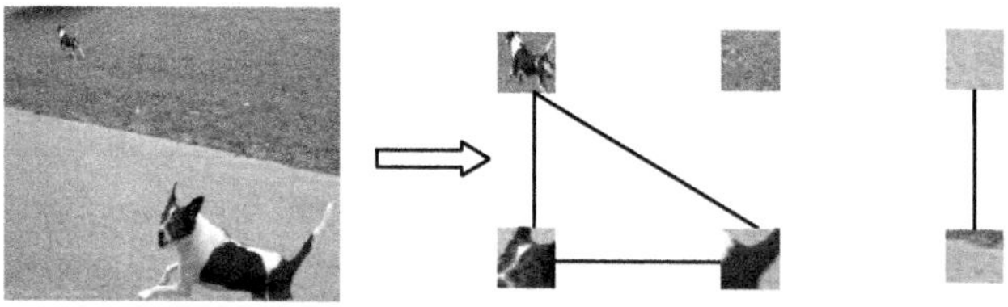

Fig. 2. An example of FCG built for an image

In this paper, we fuse different features into a uniform *Feature Correlation Graph (FCG)* $G^T=(S, E)$ by integrating various features through their correlation, where S is a set of nodes which represent both of the semantic features and visual features and E is the set of edges which denote the correlations between feature nodes. Then, the relevance of a tag to an image can be measured by the joint distribution probability of the tag and the visual features of this image on the graph. However, for a large dataset, it is complexity to build a global graph which contains all of the features to evaluate tag relevance, and it is also not necessary to build such a global graph because that most of the features have no correlation with the given tag. Thus, a local *FCG* G_i is built for each tag t_i to fuse its correlated tags with all of the correlated visual features through their correlations as shown in Fig. 1. For each query image, a *FCG* G_q is built to integrate its visual features and their correlations as shown in Fig. 2. Then, we define the problem of image tagging based on feature fusion as following:

For each tag t_i ($1<i<|T|$) in the training tag vocabulary, a *FCG* representation is built in the training process. Given a query image I_q, its *FCG* representation is built first. We calculate a joint distribution probability based on their *FCGs* to measure the relevance of t_i to I_q, and then all the tags are ranked by their relevance value and the top ranked tags are assigned to the query image.

4 Image Tagging

In this section, we first introduce a new structure named Feature Correlation Graph (*FCG*) to represent the candidate tags and query images, and then we design a probabilistic model to evaluate tag's relevance to the query image.

4.1 Feature Correlation Graph

The *FCG* integrates multiple features and their correlations into an undirected graph. Each feature is represented as a node in the graph. Two nodes are linked with an edge if and only if they are correlated, and the correlation between features is determined by the analysis of statistical information of the training image set. For a given tag t_i ($i=1,2,\ldots,|T|$), we take itself and its correlated tags as its semantic features, and the visual features correlated to all of those tags are taken as its visual features. Then, its *FCG* graph is built on all of these features, and the node which represents the tag t_i itself is regarded as the "virtual center" node. Fig. 1 illustrates an example of *FCG* built for tag "sea". Fig. 2 illustrates an example of *FCG* built for an query image.

There exist two kinds of correlations between features in social image environment, i.e., the inter-type and intra-type correlation. Given two nodes d_1 and d_2, we use a function *Cor*(. , .) to measure the correlation between them, and draw an edge between them if $Cor(d_1, d_2)$ is larger than a trained threshold.

Intra-type correlation represents the relation between features with the same type. In this paper, there are two types of intra-type correlation.

Correlations between semantic feature nodes: There are several approaches to determine whether two tags are correlated i.e., WordNet-based approach and the approach based on term co-occurrence. We use the combination as:

$$Cor(t_i,t_j)=\alpha * Sim(t_i,t_j)+(1-\alpha) * Co(t_i,t_j) \quad (1)$$

where $Sim(t_i, t_j)$ is the WordNet based Jiang and Conrath Measure (*JNC*) [13], and $Co(t_i, t_j)$ is the term co-occurrence correlation measure introduced in [2] and α is a smoothing parameter.

Correlation between visual feature nodes: We use the visual similarity between the two corresponding visual features. In this paper, each visual word is a 160-D feature vector, and the traditional Euclidean distance function is used to measure the similarity between two visual words.

Inter-type correlation represents the relation between features with different types. For each pair of nodes which represent different types of features, i.e., d_1 and d_2, they can be represented with vectors $\vec{d_1}$ and $\vec{d_2}$. Each dimension of those vectors represents a corresponding image in the training set, and its value equals to the frequency of d_1 or d_2 appearing in this image. The inter-type correlation strength can be estimated by calculating the cosine similarity between their vectors:

$$Cor(d_1, d_2) = \frac{\vec{d_1} \cdot \vec{d_2}}{|\vec{d_1}| \times |\vec{d_2}|} \tag{2}$$

4.2 Probabilistic Model for Tag Relevance Evaluation

Given a query image I_q, we can build a *FCG* G_q for it, and all its feature nodes are represented as $<d_1, d_2, ..., d_n>$. For each candidate tag t_i $(i=1,2,...,|T|)$ in the tag vocabulary, it has its own *FCG* representation G_i, and a temp *FCG* representation G is also built by linking all of the nodes $<d_1, d_2, ..., d_n>$ in G_q with an additional node representing the candidate tag t_i. Since the *FCG* structure is an undirected graph, the calculation of joint distribution probability of G can be solved by treating this graph as a Markov Random Field (MRF) [3]:

$$P(G) = P(t_i, d_1, d_2, ..., d_n) = P(t_i, I_q) = \frac{1}{Z_\Lambda} \prod_{c \in C(G)} \varphi(c; \Lambda) = \frac{1}{Z_\Lambda} \prod_{c \in C(G)} \exp(\lambda_c f(c)) \tag{3}$$

where $C(G)$ is the set of cliques in G, and each $\varphi(\ ;\ )$ is a non-negative potential function over cliques, and $f(c)$ is a real-valued feature function over clique c and λ_c is a parameter which is the weight of that particular feature function. Z_Λ normalizes the distribution. In this paper, the term "clique" refers to the complete subgraph of G. We constrain that each clique is a complete subgraph of G with the tag node and at least one feature node d_i, and each two "cliques" have no other intersecting nodes except the tag node. We use the posterior probability $P(t_i/I_q)$ to interpret the relevance of tag t_i to query image I_q:

$$\begin{aligned} P(t_i | I_q) &= \frac{P(t_i, I_q)}{P(I_q)} \\ &\propto \log P(t_i, I_q) - \log P(I_q) \\ &\propto \sum_{c \in c(G)} \log \varphi(c; \Lambda) = \sum_{c \in c(G)} \lambda_c f(c) \end{aligned} \tag{4}$$

It indicates that the joint-distribution probability of G is proportional to $P(t_i/I_q)$. Thus we can use the joint-distribution probability to rank the candidate tags. This joint distribution probability relies on the structure of graph G and the potential function $\varphi(c; \Lambda)$ and the parameter set Λ. We rewrite this formula as following:

$$P(t_i \mid I_q) \propto \sum_{c\in C(G)} \lambda_c f(c) = \sum_{c\in C(G)} \phi(c) \tag{5}$$

This formulation indicates that the posterior probability can be calculated by the sum of $\phi(c)$ over all the cliques in G. For presentation simplicity, we also name $\phi(c)$ as the potential function of clique c. The intuition behind this approach can be concluded from following aspects. If there exists correlation between a subset of visual features in the query image I_q, the graph G will has a corresponding clique c. When calculating the potential function value of c, the features and their correlation are all taken into consideration. c contributes to the relevance of a candidate tag t_i if it also appears in G_i or it has relation with other nodes in G_i. However, we don't aim to model the exact joint distribution $P(G)$ here. Instead, we attempt to approximate the distribution using a generalized form and use it for tag ranking.

These functions can be regarded as compatibility functions, and high values should be assigned to the clique settings that are more "compatible" with each other under the given distribution. In this paper, these functions attempt to abstract the idea of feature correlation, and they are used to measure the probability of the cliques appearing in feature set of the candidate tag. Given a clique $c=\{d_1, d_2,\ldots, d_{|c|-1}, t_i\}$ in G, where d_i represents nodes decomposed from query image I_q, and t_i refers to the candidate tag. The potential function need not only consider the features in the clique, but also their correlation with other features in G_i which is built for the tag t_i. Thus, the corresponding potential function is defined as following:

$$\phi(c) = \lambda_c f(c) = \lambda_c \cdot W(c) \cdot P(d_1, d_2, \ldots, d_{|c|-1} \mid t_i) \tag{6}$$

where $W(c)$ is the weight of clique c itself, and it is independent of the candidate tag t_i. $P(d_1, d_2,\ldots, d_{|c|-1}/t_i)$ is used to estimate the probability that those feature nodes in clique c also appear together in G_i:

$$\begin{aligned} P(d_1, d_2, \ldots, d_{|c|-1} \mid t_i) &= \eta \frac{Coc(d_1, d_2, \ldots, d_{|c|-1}, t_i)}{N(t_i)} \\ &+ (1-\eta) \frac{\sum_{d_k\in c} \sum_{d_j\in(G_i-c)} Cor(d_k, d_j) \times Cor(d_j, t_i)}{(|c|-1)\times(|G_i - c|)} \end{aligned} \tag{7}$$

where $N(t_i)$ is the sum of the co-occurrence frequencies of t_i with other features in G_i. Obviously, $P(d_1, d_2, \ldots, d_{|c|-1}/ t_i)$ is composed of two components. The first component is $Coc(d_1, d_2, \ldots, d_{|c|-1}, t_i)$ which refers to the co-occurrence frequency of features $(d_1, d_2, \ldots, d_{|c|-1})$ and t_i. The second component measure the contribution of the rest features in G_i, which multiply their correlations with the features in clique c by their correlations with the "virtual center" node t_i. The function $Cor(.)$ is used to evaluate the correlation between two features, which may vary from different types of features, e.g., inter-type and intra-type. The smoothing parameter η is used to trade-off the effects of two components.

Usually, the clique itself has different importance, as the strong correlation between features in a clique yields more semantic information. Thus we multiply the probability with a weight *W(c)* which capture the semantic information of the clique *c* as formula (8). A strong connection between a set of features reflect in that those features co-occur frequently in the images which contain the objects related to them, while a weak connection between a set of features reflect in that the occurrences of those features are relatively independent. Thus we estimate the correlation strength of the features in a clique as following:

$$W(c) = \frac{\sum_{i=1}^{|I|} \frac{fre(d_1, d_2, ..., d_{|c|-1} \mid I_i)}{ave(d_1, d_2, ..., d_{|c|-1} \mid I_i)}}{\log N_{oc}(d_1, d_2, ..., d_{|c|-1}) + 1} \tag{8}$$

where $|I|$ is the total number of training social images, $fre(d_1, d_2, ..., d_{|c|-1}/ I_i)$ is the co-occurrence frequency of features $(d_1, d_2, ..., d_{|c|-1})$ in image I_i, $ave(d_1, d_2, ..., d_{|c|-1}/ I_i)$ is the average of the appearance frequencies of those individual features in image I_i, $N_{oc}(d_1, d_2, ..., d_{|c|-1})$ is total number of images which contain features $(d_1, d_2, ..., d_{|c|-1})$ within the image set.

Given the parameterized joint distribution and a set of potential functions, the final step is to set the parameter values. As discussed above, the parameter λ_c is varied for different clique. In the traditional generative models, the maximum likelihood and maximum a posteriori estimation are used to estimate the parameters. However, the number of cliques in our literature may be very large in social image environment because of various visual content and high-dimensionality of features with their rich correlations, and also the event space is extremely large or even infinite depending on how it is defined. Therefore, training λ_c of MRF for each individual clique is nearly infeasible in reality due to the computation and sample limitation. Usually, λ_c is employed with some constraints in the literature. In our circumstance, cliques are different in their number of nodes and their features with the corresponding correlations. We train the parameter set Λ using the training strategy introduced in [13], in which the model is trained by directly maximizing mean average precision adopting a simple coordinate-level hill climbing search.

5 Evaluations

To validate the effectiveness of our proposed approach on real-world social image datasets, we conduct extensive experiments on the image dataset NUS-WIDE [15], and the superiority of our method is demonstrated by comparing it with other baseline algorithms.

5.1 Dataset and Baselines

The NUS-WIDE dataset [15] contains 269,648 images and the associated 5,018 unique tags all of which exist in WorldNet. All of the images in this dataset are crawled from the photo sharing website Flickr by using its public API. For evaluation, we split the dataset of image into two disjoint partitions: a training set with 150,000 images and a test set with the remaining images. The underlying image diversity and complexity make these dataset a good testbed for real-world real life social image tagging experiments. To get the "visual words", we divide each image into uniformly distributed 16×16 patches, and raw visual features are extracted for each patch. The raw visual feature include 32-D color histogram and 128-D SIFT descriptor. Then raw visual features are converted to 4024 visual words by k-means clustering. Thus, the visual content of each image can be represented by a group of visual words.

We name our proposed algorithm FCGTag, and the other four algorithms i.e., MMCRF[3], DCMRM [8], NVTag [9] and RankTag [11] are evaluated for comparison. The MMCRF use the Conditional Random Field to model semantic context and treat an image as a whole observation. Image tagging is taken as an inference problem in CRFs which is to find the most compatible configuration of the sites with a given observation. The DCMRM proposes a dual cross-media relevance model for automatic image tagging, which estimates the joint probability by the expectation over words in a pre-defined lexicon. With respect to their prior distribution among all images, NVTag assign the tags which frequently appear among the nearest neighbors to the query image. The RankTag performs a random walk over a tag similarity graph to refine the relevance scores, and the top ranked tags associated with the visually similar images are assigned to the query image. Traditional performance evaluation criteria i.e., F1 scores, average precision *APT* and recall *ART* of per-tag are used to evaluate tagging performance. Also the average precision *API@n* and recall *ARI@n* of per-image with *n* result tags are used.

5.2 Experiment Results

Fig. 3 and Fig. 4 show the *API@n* and *ARI@n* with *n* varied from 1 to 6. To compare the average precision and recall of tag, we fix the number of result tags for each query image with 6 and obtain the *APT* and *ART* by averaging the precision and recall on the tags associated with the query images. Fig. 5 shows the average precision and recall of tag. From these figures, we can see that FCGTag archives encouraging improvements on these measures, in which the average per-image precision and recall are greatly improved. Several conclusions can be drawn from the figures. First, on the large-scale real-world image dataset, our proposed algorithm FCGTag outperforms the other algorithm significantly on different number of result tags per query image. For example, with 6 result tags, FCGTag has an improvement about 15% over MMCRM and DCMRM , 20% over RankTag and NVTag. The improvement is supposed to stem from the fact that our algorithm integrates multi features and their correlations which can select the more relevant tags. However, other methods are mainly based on the visual similarity or tag correlation. Second, with the increasing number of result tags per image, the precision of all algorithms decease. This is because that the probability of including noisy tags increase with the number of result tags increasing. However, our proposed algorithm keeps higher average precision and recall value than other four algorithms over all numbers of result tags. Fig. 15 gives some real examples of image tagging on NUS-WIDE.

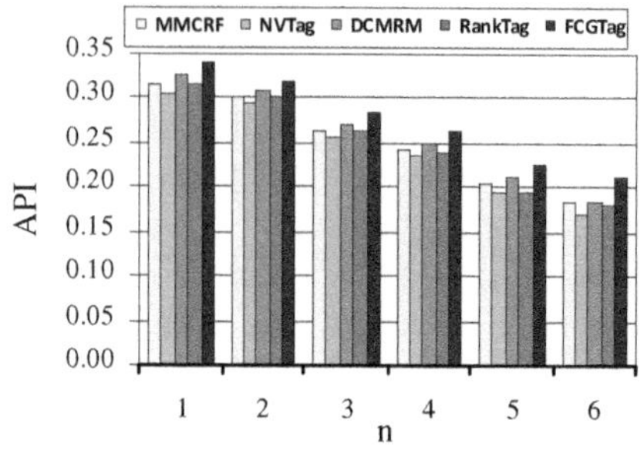

Fig. 3. Comparison of API@n

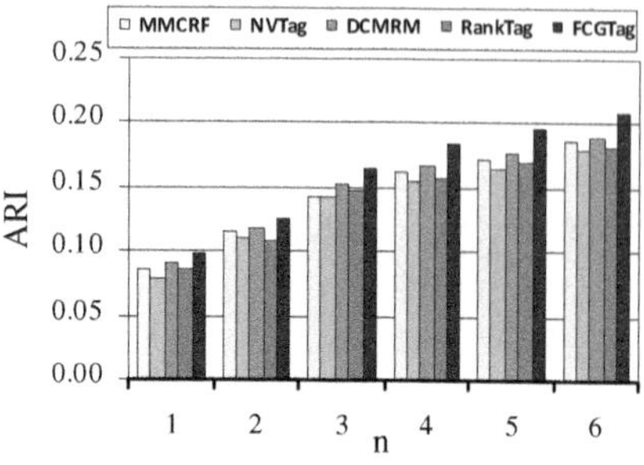

Fig. 4. Comparison of ARI@n

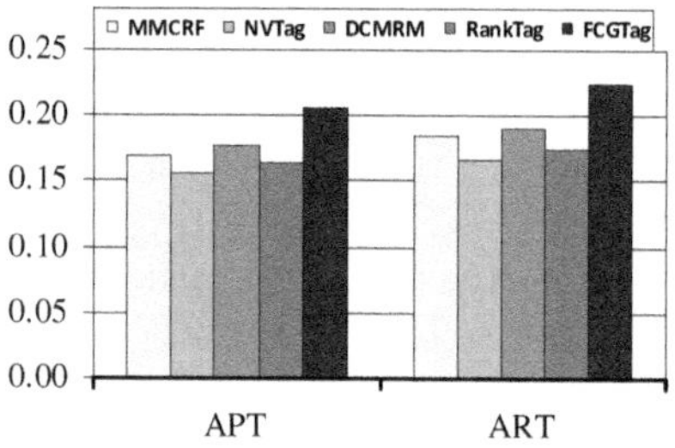

Fig. 5. Comparison of APT and ART

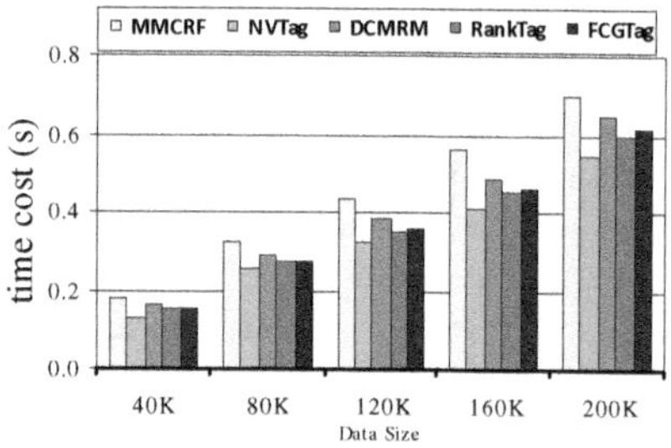

Fig. 6. Comparison of time efficiency

Fig. 6 illustrates the time cost per query image with varied scale of training set from 40K to 200K. According to the experiment results, the NVTag is the most efficient because it directly selects the tags from those associated with the neighbors. The second efficient one is RankTag which applies RWR on the tags associated with the neighbors. In the realization, we first retrieve a set of tags which have correlation with the cliques of the query image, which decrease the number of candidate tags needed to be processed. Moreover, the evaluation of tag relevance is linear complexity given the number of cliques within the query image. Thus, our algorithm is more efficient than the other two algorithms and is only a little less efficient than RankTag. We think this trade-off is worthwhile because the time efficiency can be potentially increased by adopting parallel algorithm and distributed architectures. The forth efficient one is DCMRM. It is quadratic complexity to estimate the joint probability by the expectation over the whole lexicon. While the model-based algorithm MMCRF has to infer tags from the whole training dataset, which makes it the least efficient.

6 Conclusion

In this paper, we propose and validate a new algorithm introduce a new structure Feature Correlation Graph FCG to fuse multiple feature with their correlation. We model each tag and query image as a feather correlation graph representation FCG which can integrate multiple features through the correlations among them. Using this model, we propose a revised probabilistic model based on MRF to effectively measure the relevance of a tag to the query image. We compare our approach with other existing approaches on two large real-world datasets. The experiment results on both of the dataset demonstrate the advantage of our approach over existing approaches.

References

1. Naaman, M., Song, J., et al.: Automatic Organization for Digital Photographs with Geographic Coordinates. In: Proc. ACM/IEEE JCDL (2004)
2. Shen, Y., Fan, J.P.: Leveraging Loosely-Tagged Images and Inter-Object Correlations for Tag Recommendation. In: Proc. ACM Multimedia (2010)
3. Xiang, Y., Zhou, X.D., Liu, Z.T., Chua, T.-S., Ngo, C.-W.: Semantic Context Modeling with Maximal Margin Conditional Random Fields for Automatic Image Annotation. In: Proc. CVPR (2010)
4. Kindermann, R., Laurie Snell, J.: Markov Random Fields and Their Applications. American Mathematical Society, Providence (1980)
5. Wu, F., Han, Y.H., Tian, Q., Zhuang, Y.T.: Multi-label Boosting for Image Annotation by Structural Grouping Sparsity. In: Proc. ACM Multimedia (2010)
6. Bailloeul, T., Zhu, C.Z., Xu, Y.h.: Automatic Image Tagging As A Random Walk With Priors On The Canonical Correlation Subspace. In: Proc. ACM MIR (2008)
7. Lei, W., Linjun, Y., Nenghai, Y., Hua, X.S.: Learning to tag. In: Proc. ACM WWW (2009)
8. Liu, J., Wang, B., Li, M.J., Li, Z.W., Ma, W.Y., Lu, H.Q., Ma, S.D.: Dual Cross-Media Relevance Model for Image Annotation. In: Proc. ACM Multimedia (2007)
9. Li, X., Snoek, C.G.M., Worring, M.: Learning social tag relevance by neighbor voting. IEEE Transactions on Multimedia 11(7), 1310–1322 (2009)
10. Siersdorfer, S., San Pedro, J., Sanderson, M.: Automatic Video Tagging using Content Redundancy. In: Proc. ACM SIGIR (2009)
11. Liu, D., Wang, M., Hua, X.S., Zhang, H.J.: Tag Ranking. In: Proc. ACM WWW (2009)
12. Wu, L., Li, M.J., Li, Z.W., Ma, W.Y., Yu, N.H.: Visual language modeling for image classification. In: Proc. MIR (2007)
13. Jiang, J., Conrath, D.: Semantic similarity based on corpus statistics and lexical taxonomy. In: Proc. International Conference on Research in Computational Linguistics (1997)
14. Metzler, D., Bruce Croft, W.: A markov random field model for term dependencies. In: Proc. ACM SIGIR (2005)
15. Chua, T., Tang, J., Hong, R., Li, H., Luo, Z., Zheng, Y.: Nus-wide: A real-world social image database from national university of singapore. In: Proc. CIVR (2009)

Semi-supervised Bibliographic Element Segmentation with Latent Permutations

Tomonari Masada[1], Atsuhiro Takasu[2], Yuichiro Shibata[1], and Kiyoshi Oguri[1]

[1] Nagasaki University, 1-14 Bunkyo-machi, Nagasaki-shi, Nagasaki, Japan
{masada,shibata,oguri}@nagasaki-u.ac.jp
[2] National Institute of Informatics, 2-1-2 Hitotsubashi, Chiyoda-ku, Tokyo, Japan
takasu@nii.ac.jp

Abstract. This paper proposes a semi-supervised bibliographic element segmentation. Our input data is a large scale set of bibliographic references each given as an unsegmented sequence of word tokens. Our problem is to segment each reference into bibliographic elements, e.g. authors, title, journal, pages, etc. We solve this problem with an LDA-like topic model by assigning each word token to a topic so that the word tokens assigned to the same topic refer to the same bibliographic element. Topic assignments should satisfy contiguity constraint, i.e., the constraint that the word tokens assigned to the same topic should be contiguous. Therefore, we proposed a topic model in our preceding work [8] based on the topic model devised by Chen et al. [3]. Our model extends LDA and realizes unsupervised topic assignments satisfying contiguity constraint. The main contribution of this paper is the proposal of a *semi-supervised* learning for our proposed model. We assume that at most one third of word tokens are already labeled. In addition, we assume that a few percent of the labels may be incorrect. The experiment showed that our semi-supervised learning improved the unsupervised learning by a large margin and achieved an over 90% segmentation accuracy.

1 Introduction

Bibliographic element segmentation is an important problem when we build a large publication database from "raw" references, i.e., the references that are not segmented into bibliographic elements, e.g. author names, paper title, journal name, pages, publication year, etc. By scanning reference sections of printed articles or by crawling publication data from researchers' Web sites, we can obtain raw references as unlabeled word token sequences. Our problem is to segment each raw reference into bibliographic elements. Once bibliographic elements are identified, relevant tasks, e.g. reference alignment, reference deduplication, etc, will become easy. Figure 1 shows a segmentation example our method provides.

In a preceding paper [8], we proposed a completely new approach for bibliographic element segmentation. Existing works solve this problem by modeling the appearance order of bibliographic elements as *state transition* [2,7,10,11,6,5]

C. Xing, F. Crestani, and A. Rauber (Eds.): ICADL 2011, LNCS 7008, pp. 60–69, 2011.

```
A S Williams S V Ponchillia | Psychosocial sequelae of visual loss in diabetes. The | Diabetes educator | 675-6
T May | 1998 | Assessing competency without judging merit. The Journal of | clinical ethics | 247-57
D F Fiorino D Treit J Menard L Lermer A G Phillips | 1998 | Is barakol | anxiolytic? Behavioural pharmacology | 375-8
I Perera B K Yeo S M Ko E H Kua | 1998 | Telephone counselling in | psychiatry. Singapore medical journal | 488-90
H Kojima R Blake | 1998 | Role of spatial and temporal coincidence in depth organization. | Perception | 541-52
D Hussain J D Gass | 1998 | Idiopathic central serous chorioretinopathy. Indian | journal of ophthalmology | 131-7
T Hain | 1998 | Working in harmony: the role of a musician in residence. | Paediatric nursing | 28-9
B A Krumme | 1998 | Experiences with humanitarian interventions in crisis areas | Krankenpflege Journal | 486-91
J Kellett | 1998 | Reflections on the practice of acute Irish hospital medicine. | The hospitalist | 4
Histoid | leprosy with episcleral nodule--after MDT-MB. | Indian journal of leprosy | 411-2 | M A Rajan
J G Barranco | 1998 | Glucose control guidelines: current concepts. Clinical | nutrition (Edinburgh, Scotland) | 7-17
G Sermonti L Di Bella | 1998 | Di Bella--candidate failed before the exam? | Rivista di biologia 363-5, | 367-9
9 The possessive form for a plural compound noun. | Nurse author & editor | A Taylor S Y Chao | 1998
R Yelsangikar | 1998 | Status of poliomyelitis after pulse polio immunization. Indian | pediatrics | 480-1
Decision making by emergency nurses in triage assessments. | 184-91 | Accident and emergency nursing | 1998 | J Cioffi
A A de Sousa | 1998 | Carotid endarterectomy under regional anesthesia. | Neurologia medico-chirurgica | 279-83
By-laws | of the Mexican | Association of Gastroenterology Revista de gastroenterología de México | 234-49
1587-8 | Interleukin 10 in febrile patients and patients with sepsis. | 1998 | A Bouchama M Hammami | Lancet
J C Patel | 1998 | Melatonin. Pineal gland hormone--a brief review. Indian | journal of medical sciences | 567-8
P Mårin S Arver | 1998 | Androgens and abdominal obesity. Baillière's clinical | endocrinology and metabolism | 441-51
E Yamada | 1998 | The clinical development of KOMI charts Sogo | kango. Comprehensive nursing, quarterly | 49-58
R M Jeresaty | 1999 | Mitral-valve prolapse. The New | England journal of medicine 1471; author reply | 1472
A A Sariev | 1999 | Acute anal fissures in puerperants Vestnik | khirurgii imeni I. I. Grekova | 80-3
S Onishi | 1999 | The entity "autoimmune cholangitis": hanging by a thread? Journal of | gastroenterology | 657-8
A Brandrup-Lukanow | 1999 | Priorities in reproductive health in eastern Europe. | Medicine and law | 167-75
Medicine and law | 483-6 | 1999 | Internet websites for reproductive and sexual health law and ethics.
G Alvarez | 1999 | Bring your own sutures--Peruvian mission. Nursing spectrum | (D.C./Baltimore metro ed.) | 10-1
F W Verheugt | 1999 | Hotline sessions of the 21st European Congress of | Cardiology. European heart journal | 1603-6
Methods in enzymology | 1999 | T R Neu J R Lawrence | 145-52 | Lectin-binding analysis in biofilm systems.
N J Stone | 1999 | Hyperlipidaemia and cardiovascular disease. Current opinion in | lipidology | 479-81
Annals of plastic surgery | 1999 | P Sylaidis A Logan Re: | Epinephrine in digital blocks: revisited. | 572
A B Cairns | 1999 | Spirituality and religiosity in palliative care. | Home healthcare nurse | 450-5
```

Fig. 1. A segmentation example of our method for M20 dataset (cf. Section 4), where the bibliographic elements to be segmented are `authors`, `year`, `title`, `journal`, and `pages`. Each line gives a different reference. The symbol "|" shows the segmentation.

with hidden Markov models or conditional random fields. In contrast, we proposed a Bayesian probabilistic model as an extension of latent Dirichlet allocation (LDA) [1] and modeled the appearance order as *topic permutation*.

We regard each raw reference as a document and assign each word token to a topic so that the word tokens assigned to the same topic refer to the same bibliographic element. However, the topic assignments obtained by LDA fails to solve our problem, because the assignments do not satisfy *contiguity constraint*, i.e., the constraint that the same bibliographic element should be referred to by contiguous word tokens. In the topic assignments given by LDA, the word tokens assigned to the same topic are in general not contiguous.

Therefore, in [8], we borrowed a remarkable idea proposed by Chen et al. [3] and modified LDA as follows: For each document, 1) draw as many topics as word tokens; 2) draw a topic permutation from a probability distribution defined by generalized Mallows model (GMM) [4] over all topic permutations; and 3) sort the drawn topics according to the drawn permutation (cf. Figure 2). The idea to use GMM in topic models was originally proposed in [3] to solve *document structure learning*. However, this problem is widely different from ours. Therefore, we modified their method and gave an unsupervised method for bibliographic element segmentation. In this paper, we give a fixed name *PaiFen*[1] to our segmentation method. The main contribution of this paper is to propose a *semi-supervised learning* for PaiFen and to improve the segmentation accuracy up to over 90%. We call the semi-supervised version of PaiFen *BanPaiFen*.[2]

[1] "Pai" is the pronunciation of the Greek character π often used to denote permutations, and "Fen" is the pronunciation of the Chinese character meaning segmentation.

[2] "Ban" is the pronunciation of the Chinese character standing for "semi."

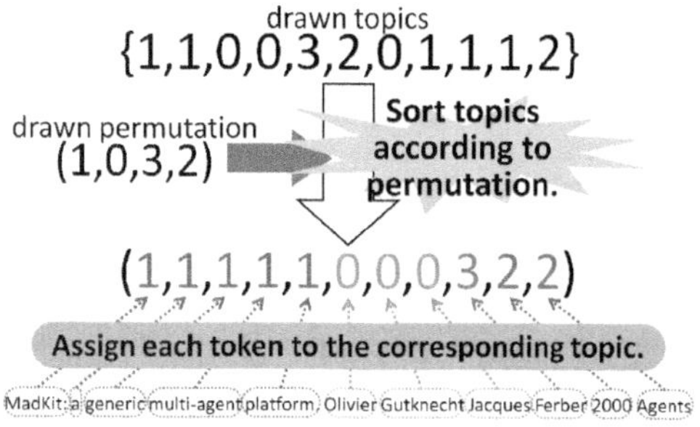

Fig. 2. How to make topic assignments satisfy contiguity constraint

BanPaiFen accepts an input set of references where some word tokens are already labeled. In our experiment, at most one third of word tokens are labeled. However, these supervised labels are allowed to be imperfect, because we assume that the labels are given not by careful hand-labeling but by automated labeling. In our experiment, around 3% of the labels were incorrect. However, BanPaiFen could achieve an over 90% segmentation accuracy.

The rest of the paper is organized as follows. Section 2 describes the model PaiFen in a generative manner. Section 3 explains how we modify PaiFen for semi-supervised learning. Section 4 includes the settings and the results of our experiment. Section 5 concludes the paper with discussions and future work.

2 Topic Modeling for Segmentation

As is widely known, topic modeling trend in text mining was inaugurated by LDA [1]. By regarding references as documents and bibliographic elements as topics, we can interpret the topic assignments of LDA as bibliographic element labelings. However, the topic assignments by LDA do not satisfy *contiguity constraint*, i.e., the constraint that the word tokens assigned to the same topic should appear contiguously. This constraint is required for our problem, because each bibliographic element should be referred to by contiguous word tokens. Here we can borrow an intuition from the work of Chen et al. [3] and can use generalized Mallows model (GMM) for meeting contiguity constraint. Since GMM defines a probability distribution over topic permutations, we can model the topic appearance order in each document by a draw from this distribution.

However, the problem envisioned in [3] was *document structure learning*, a problem widely different from ours. Therefore, in [8], we proposed a modified version of their method. We call our method *PaiFen* in this paper. Chen et al. assumed *sparse topic distribution* in [3]. That is, they assumed that only a small number of topics appear in each document. This assumption is not appropriate for our problem, because all topics, i.e., all bibliographic elements, are basically expected to appear in every references. Therefore, while Chen et al. draw topic multinomial distributions from the Dirichlet prior to achieve topic sparseness, we replace all topic multinomials by the uniform distribution in PaiFen.

Due to space limitation, we skip the mathematical details of PaiFen and refer the details to [8]. We only repeat the generative description of PaiFen as below:

1. Draw a word multinomial parameter $\phi_k = (\phi_{k1}, \dots, \phi_{kW})$, $k = 1, \dots, K$, from the symmetric Dirichlet prior $\text{Di}(\beta)$ for each of the K latent topics.
2. Draw a parameter $\rho = (\rho_1, \dots, \rho_{K-1})$ of generalized Mallows model $\text{GMM}(\rho)$ from the conjugate prior, which is defined by Equation (2) in [3].
3. Let n_j be the number of word tokens contained in the jth reference d_j. Draw d_j as an ordered word token sequence $\mathbf{x}_j = (x_{j1}, \dots, x_{jn_j})$ as follows:
   (a) Draw n_j topics $\mathbf{t}_j = \{t_{j1}, \dots, t_{jn_j}\}$ uniformly from the set of K topics.
   (b) Draw a permutation π_j of the K topics from $\text{GMM}(\rho)$.
   (c) Sort the drawn topics $\mathbf{t}_j$ according to the drawn permutation π_j and obtain an ordered multiset of topics $\mathbf{z}_j = (z_{j1}, \dots, z_{jn_j})$ (cf. Figure 2).
   (d) For x_{ji}, draw a word w from the multinomial $\text{Multi}(\phi_{z_{ji}})$ and set $x_{ji} = w$.

3 Semi-supervised Inference

Basically, we adopt the MCMC inference described in [8]. However, we realize a semi-supervised learning by modifying the conditional posteriors in a *scheduled* manner. We need to generate MCMC samples for the following three types of variables: GMM parameters ρ, inversion counts $\mathbf{v}_j$, and topic draws $\mathbf{t}_j$. GMM parameters are sampled in the same manner as [8]. Therefore, we explain how inversion counts and topic draws are sampled in our semi-supervised inference.

We present necessary notations. *Inversion counts* are introduced to specify a topic permutation in a one-to-one manner. For each reference d_j, a set of $K-1$ inversion counts $\mathbf{v}_j = (v_{j1}, \dots, v_{jK-1})$ are prepared as latent variables. Each v_{jk} is an integer satisfying $0 \le v_{jk} \le K-1$. Let $\boldsymbol{\pi}_j = (\pi_{j1}, \dots, \pi_{jK})$ be the topic permutation used for sorting the topic draws $\mathbf{t}_j$ for reference d_j. We denote the position of the kth topic in $\boldsymbol{\pi}_j$ by $\boldsymbol{\pi}_j^{-1}(k)$. $\boldsymbol{\pi}_j$ can be specified by $\mathbf{v}_j$ in a one-to-one manner via the following relationship: $v_{jk} \equiv |\{k' : k' > k \text{ and } \boldsymbol{\pi}_j^{-1}(k') < \boldsymbol{\pi}_j^{-1}(k)\}|$. That is, v_{jk} is the number of topics whose indices are larger than k and whose positions in $\boldsymbol{\pi}_j$ are earlier than the kth topic. For example, $\boldsymbol{\pi}_j = (4, 3, 1, 5, 2)$ can be uniquely specified by $\mathbf{v}_j = (2, 3, 1, 0)$.

In the unsupervised inference described in [8], we sample the inversion count v_{jk} for each j and k from the full conditional posterior $P(v_{jk}|\cdots)$. Further, we sample the latent topic t_{ji} for each j and i from the full conditional posterior $P(t_{ji}|\cdots)$. The mathematical details of these posteriors are referred to [8]. In BanPaiFen, these posteriors are modified by using the two types of penalties calculated for each reference d_j. The one is R_j^{mis}, the number of mismatches between the supervised label and the inferred topic. The other is R_j^{red}, the number of *redundant assignments* that are explained as follows. When more than one word tokens are assigned to the topics corresponding to the bibliographic elements that should be referred to by one word token, we call such assignments *redundant*. For example, when three word tokens are assigned to the topic corresponding to publication year, we set $R_j^{red} = 2$, because publication year should

Table 1. Specifications shared by the three DBLP datasets, i.e., D0, D20, and D50

#references = 944,755, #different words = 685,799, #word tokens = 17,408,876					
#labels	authors	title	journal	year	total
	2,851,470	1,761,034	126,420	950,721	5,689,645
(/ #word tokens)	(16.4%)	(10.1%)	(0.73%)	(5.5%)	(32.7%)
#incorrect labels	authors	title	journal	year	total
	7,549	160,325	1,218	5,966	175,058
(/ total #labels)	(0.13%)	(2.8%)	(0.02%)	(0.10%)	(3.1%)

be referred to by one word token, and thus two among the three assignments are redundant. Then, $P(v_{jk}|\cdots)$ and $P(t_{ji}|\cdots)$ are modified by R_j^{mis} and R_j^{red} as

$$P(v_{jk}|\cdots)^{\exp(R_j^{mis}I/M)} \text{ and } P(t_{ji}|\cdots)^{\exp\{(R_j^{mis}+R_j^{red})I/M\}}, \tag{1}$$

respectively. In Eq. (1), I is the number of iterations, and M is the parameter that controls how fast the effect of the supervised labels becomes strong as iterations proceed. The modified probabilities are used after being normalized. As Eq. (1) shows, when we sample inversion counts, we only consider R_j^{mis}. This is because R_j^{red} does not change by any update of $\mathbf{v}_j$. While we tested various ways to modify the posteriors, Eq. (1) gave the best result.

We can explain what Eq. (1) means as follows. Suppose that we terminate the inference after 300 iterations. Further, suppose that we choose M as 100. Then, for earlier iterations, $R_j^{mis}I/M$ and $(R_j^{mis} + R_j^{red})I/M$ are nearly equal to zero, because $I \ll M$ holds. Therefore, the penalty scores only have a small effect. This means that we conduct an almost unsupervised inference. However, as iterations proceed, $R_j^{mis}I/M$ and $(R_j^{mis} + R_j^{red})I/M$ become larger, because I becomes larger. Consequently, the posterior probabilities become closer to zero for the sample values making penalties positive. The inference then comes to sharply avoid such sample values. For example, if some sample value for t_{ji} increases the number of the word tokens referring to publication year from one to two, the inference after hundreds of iterations sharply avoids this sample value.

We can find a previous work [9] that utilizes supervised labels by introducing response variables into topic models. However, we use the labels to directly modify the "shape" of the posterior distributions. We think that our method is more intuitive and efficient, though not mathematically elegant.

4 Evaluation Experiment

4.1 Dataset Composition

We composed the datasets by using DBLP database[3] and MEDLINE/PUBMED database[4]. With respect to DBLP database, we composed the three datasets D0, D20, and D50 based on the file `dblp.xml` dated February 8, 2010 as follows:

[3] `http://dblp.uni-trier.de/xml/`

[4] MEDLINE®/PUBMED®, a database of the U.S. National Library of Medicine.

Table 2. Specifications shared by the MEDLINE datasets, i.e., M0, M20, and M50

#references = 3,001,207, #different words = 2,168,061, #word tokens = 87,085,708						
#labels	`authors`	`year`	`title`	`journal`	`pages`	total
	4,983,698	2,770,489	483,878	294,330	2,710,264	11,242,659
(/ #word tokens)	(5.7%)	(3.2%)	(0.56%)	(0.34%)	(3.1%)	(12.9%)
#incorrect labels	`authors`	`year`	`title`	`journal`	`pages`	total
	60,536	13,951	179,958	4,570	8,198	267,213
(/ total #labels)	(0.54%)	(0.12%)	(1.6%)	(0.04%)	(0.07%)	(2.4%)

- We collected the references having publication years ranging from 2000 to 2009 and extracted the bibliographic elements `authors`, `title`, `booktitle`, `journal`, and `year`. We identified `booktitle` with `journal`, because these two elements could be regarded as playing the same role for our problem.
- We defined the *canonical order* of bibliographic elements as `authors` < `title` < `journal` < `year` and sorted the bibliographic elements in this order for all references. The canonical order determines the correspondence between the bibliographic elements and the latent topics. That is, any inferred segmentation is expected to assign word tokens to topic 1 (resp. 2, 3, and 4) when those tokens refer to `authors` (resp. `title`, `journal`, and `year`).
- We randomly selected Q% of the references and randomly shuffled the order of the bibliographic elements. We did not change the word token order in each bibliographic element. By forgetting bibliographic elements information, we obtained each reference as a "raw" word token sequence. D0, D20, and D50 are the datasets obtained when $Q = 0, 20,$ and 50, respectively.

We call D0, D20, and D50 *DBLP datasets.* Table 1 summarizes the specifications. Note that these specifications are shared by D0, D20, and D50, because only the order of the bibliographic elements is different among these datasets.

With respect to MEDLINE/PUBMED database, the 100 files whose names ranged from `medline09n0400.xml` to `medline09n0499.xml` were used to compose the three datasets M0, M20, and M50 by applying the same procedure with D0, D20, and D50, respectively. We call the three datasets *MEDLINE datasets*, whose specifications are shown in Table 2. For MEDLINE datasets, we extracted the five bibliographic elements: `authors`, `year`, `title`, `journal`, and `pages`, and regarded this order as the canonical order. Further, we eliminated the parentheses `[` and `]` at the head and the tail of every title, because they are artifacts. Except this, we applied no preprocessing like stemming, punctuation removal, and stop word elimination to MEDLINE datasets and also to DBLP datasets.

4.2 Automated Labeling

We applied the following automated procedures to obtain supervised labels:

1. We extracted the references having publication years earlier than 2000 from DBLP database. When a word appeared as a part of some bibliographic

element and never appeared as a part of the other bibliographic elements in the extracted references, we labeled all tokens of that word by the corresponding bibliographic element. The same procedure was applied also for MEDLINE datasets by using the references contained in the 100 files from `medline09n0000.xml` to `medline09n0099.xml` of MEDLINE database. This automated labeling was applied only for `authors`, `title`, and `journal`.
2. We labeled all tokens of the words giving an integer in the interval [1900, 2012] as `year`. Further, we labeled all tokens of the words matching the regular expression `[1-9][0-9]*\-[1-9][0-9]*` as `pages`.

Since the first procedure gave many incorrect labels, we removed the labels of `authors` and `journal` from all tokens of the words included in the SCOWL word list[5]. Consequently, the tokens of the words included in the SCOWL list cannot have any labels other than `title`. The label statistics are shown in Table 1 for DBLP datasets and in Table 2 for MEDLINE datasets. For DBLP datasets, 32.7% of the word tokens are labeled, but 3.1% of them are incorrectly labeled. For MEDLINE datasets, only 12.9% of the word tokens are labeled, and 2.4% of them are incorrect.

4.3 Implementation

We refer the details of the inference implementation to [8]. The number of iterations were 300 for every experiment setting. Therefore, I in Eq. (1) is at most 300. We ran the inference 10 times for every experiment setting by starting from a random initialization. The segmentation quality was measured by the F-score defined in [3], which can be roughly viewed as the proportion of the correctly segmented word tokens. Based on the 10 runs, we calculated the mean and standard deviation of the corresponding 10 F-scores. However, we do not report the standard deviation here, because it was always negligibly small.

4.4 Evaluation Results for DBLP Datasets

Figure 3 summarizes the evaluation results for DBLP datasets. We set $M = 10000$ in Eq. (1), because this setting gave fairly good results. The wall clock time of 300 iterations was 4,400 seconds on a Fedora 14 PC equipped with Intel Core i7 970 at 3.20GHz. Each bar in Figure 3 represents the mean F-score.

In Figure 3, the charts in the top left, top right, and bottom panels present the mean F-scores for D0, D20, and D50, respectively. The bottom black bar in each chart gives the baseline F-score, i.e., the F-score achieved by PaiFen. The other bars give the F-scores achieved by BanPaiFen under various settings. The tag of each bar shows the labels used in semi-supervised inference. A, T, J, and Y mean `authors`, `title`, `journal`, and `year`, respectively. For example, the tag ATY means that we used the supervised labels of `authors`, `title`, and `year`.

[5] http://wordlist.sourceforge.net/

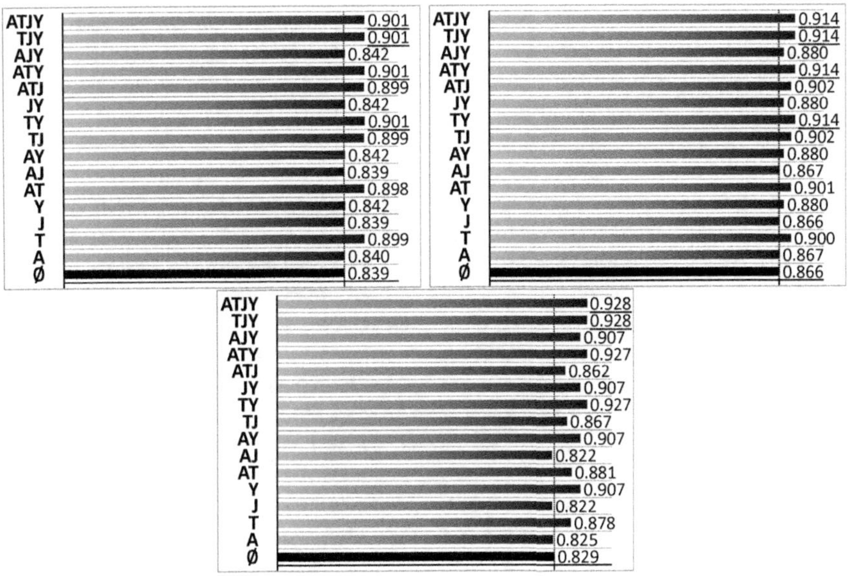

Fig. 3. Evaluation results for DBLP datasets. The charts in the top left, top right, and bottom panels present the F-scores for D0, D20, and D50, respectively. Each F-score is the mean calculated over 10 runs of the inference. The best F-scores are underlined.

The best F-scores in each chart are underlined. We can observe that the settings TJY and ATJY gave the best F-scores for all D0, D20, and D50. Interestingly, we could achieve better F-scores for D50 than for D0 and D20. In D50, 50% of the references contain the bibliographic elements in a random order. It seems that the existence of a wide variety of orderings of bibliographic elements helped the inference to reach better local optima at least for DBLP datasets. Obviously, Figure 3 shows that BanPaiFen improved PaiFen by a large margin.

4.5 Evaluation Results for MEDLINE Datasets

Figure 4 presents the evaluation for MEDLINE datasets. We set $M = 100$ in Eq. (1), because this setting gave better results than $M = 10000$. 42,000 seconds were required for 300 iterations on a Fedora 14 PC equipped with Intel Core i7 970 at 3.20GHz. The tag characters A, T, J, Y, and P attached to each bar mean `authors`, `title`, `journal`, `year`, and `pages`, respectively. We only show the results for the baseline case tagged as $\emptyset$ and the following eight cases: A, T, J, Y, P, YP, ATJ, and ATJYP, because other cases gave no better results.

Figure 4 shows that the combined use of the supervised labels of `year` and `pages` achieved the best F-score for all of M0, M20, and M50. Especially, we can observe that the supervised labels of `year` played an important role. With no `year` labels, we could not improve PaiFen. This may be because `year` is placed between `authors` and `title` in the canonical order of bibliographic elements

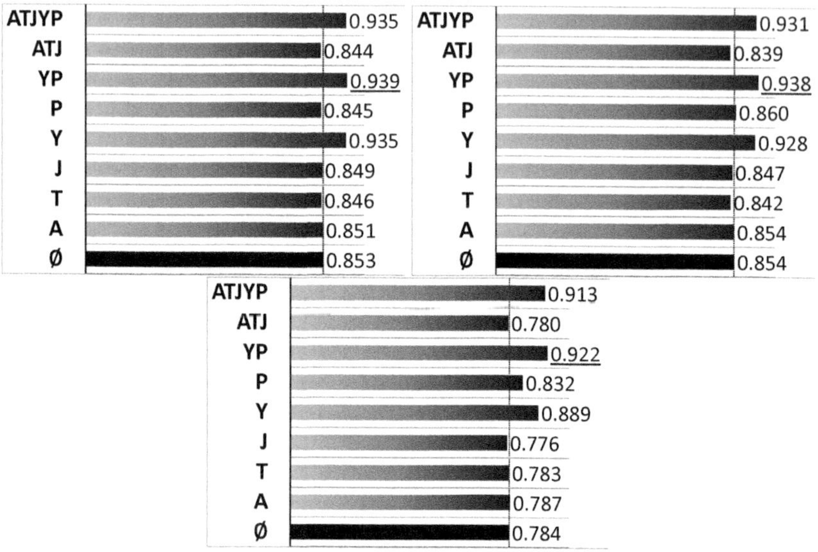

Fig. 4. F-scores for MEDLINE datasets. The charts in the top left, top right, and bottom panels present the F-scores for M0, M20, and M50, respectively. Each F-score is the mean calculated over 10 runs of the inference. The best scores are underlined.

for MEDLINE datasets, and thus the correct segmentation of `year` drastically contributed to the correct segmentation of both `authors` and `title`.

Further, we can observe that the segmentation quality was worse for M50 than for M0 and M20, though we obtained better F-scores for D50 than D0 and D20 in case of DBLP datasets. While there are only $4! = 24$ permutations of bibliographic elements in total for DBLP datasets, we have $5! = 120$ permutations for MEDLINE datasets. Therefore, the inference of the correct permutation for each reference is far more difficult in case of MEDLINE datasets. This may be the reason we obtained worse F-scores for M50 than M0 and M20. However, BanPaiFen improved PaiFen by a large margin for all of M0, M20, and M50.

5 Conclusion

We proposed a semi-supervised bibliographic element segmentation called *BanPaiFen* by modifying our unsupervised method PaiFen [8]. We modified posterior probabilities in PaiFen so that the inference can avoid the sample values leading to mismatches between supervised labels and inferred topics or leading to redundant topic assignments, where the word "redundant" means that we assign more than one word tokens to the topic corresponding to the bibliographic element that should be referred to by one word token, e.g. publication year. Our experiment showed that BanPaiFen could improve PaiFen by a large margin.

In a more realistic situation, OCR errors may be included in the references obtained from the scanned articles, and crawled Web pages may include typos.

GPSR Compliance

The European Union's (EU) General Product Safety Regulation (GPSR) is a set of rules that requires consumer products to be safe and our obligations to ensure this.

If you have any concerns about our products, you can contact us on ProductSafety@springernature.com

In case Publisher is established outside the EU, the EU authorized representative is:

Springer Nature Customer Service Center GmbH
Europaplatz 3
69115 Heidelberg, Germany

Batch number: 10844627

Printed by Printforce, the Netherlands

Therefore, it is an important future work to incorporate an error correction at the word token level into our model, as a preceding work did for hidden Markov models [10]. One promising research direction is to propose a modeling of each word token as a character string for calculating the penalties used in Eq. (1).

We know that existing successful databases regard bibliographic element order as *state transition* and use hidden Markov models or conditional random fields. However, such approaches achieve their superiority not only with such data modelings but also with practical tunings. While our approach also requires additional practical tunings, we think that our approach can be an alternative to existing approaches as a new style of segmentation based on *permutation*, not on transition, because we have already achieved an over 90% accuracy without getting into any details about how each word token is composed as a string.

Acknowledgement. This work was supported in part by Japan Society for the Promotion of Science Grant-in-Aid for Young Scientists (B) 60413928.

References

1. Blei, D.M., Ng, A.Y., Jordan, M.I.: Latent Dirichlet Allocation. Journal of Machine Learning Research 3, 993–1022 (2003)
2. Connan, J., Omlin, C.W.: Bibliography Extraction with Hidden Markov Models. Technical Report US-CS-TR-00-6, University of Stellenbosch (2000)
3. Chen, H., Branavan, S.R.K., Barzilay, R., Karger, D.R.: Global Models of Document Structure Using Latent Permutations. In: Proc. of North American Chapter of the Association for Computational Linguistics - Human Language Technologies (NAACL HLT) 2009 Conference, pp. 371–379 (2009)
4. Fligner, M.A., Verducci, J.S.: Distance Based Ranking Models. Journals of the Royal Statistical Society B 48(3), 359–369 (1986)
5. Hetzner, E.: A Simple Method for Citation Metadata Extraction Using Hidden Markov Models. In: Proc. of the 8th ACM/IEEE-CS Joint Conference on Digital Libraries, pp. 280–284 (2008)
6. Kramer, M., Kaprykowsky, H., Keysers, D., Breuel, T.M.: Bibliographic Meta-Data Extraction Using Probabilistic Finite State Transducers. In: Proc. of the 9th International Conference on Document Analysis and Recognition, pp. 609–613 (2007)
7. Lafferty, J.D., McCallum, A., Pereira, F.C.N.: Conditional Random Fields: Probabilistic Models for Segmenting and Labeling Sequence Data. In: Proc. of the Eighteenth International Conference on Machine Learning, pp. 282–289 (2001)
8. Masada, T., Shibata, Y., Oguri, K.: Unsupervised Segmentation of Bibliographic Elements with Latent Permutations. International Journal of Organizational and Collective Intelligence 2(2), 49–62 (2011)
9. Sharifi, M.: Semi-supervised Extraction of Entity Attributes Using Topic Models. Master's Thesis, Carnegie Mellon University (2009)
10. Takasu, A.: Bibliographic Attribute Extraction from Erroneous References Based on a Statistical Model. In: Proc. of the 3rd ACM/IEEE-CS Joint Conference on Digital Libraries, pp. 49–60 (2003)
11. Yin, P., Zhang, M., Deng, Z.-H., Yang, D.-Q.: Metadata Extraction from Bibliographies Using Bigram HMM. In: Proc. of the 7th International Conference on Asian Digital Libraries, pp. 1–14 (2004)

A Discretization Algorithm of Numerical Attributes for Digital Library Evaluation Based on Data Mining Technology

Yumin Zhao, Zhendong Niu, Xueping Peng, and Lin Dai

School of Computer Science and Technology, Beijing Institute of Technology
Beijing 100081, China
{ymzbit,bitd2000}@gmail.com, {pengxp,zniu}@bit.edu.cn

Abstract. We present here a discretization algorithm for numerical attributes of digital collections. In our research data mining technology is imported into digital library evaluation to provide a better decision-making support. But data prediction algorithms work not well based on the traditional discretization method during the data mining process. The reason is that numerical attributes of digital collections are complicated and not in the same scale of distribution distance. We study the characteristic of numerical attributes and put forward a discretization method based on the Z-score idea of mathematical statistics. This algorithm can reflect the dynamic semantic distance for different numerical attributes and significantly enhance the precision rate and recall rate of data prediction algorithms. Furthermore a 'nonlinear conditional relationship' among attributes of digital collections is discovered during the study of discretization algorithm and impacts the actual application result of traditional data mining algorithms.

Keywords: Discretization algorithm, data mining, digital library evaluation.

1 Introduction

At present there are two main ideologies in digital library evaluation [1]. One is expert-centric evaluation which bases on expert's professional way and the other is user-centric evaluation which bases on the user experience of using digital libraries [2] [3]. Both methods are mainly based on evaluation index and user survey. There lacks advanced idea of computer intelligent data processing, analysis and decision support. Nowadays evaluation methods face many problems difficult to solve with the rapid development of digital library, such as mass information processing and utilization, complex data relationship, etc. We import the data mining technology into digital library evaluation [4]. This way can not only utilize the mass information well, but provide a natural pattern for digital library evaluation to integrate the expert-centric data and user-centric data. In our research data is divided into two types, one is objective data, the other is subjective data. We integrate two types of data to build a dataset for digital collections evaluation.

The subjective data come from the website named DOUBAN.com. The website provides the communication platform for personal feeling information about everything in which people are interesting. Many people pose very interesting reviews,

C. Xing, F. Crestani, and A. Rauber (Eds.): ICADL 2011, LNCS 7008, pp. 70–76, 2011.

discussions and evaluation information about digital collections presented in website, such as books, movies and music they have read, looked and heard. This kind of information about digital collection is gaining in popularity and now DOUBAN.COM has became one of the most popular SNS websites in China. DOUBAN.com is maintaining the biggest users' feeling information database about billions of book, such as score, number of people who have read, number of people who are reading, number of people who want to read, etc. We take these kinds information as the important subjective data in digital library evaluation.

In our dataset there are many attributes such as *author*, *nameLength*, *numberAuthor*, *haveTranslator*, *numberTranslator*, *pages*, *price*, *time*, *numberEvaluation*, *numberReading*, *numberHaveread*, *numberWantread*, *numberThree*, *score*, etc. We import data mining technology to build a prediction model of score level about single digital collection. This prediction model can provide a score level for a new book, movie or music without users' evaluation. This work is very meaningful. We can improve individuation recommendation system of an information service website or a digital library based it. We can also provide a comparative analysis for old digital collection which has been evaluated and let users find the value of digital collection better. For the evaluation of digital library, traditional evaluation work is often based on the scale of digital collections, now we can get the value of digital library collections based this prediction model. So we can evaluate digital library more rationally through merging the scale evaluation and value evaluation.

Decision tree algorithm is one of the most widely used data mining prediction technologies [5] [6]. We adopt and research the algorithm for our dataset. Because there exists many numerical attributes, such as *score*, *nameLength*, *pages*, *price*, *time*, *numberHaveread*, *numberWantread*, *numberReading*, etc. at first we should disperse the data of numerical attributes. Consequently we have developed a new discretization algorithm for numerical attribute and a new decision tree algorithm to build the prediction model. This paper focuses on the study about discretization algorithm.

2 A Parallel Discretization Algorithm Based on Z-Score Idea

Through research on attributes' characteristic of digital library collections, a parallel discretization algorithm based on Z-score idea (PDOZ algorithm) is put forward in our research.

We found almost all the numerical attributes in our dataset obey the normal distribution, such as *score*, *nameLength*, *pages*, *price*, *time*, especially *score* is the CLASS attribute. Z-score of mathematical statistics listed in Equation (1) is to provide a standardization comparison platform among different normal distributions and has the probability distribution significance. We think utilizing this kind of probability distribution points to disperse data can reflect the dynamic semantic distance for different numerical attributes in dataset of digital collections.

$$z_i = \frac{(X_i - Mean)}{s} \tag{1}$$

Process of PDOZ algorithm is as follow.

1. Normalization for origin data distribution

Select the representative value in turn form mean, median and mode about origin data of one numerical attribute, and adjust data range to get three approximate normal distributions based on the representative value for origin data. This step would eliminate exceptional data point and make the initial data clean.

2. Design weight algorithm to determine the relationship among three approximate normal distributions

Our goal is to get a normal distribution as accurate as possible for characteristics description about origin data. Skewness is an indicator in statistical subjects to measure how far the distribution is away from absolute normal distribution. So we take skewness as the base to compare the three approximate normal distributions.

The concrete compution process is Equation (2), (3) and (4). Because smaller skeness means closer absolute normal distribution, the distribution with minimum skeness should have the maximum weight.

$$skewness = Mean - Median \tag{2}$$

$$sSD = \frac{|Mean - Median|}{s} \tag{3}$$

$$w_{\min\_skewness} = \frac{sSD_{\max}}{\sum_{i=1}^{n} sSD_i}, w_{med-skewness} = \frac{sSD_{med}}{\sum_{i=1}^{n} sSD_i}, w_{\max\_skewness} = \frac{sSD_{\min}}{\sum_{i=1}^{n} sSD_i} \tag{4}$$

3. Compute mean-weighted and standard-deviation-weighted

We get the weight from step 2 and compute the mean-weighted and standard-deviation-weighted with Equation (5) and (6).

$$Mean_{weighted} = w_1 Mean_1 + w_2 Mean_2 + w_3 Mean_3 \tag{5}$$

$$s_{weighted} = w_1 s_1 + w_2 s_2 + w_3 s_3 \tag{6}$$

4. Determine the discretization point based on the proper Z value

Here we should put the mean-weighted and standard-deviation-weighted in the Z-value Equation (7) to get the discretization point X_i.

$$z_i = \frac{(X_i - Mean_{weighted})}{s_{weighted}} \tag{7}$$

Selection of Z value should rely on the Table 1. The value of this table comes from the relationship table between Z value and normal curve area which can be found in any statistics books. The area point means the distribution probability. We rely on the probability distribution value to disperse the number attributes' intervals. Because we choose ten intervals in experience, nine probability distribution points are listed in Table 1.

Table 1. Area point under nomal curve

area	Z	area	Z	area	Z
-40%	-1.28	-10%	-0.25	20%	0.535
-30%	-0.84	0	0	30%	0.84
-20%	-0.535	10%	0.25	40%	1.28

We conduct the experiment in different traditional decision tree algorithm to build the score-level prediction model. The J48 is the best one for the origin data, and we compare precision rate and recall rate under J48 algorithm.

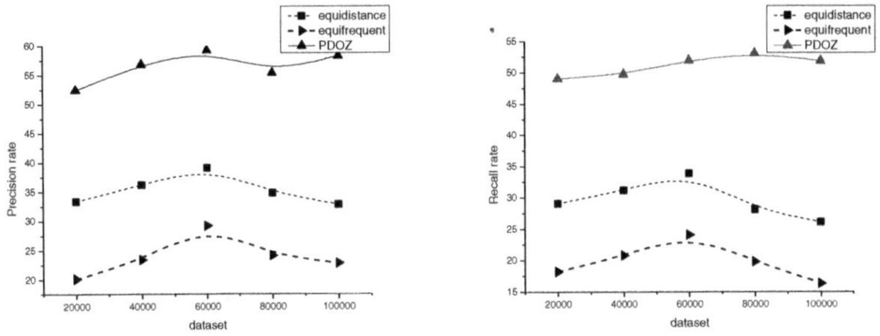

Fig. 1. Precision rate and recall rate comparison under J48 algorithm

We compare the equidistance, equifrequent and PDOZ methods for data discretization of numerical attributes. From Fig.1 we can find precision rate and recall rate both increase obviously in different data scale when we adopt the PDOZ algorithm for numerical attributes. The result of PDOZ raised nearly 100% than equidistance method, and raised nearly 300% than equifrequent method. This shows PDOZ discretization algorithm fit the dataset of digital collections and facilitate the data mining application.

3 Nonlinear Conditional Attributes Relationship

From the above experience we found traditional decision tree algorithm doesn't work well even under the improvement of the PDOZ discretization algorithm. Precision rate and recall rate is about 55%. The rate is too low to apply. Why is there such a situation?

During the study of discretization algorithm a complicated relationship among attributes of digital collections is discovered and impacts the actual application result of traditional data mining algorithms. We call it 'nonlinear conditional attributes relationship'.

'Nonlinear conditional attributes relationship' defined by this paper is that there is no linear relationship or other explicit mathematical relationship among attributes. We can suppose an attributes-set of a dataset is $\{A,B,C,\cdots,Z\}$. Attribute Z is the class attribute. Attribute B is a nonlinear conditional attribute for attribute A. There is a phenomenon that correlation coefficients are all zero between A and B, A and Z, B and Z. We directly apply the traditional decision tree algorithm in this dataset, A would be treat as a weak ability attribute for distinguishing data. But when origin dataset is splitted by the different values of attribute B, A has a strong ability to distinguish data. Namely A would play an important role in subset of origin data.

We make an experience to prove this relationship as follows.

At first we draw the relationship picture between Attribute *price* and Attribute *score* with the fitted lines based on the origin dataset presented in Fig.2.

Then we split the origin dataset into three subsets according to the value of Attribute *time* and draw the relation picture between *Attribute price* and *Attribute score* with the fitted lines based on the subsets of origin data presented in Fig.3.

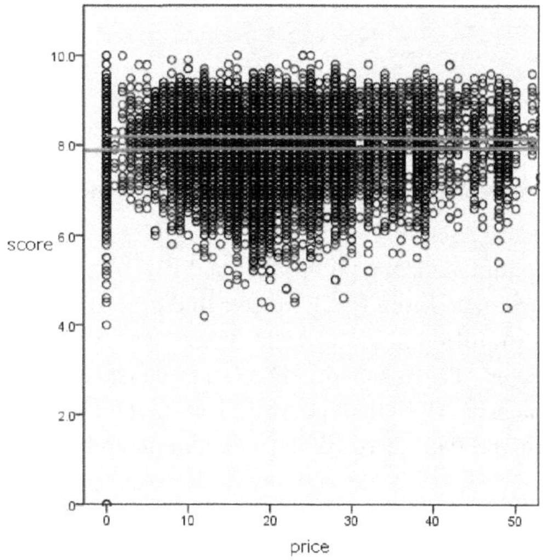

Fig. 2. The relationship between price and score in origin dataset

From Fig.2 we can find the fitted lines demonstrate the relationship between price and score is irrelevant.

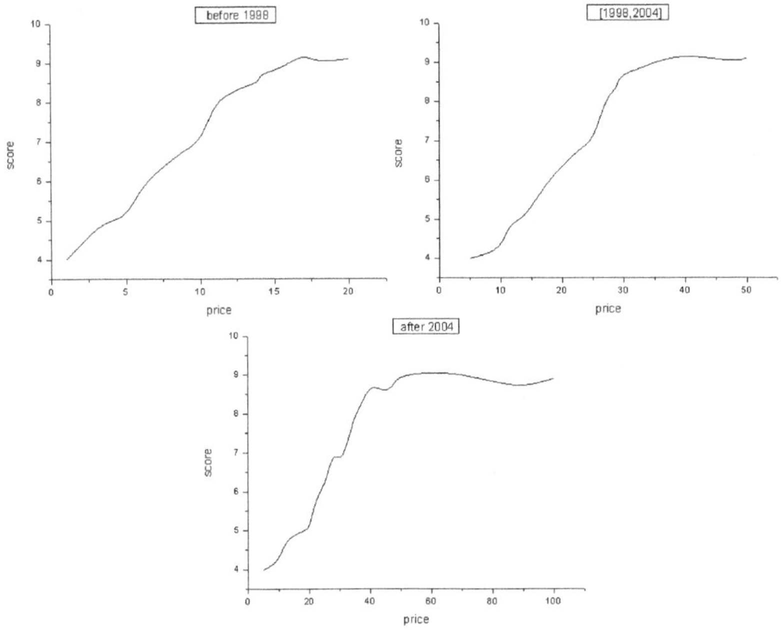

Fig. 3. The relationship between price and score in three subsets of origin dataset

From Fig.3 we can find the relationship between price and score become very evident in three subsets of origin dataset.

To solve this problem we have developed a stratified decision tree algorithm, stratified decision tree algorithm would be presented in future paper. Here we confirm this relationship is to explain why the rate is too low to apply even when the PDOZ algorithm raises the rate of precision and recall obviously.

4 Conclusions

This paper imports the subjective data which express users' feel and evaluation information into digital collections dataset, and then tries to build a prediction model of score level about digital collection as a solution quantizing collection's quality. We study the data characteristics of numerical attributes and propose a parallel discretization algorithm based on z-score idea. This discretization algorithm for numerical attributes uses dynamic range that reflects the significance of probability distributions instead of traditional equidistant and equifrequent discretization methods. It can reflect the dynamic semantics information of different attribute, discover attribute relation more efficiently and raise the rates of precision and recall obviously. At last we present a complicated condition attribute relation during the study of PDOZ algorithm. This attribute relationship explains why traditional decision tree algorithms can't work well even under the improvement of PDOZ algorithm. This problem should be solved in future paper.

Acknowledgement. This work is supported by the Natural Science Foundation of China (grant no. 60803050).

References

1. Hsieh-Yee, I.: Digital Library Evaluation: Progress & Next Steps. In: 2005 Annual Meeting of the American Society for Information Science and Technology Charlotte, N.C. (2005)
2. Snead, J.T., Bertot, J.C., Jaeger, P.T., McClure, C.R.: Developing multi-method, iterative, and user-centered evaluation strategies for digital libraries: Functionality, usability, and accessibility. Proceedings of the American Society for Information Science and Technology 42 (2005)
3. Kyrillidou, M., Giersch, S.: Developing the DigiQUAL protocol for digital library evaluation. In: Proceedings of the 5th ACM/IEEE-CS Joint Conference on Digital Libraries, Denver, CO, USA (2005)
4. Zhao, Y., Niu, Z., Dai, L.: Evaluation algorithm about digital library collections based on data mining technology. In: Proceedings of Role of Digital Libraries in a Time of Global Change - 12th International Conference on Asia-Pacific Digital Libraries (2010)
5. Han, J., Kamber, M.: Data mining: concepts and techniques, 3rd edn. Morgan Kaufmann Pub., San Francisco (2011)
6. Witten, I.H., Frank, E.: Data Mining: Practical machine learning tools and techniques. Morgan Kaufmann Pub., San Francisco (2005)

Sentence-Level Sentiment Polarity Classification Using a Linguistic Approach

Luke Kien-Weng Tan[1], Jin-Cheon Na[1], Yin-Leng Theng[1], and Kuiyu Chang[2]

[1] Wee Kim Wee School of Communication and Information
Nanyang Technological University, Singapore
{w080078,tjcna,tyltheng}@ntu.edu.sg
[2] School of Computer Engineering, Nanyang Technological University, Singapore
askychang@ntu.edu.sg

Abstract. Recent sentiment analysis research has focused on the functional relations of words using typed dependency parsing as this provides a refined analysis on the grammar and semantics of the textual data, which could improve performance. However, typed dependencies only provide the grammatical relationships between individual words while there exist more complex relationships between words that could influence a sentence sentiment polarity. In this paper, we propose a linguistic approach, called Polarity Prediction Model (PPM), that combines typed dependencies and subjective phrase analysis to detect sentence-level sentiment polarity. Our approach also considers the intensity of words and domain terms that could influence the sentiment polarity output. PPM is shown to provide a fine-grained analysis for handling and explaining the complex relationships between words in detecting a sentence sentiment polarity. PPM was found to consistently outperform a baseline model by 5% in terms of overall F1-score, and exceeding 10% in terms of positive F1-score when compared to a Typed-dependency only approach.

Keywords: Sentiment analysis, polarity classification, linguistic approach.

1 Introduction

The proliferation of Web 2.0 and the easy access and availability of blog sites (e.g., www.blogspot.com) have encouraged web users to change from consumers to providers of information. A Weblog or Blog allows an individual or group of individuals to express their thoughts, voice their opinions, and share their experiences and ideas, creating a huge repository of online content that contains opinion expressions, which could be mined for sentiment orientation to determine its possible influence on the readers. This has led to new opportunities and challenges for researchers to extract opinionated information and sentiments from digital content.

Recent sentiment analysis studies used a linguistic approach that leveraged on the semantic dependencies between words to predict sentiments. Wilson et al. [13] observed that word patterns along with its modifiers could be used to determine sentiment polarity. For example, in the sentence "*I really love the nice phone*", the adverbial modifier "*really*" intensifies the positive sentiment expressed by "*love*",

C. Xing, F. Crestani, and A. Rauber (Eds.): ICADL 2011, LNCS 7008, pp. 77–87, 2011.

while the adjectival modifier "*nice*" gives a positive sentiment polarity to the noun object "*phone*". Therefore, the discovery of word patterns within the typed dependencies, which contains the modifier relationship between words, could potentially reveal clues to identifying sentiment polarity. A common approach to derive the dependencies between words is through typed dependency parsing [2, 10, 11], achieved via a natural language parser, which is a program that works out the grammatical structure of sentences. An example is the Stanford Parser[1], which generates Stanford typed dependencies output as shown in Figure 1 for the above example sentence. From Figure 1, the grammatical relations for the adverbial modifier "really" and the adjectival modifier "*nice*" are described by the typed dependencies advmod(*love*, *really*) and amod(*phone*, *nice*) respectively.

nsubj(love, I)
advmod(love, really)
det(phone, the)
amod(phone, nice)
dobj(love, phone)

Fig. 1. Generated typed dependencies

det(design, A)
nsubj(stands, that)
ccomp(design, **stands**)
det(test**, the**)
dobj(stands, **test**)
prep(test, **of**)
pobj(of, **time**)

Fig. 2. Subjective phrase across typed dependencies

Although typed dependencies based approach for sentiment prediction generally performed well for cases such as movie reviews [10], it was observed in Polanyi and Zaenen [6] and Wilson et al. [11] that there exist more complex relationships between words that could influence the sentiment polarity of sentences. Examples of these complex relationships include negation (e.g., "not" in "*not good*"), subjective terms found in domain terms (e.g., "hard" in "*hard disk*"), subjective terms appearing as neutral (e.g., "fiction" in "*science fiction*"), neutral terms appearing as subjective (e.g., "red" in "*red carpet*"), and intensifier and mitigation terms. Intensifiers such as "very" intensifies the sentiments of its adjacent term (e.g., "*very good*"), while mitigators such as "few" reduces the sentiment (e.g., "*few support*"). As typed dependencies contain only the grammatical relationship between two words, they could not detect the subjectivity within multiple words in a phrase. For example, in the sentence "*A design that stands the test of time*" as shown in Figure 2, typed dependency analysis alone could not detect the positive expression "*stands the test of time*". A multi-word analysis is required to identify the subjective phrases found in sentences. In this paper, we define subjective phrases as multi-word expressions that could show sentiment polarity. In previous studies, Wilson et al. [11] used a machine-learning approach by identifying the appropriate subjective phrase features to predict the phrase sentiment polarity, while Moilanen and Pulman [5] and Shaikh et al. [9] analyzed the words constituent within phrases to detect sentiment polarity. However, none had really analyzed the subjectivity that is found in multi-word phrases to provide a fine-grained analysis of the relationships between words that could affect the sentence sentiment polarity.

[1] `http://nlp.stanford.edu/software/lex-parser.shtml` retrieved on 14 June 2011.

The contributions of this paper are as follows. First, we propose a linguistic approach to evaluate typed dependencies polarity rules. Second, we consider various complex relationships between words, which improves sentiment polarity prediction performance as well as provide a better explanation of the grammatical relationships leading to the identified sentiment polarity. Third, we systematically evaluate our proposed model on a real-life sentiment dataset against two other approaches, namely a baseline counting-based approach and a typed-dependency approach.

2 Related Work

A common approach to extract patterns within word dependencies for sentiment detection was through heuristic means [5, 9, 10]. Moilanen and Pulman [5] proposed a sentiment composition model based on the concept that the global polarity of a sentence is a function of the polarities of its parts. For example, a rule ($OUT^{\alpha ij} \rightarrow SPR^{\alpha i} + SUB^{\alpha j}$) includes the consequent (OUT), the superordinate (SPR), which is the stronger of the input constituents, and the dominated constituent subordinate (SUB). The polarity (α) of the OUT constituent was determined by the SPR constituent and the compositional processes executed by the SPR constituent on the SUB constituent. Shaikh et al. [9] considered the semantic relationship between textual components in a sentence and the computation of contextual valence of the words to create word pattern rules. Examples of rules are [ADJ_{pos}+(CON_{neg} | NE_{neg})→(negative output)] (e.g., "*strong cyclone*) and [ADJ_{pos}+(CON_{pos} | NE_{pos})→(positive output)] (e.g., "*brand new car*") where the adjectives (ADJ), concepts (CON), and named entities (NE) were assigned valence groupings. Thet et al. [10] proposed the use of heuristic rules to compute the sentiment of a clause from the prior sentiment score assigned to individual words, taking into consideration the clause grammatical relations. The sentiment scores of each clause was then inferred with heuristic rules by using the grammatical dependencies and prior sentiment scores of the sentiment and feature terms. An example rule is [(ADJ_{pos}+N_{pos})→(positive output)] (e.g., "beautiful art") where ADJ(adjective) and N(noun) are the respective feature term prior sentiment scores. The use of typed dependency pattern rules alone may not predict the sentiment polarity accurately especially for long sentences. It was noted by Kennedy and Inkpen [4] that parsing the text can also serve as a basis for modeling *valence shifters* such as negation, intensifiers, and diminishers. Collocations and more complex syntactic patterns have also been found to be useful for subjectivity detection as reported in Riloff and Wiebe [8]. Polanyi and Zaenen [6] observed that there are contextual valence shifters that could influence the sentiment polarity of sentences. These include negators (e.g., "*is not*"), intensifiers (e.g., "*deeply*"), presuppositional items (e.g., "*barely*"), modals (e.g., "*might*"), and irony. Modal operators set up a context of possibility or necessity which do not necessarily reflect the author's attitude towards those entities in an actual situation under discussion. Modal and irony analysis require contextual information, which we are not considering in this study.

3 Linguistic Approach to Sentiment Polarity Detection

Our Polarity Prediction Model (PPM) uses a set of typed dependency polarity pattern rules and phrase-level analysis to predict the sentiment polarity of sentences. First, we parse the sentence dataset using the Stanford parser to generate the typed dependencies for the words. Secondly, we match the words with a list of lexicons and adjust the polarity. Thirdly, the rules are recursively evaluated to give the output polarity of the corresponding level of the phrase structure tree. Fourthly, at each level, the words in the phrase are matched against a subjective phrase lexicon to extract multi-word expressions that are subjective. In the final recursion, the sentence's sentiment polarity is predicted. An overview of PPM is shown in Figure 3.

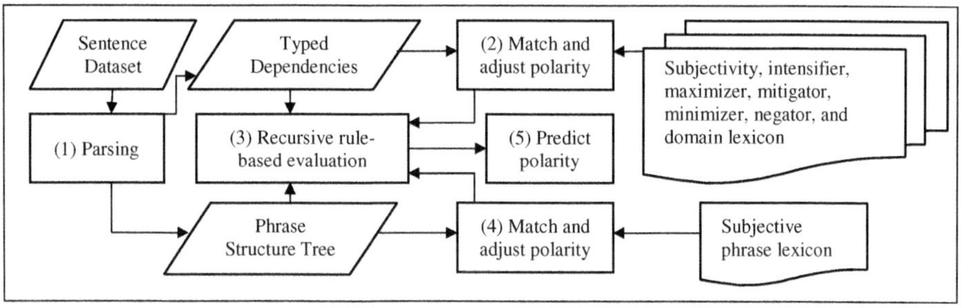

Fig. 3. Overview of the rule-based Polarity Prediction Model (PPM)

We use the blog dataset available from Zafarani and Liu [14] where long sentences (10 words or longer) were extracted as we attempt to handle the more complex relationship between words cases. A total of 402 sentences were used for lexicon building, and another 1000 sentences were used as the test data, which contains a balanced distribution of 300 positive and 300 negative sentences, and 400 neutral sentences. Two coders manually classified the sentiment orientation of each sentence into positive (+), negative (-), or neutral (0). The intercoder agreement Cohen's kappa index [1] is 0.78, which is in the acceptable range. The conflicting tags by the two coders were reviewed and then manually re-classified and used as the answer keys.

3.1 Typed Dependency Rules

The typed dependency polarity pattern rules used in our model, as shown in Tables 1 and 2, are adopted from Thet et al. [10]. The adjectival modifier (AMOD) typed dependency and adverb modifier (ADVMOD) typed dependency are evaluated as these are deemed to contain the vast majority of opinionated expressions [3, 13]. The AMOD of a noun phrase is any adjectival phrase that modifies its meaning. Similarly, an ADVMOD of a word is an adverb or adverbial phrase that modifies its meaning. The governor is a term or phrase in the typed dependency bigram that is modified by the dependent. In addition, the direct object (DOBJ) typed dependency and nominal subject (NSUBJ) typed dependency, as shown in Tables 3 and 4, are evaluated to handle the predicate (VP-to-NP) relation and the relation between noun phrase (NP) and verb phrase (VP), respectively in a sentence. Lastly, the negation (NEG) typed dependency is evaluated to detect negation in the sentence, where the output is the reverse of the recursion polarity.

Table 1. AMOD typed dependency rules

(G)overnor	(D)ependent	Sentiment Score	Examples
+ / 0*	+ / 0	+(|G|+((1-G|)*|D|)	*Fantastic*(D)[1.0]** *Product*(G)[0.0] → [1.0]
+/0	-	-|D|	*Worst*(D)[-1.0] *Scenario*(G)[0.0] → [-1.0]
-	+/0	-(|G|+((1-|G|)*|D|)	*Strong*(D)[0.5] *Complaint*(G)[-1.0] → [-1.0]
-	-	-(|G|+((1-|G|)*|D|)	*Little*(D)[-1.0] *Awkward*(G)[-1.0] → [-1.0]
*: 0 indicates neutral, **: [ ] denotes sentiment polarity score			

Table 2. ADVMOD typed dependency rules

(G)overnor	(D)ependent	Sentiment Score	Examples
+ / 0	+ / 0	+(|G|+((1-G|)*|D|)	*Enthusiastically*(D)[1.0] *Responded*(G)[0.0] → [1.0]
+/0	-	-|D|	*Glaringly*(D)[-1.0] *Stared*(G)[0.0] → [-1.0]
-	+/0	-(|G|+((1-|G|)*|D|)	*Probably*(D)[0.0] *Disappointed*(G)[-1.0] → [-1.0]
-	-	-(|G|+((1-|G|)*|D|)	*Strictly*(D)[-1.0] *Prohibited*(G)[-.05] → [-1.0]

Table 3. DOBJ typed dependency rules

(G)overnor	(D)ependent	Sentiment Score	Examples
+ / 0	+ / 0	+(|G|+((1-G|)*|D|)	*loved*(G)[1.0] the *phone*(D)[0.0] → [1.0]
+/0	-	-|D|	*face*(G)[0.0] an *onslaught*(D)[-0.5] → [-0.5]
-	+/0	-(|G|+((1-|G|)*|D|)	*killed*(G)[-0.5] *servers*(D)[0.0] → [-0.5]
-	-	-(|D| + (1-|D|)*|G|)	*destroyed*(G)[-0.5] the *virus*(D)[-1.0] → [-1.0]

Table 4. NSUBJ typed dependency rules

(G)overnor	(D)ependent	Sentiment Score	Examples
+ / 0	+ / 0	+(|G|+((1-G|)*|D|)	The *discussions*(D)[0.0] *resolve the problems*(G)[0.5] → [0.5]
+/0	-	If |D|>|G| output -|D| Else output |G|	The *failures*(D)[-1.0] *changes the power source*(G)[0.0] → [-1.0]
-	+/0	-|G|	The *thing*(D)[0.0] *really bugged me*(G)[-0.5] → [-0.5]
-	-	-(|G|+((1-|G|)*|D|)	The *virus*(D)[-1.0] *destroyed the codes*(G)[-0.5] →[-1.0]

We used the subjectivity lexicon from [12] to tag the prior polarity of the typed dependency records as denoted in the relation "typed-dependency(governor-term:[polarity], dependent-term:[polarity])→(polarity)". We convert the weak-subjective and strong-subjective tags found in the original lexicon to a polarity intensity score of ±0.5 and ±1 for the corresponding positive and negative terms. Neutral terms are tagged with a score of 0. From the example sentence, "*I really love the nice phone*" (Figure 1), the PPM recursively evaluate the respective typed dependency polarity pattern rules. From the recursive evaluation, the output sentiment polarity of each preceding typed dependency will be used as the governor polarity of the subsequent typed dependency. For example, while evaluating AMOD(*phone*[+], *nice*[+]) → [+] followed by DOBJ(*love*[+], *phone*[0]) → [+], the positive output of AMOD is input as the governor ("*love*") polarity of the subsequent DOBJ typed dependency pattern rule. The DOBJ output is then recursively input as the governor polarity of the subsequent typed dependencies, ADVMOD(*love*[+], *really*[intensify]) → [+], followed by NSUBJ(*really love the nice phone*[+], I[0]) → [+]. We could then detect the positive polarity of the relations between the NP ("*I*") and VP ("*really love the nice phone*") given by the typed dependency NSUBJ(*really love the nice phone*[+], I[0]) → [+].

3.2 Handling Complex Relationships

It is common to find more complex relationships between words in a long sentence. For example, the sentence "*While interesting, this phone is hardly in the same league with the iphone*", contains negation ("*hardly*") and a positive subjective phrase ("*in the same league*") giving the sentence an overall negative sentiment. If we only consider the adjective "*interesting*" found in the subjectivity lexicon, the prediction would yield an incorrect positive result. Phrases could contain subjectivity that influences the sentiment polarity of the sentence as seen in the positive phrase "*in the same league*" containing only neutral words. Other subjective phrases include subjective phrases that have neutral sentiment orientation such as the negative word "*close*" in "*close up*" as in "*take a close up look*". Subjective phrases could also result in opposite polarity such as the negative word "*break*" in the positive phrase "*break through*". We analyze the multi-word expressions at phrase-level in order to detect phrase subjectivity and adjust the output polarity to the correct polarity. A subjectivity phrase lexicon containing 5330 phrase-lexicon terms was built from the training dataset and a list of manually annotated sentiment polarity phrasal verbs and idioms. To handle the more complex relationships between words, additional lexicons and polarity rules, as shown in Table 5, are incorporated into the model. The idea is to identify key expressions that could alter the intensity of phrase intensity.

Table 5. Lexicon Based Rules

(M)odifier	(T)erm	Sentiment Score	Examples
Negation	T	-1*T	Hardly good
Intensify	T	(2*T), max value = ±1	Highly recommended
Mitigate	T	(0.5* T)	Slightly better
Maximize	T	1*Polarity(T)	Totally bad
Minimize	T	0.25* T	Least effort

These lexicons include negators (10 terms), intensifiers (42 terms), mitigators (45 terms), maximizers (43 terms), and minimizers (10 terms). The lexicon terms are collected from the training dataset and the given lists from Quirk et al. [7]. The intensity of sentiment polarity scores could also affect the eventual polarity of the sentence. For example, in the sentence, "*This is quite a good idea still in its infancy*" , the mitigating effect of the words "*quite*" and "*still*" reduce the intensity of the positive word "*good*", which results in a less positive sentence sentiment polarity. In the negation rule, we reverse the polarity score of the word that is modified by the negator. In the intensify rule, the polarity score of the modified word is doubled, but limited to a value of ±1 in consideration for the increasing effect of the intensifier. Conversely, the polarity of the modified word is halved in the mitigate rule. The polarity score is maximized to ±1 in the maximize rule to denote the upper extreme of the intensity scale. In the minimize rule, the polarity score of the modified word is severely reduced to account for the lower extreme of the intensity scale. Additionally, we use a domain lexicon (40 terms) to correct domain-specific subjective words (e.g., "*hard disk*" and "*open source*"). Next, we group the typed dependencies to be evaluated under the corresponding phrase according to the phrase head word. The phrases are then matched with the subjective phrase lexicon to give the annotated phrase polarity. We use a bottom-up approach in evaluating the rules, taking into consideration the nested level of each item in the phrase structure tree to ensure that only items from the same level are evaluated and the resultant polarity propagated to subsequent levels for further evaluation. For example, in the positive expression "*no doubt, it is good*", if the nested levels are not considered, the evaluation would have incorrectly negate the neutral output of ("*good*" + "*doubt*").

4 Results and Discussion

We conducted experiments using the baseline approach, typed dependency (TD)-only approach, and PPM. The baseline uses the average count of the positive terms and negative terms in the sentence to predict the polarity of the sentence. In the TD-only approach, only the AMOD, ADVMOD, NSUBJ, DOBJ, and NEG typed dependency polarity pattern rules are used in the evaluation. Lastly, we test out PPM. All three tests are conducted on the testing dataset, and their results are compared using the F1 score measure, with the formula given as F1=(2•Precision•Recall)/(Precision+Recall) . The test results are shown in Table 6, 7, and 8 respectively.

Table 6. Baseline test results

Polarity	Precision	Recall	F1-Score	Average
Neutral	56.64%	65.00%	60.54%	
Positive	51.69%	61.00%	55.96%	54.85%
Negative	62.57%	39.00%	48.05%	

Table 7. Typed dependency (TD)-only test results

Polarity	Precision	Recall	F1-Score	Average
Neutral	53.04%	89.50%	66.60%	
Positive	64.44%	38.67%	48.33%	56.43%
Negative	83.45%	40.33%	54.38%	

Table 8. Polarity Prediction Model (PPM) test results

Polarity	Precision	Recall	F1-Score	Average
Neutral	55.94%	76.31%	64.56%	59.39%
Positive	61.27%	58.00%	59.59%	
Negative	74.71%	42.33%	54.04%	

The average F1-score for the baseline approach (54.85%) is relatively lower than the TD-only (56.43%) and the PPM (59.39%) approach. The baseline approach only considers the number of positive and negative terms and does not take into account the possible syntactic or semantic relationships between words that could influence the sentence sentiment polarity. This is seen in the case for the expression "*remove the bug*". The baseline approach would have counted two negative terms ("*remove*" and "*bug*") and output a negative sentiment polarity for the sentence. In the TD-only and PPM approach, the DOBJ(remove[-], *bug*[-])→ [+] polarity pattern rule would have given the correct positive sentiment output. The low negative F1-score (48.05%) of the baseline could be due to the baseline approach not being able to detect negation of positive sentiments resulting in a low recall of negative sentiment polarity cases.

Although the PPM performed only around 3% better than the TD-only approach in terms of overall F1-score, its positive F1-score (59.59%) is more than 10% better than that (48.33%) for the TD-only approach. This is because the test dataset contains a large proportion of long sentences, whose structure could significantly influence the sentiment polarity output; the TD-only approach is unable to account for the syntactic phrase structure of the sentence while analyzing the words' sentiment polarity. Additionally, the PPM approach considered the complex relationships between words, such as subjective phrases, intensifiers, mitigators, maximizers, minimizers, and domain terms, which could provide a more accurate representation of the sentence sentiment polarity. However, the small improvement (2.96%) in overall F1 performance could be due to the fact that the test dataset may not contain enough subjective phrases like idioms or subjective phrasal verbs that are found in the subjective phrase lexicon. An example of how the complex word relationships found in a sentence could affect the overall sentence sentiment polarity is shown in the neutral sentiment oriented sentence "*This is a quick and dirty*[-] *podcast so the production* *isn't*[negation] good [+] *but the content* makes up[+] *for that*", with the generated typed dependencies and phrase structure tree shown in Figure 4 and 5 respectively. The TD-only approach would evaluate the negative term ("*dirty*") and the negation of the positive term ("*good*") to give a negative sentiment polarity output. The TD-only approach could not detect the positive subjective phase "*makes up*" in the bigram words as "*makes*" and "*up*" are individually neutral terms. The PPM analyzed the sentence recursively from bottom-up, "*makes up*" [+1.0 expression] → intermediate output [+1.0] of the grouping (S1). Next, following the recursion for the grouping (S2), "*good*" [+1.0] and "isn't" [negation] → intermediate output [-1.0]. The grouping (S3) evaluation would yield an intermediate output [-1.0] from the negative oriented phrase "*quick and dirty*". S2's [-1.0] output would then combine with S3's [-1.0] output due to the nested hierarchy of S2 within S3. This gives the combined S2S3[-1.0] polarity output by taking the average of the sentiment scores of both S2 and S3. Eventually, the S2S3[-1.0] sentiment score and S1's [+1.0] sentiment score is averaged to give a neutral sentiment polarity prediction for the sentence. This shows that by considering the more complex relationships between words, we could conduct a more in-depth analysis on the possible influence that could affect sentence sentiment polarity. The typed dependency tree was not used in our approach as it is more difficult to identify the phrase structure from a typed dependency tree as compared with the phrase structure tree.

nsubj(podcast-7, This-1)
cop(podcast-7, is-2)
det(podcast-7, a-3)
amod(podcast-7, quick-4)
cc(quick-4, and-5)
conj(quick-4, dirty-6)
mark(good-13, so-8)
det(production-10, the-9)
nsubj(good-13, production-10)
cop(good-13, is-11)
neg(good-13, n't-12)
dep(podcast-7, good-13)
cc(podcast-7, but-14)
det(content-16, the-15)
nsubj(makes-17, content-16)
conj(podcast-7, makes-17)
prt(makes-17, up-18)
prep(makes-17, for-19)
pobj(for-19, that-20)

Fig. 4. Typed dependencies

```
(ROOT
  (S
    (NP (DT This))
    (VP (VBZ is)
      (NP
        (NP (DT a) (JJ quick)                 } S3
          (CC and)
          (JJ dirty) (NN podcast))
        (SBAR (IN so)
          (S
            (NP (DT the) (NN production))
            (VP (VBZ is) (RB n't)             } S2
              (ADJP (JJ good))))))))
    (CC but)
    (S
      (NP (DT the) (NN content))
      (VP (VBZ makes)                          } S1
        (PRT (RP up))
        (PP (IN for)
          (NP (DT that)))))))
```

Fig. 5. Phrase structure tree

However, the baseline approach can sometimes predict a correct sentiment polarity output without considering the subjective phrases as shown in the single count for "*good*" in the positive sentence, "*Now I* <u>*don't*</u>(negation) *use the term* <u>*back seat*</u>(-expression) *lightly since Macs still being in the car for the ride is a* <u>*good*</u>(+) *thing.*". Nevertheless, the PPM provides a deeper analysis of the relationships between words in predicting the output. Errors in the PPM include subjective phrases that are not identified in the subjective phrase lexicon. Examples include the negative term "cut" which is neutral in context for the expression "*you can download, print,* <u>*cut out*</u> *and fold together*". While the subjectivity lexicon contains quality annotation of subjective terms, there are still subjective words (e.g., recognize, trained, and etc.) that are not included. It is also possible to provide a wrong sentiment polarity to words due to different context. For example, the word "*little*" is considered negative in the subjectivity lexicon, but is neutral in context in the sentence "*put up a* <u>*little*</u> *gallery with a bunch of the template…*". Errors were found in the typed dependencies parsing. For example, negation was not identified in the phrase "no bargaining chips", where only the det(*chips, no*) and nn(*chips, bargaining*) typed dependencies were generated. We observed that subjective phrases not only determine the sentiment polarity of phrases, but could also alter the intensity of sentiment polarity through intensification (e.g., "long since" in "he has long since won the award"), maximization (e.g., "the one and only"), mitigation ("more or less"), minimization (e.g., "not going anywhere" in "the product is not going anywhere").

5 Conclusion

Previous studies [5, 10] have shown that typed dependencies could be used to determine the sentence-level sentiment polarity. However, the more complex relationships between words affect the sentiment polarity of sentences. Our study further analyzes the complex relationships between words in a sentence, with PPM performing better than the baseline models, achieving more than 10% better performance in identifying positive sentiments. This shows that sentiment polarity prediction performance could be improved significantly by taking into consideration the more complex relationships between words. However, our model could not detect subjectivity due to context. Our study has provided a fine-grained analysis of the word relationships which could explain the effect that these relationships have on the sentence sentiment polarity. We believe that this study would provide a sound basis for future work using typed dependencies to predict sentence-level sentiment polarity.

References

1. Cohen, J.A.: Coefficient of Agreement for Nominal Scales. Educational and Psychological Measurement 20(1), 37–46 (1960)
2. Hassan, A., Qazvinian, V., Radev, D.R.: What's with the attitude? A study of participant attitude in multi-party online discussions. In: Empirical Methods on Natural Language Processing, pp. 1245–1255. ACL (2010)
3. Jakob, N., Weber, S.H., Muller, M.C., Gurevych, I.: Beyond the stars: exploiting free-text user reviews to improve the accuracy of movie recommendations. In: Conference on Information and Knowledge Management, pp. 57–64. ACM, New York (2009)
4. Kennedy, A., Inkpen, D.: Sentiment classification of movie reviews using contextual valence shifters. Computational Intelligence 22(2), 110–125 (2006)
5. Moilanen, K., Pulman, S.: Sentiment Composition. In: Recent Advances in Natural Language Processing, pp. 378–382 (2007)
6. Polanyi, L., Zaenen, A.: Contextual Valence Shifters. Computing Attitude and Affect in Text: Theory and Applications 20, 1–10 (2006)
7. Quirk, R., Greenbaum, S., Leech, G., Svartvik, J.: A Comprehensive Grammar of the English Language. Longman, Redwood City (1985)
8. Riloff, E., Wiebe, J.: Learning extraction patterns for subjective expressions. In: Empirical Methods on Natural Language Processing, pp. 105–112. ACL (2003)
9. Shaikh, M.A.M., Prendinger, H., Ishizuka, M.: Sentiment Assessment of Text By Analyzing Linguistic Features And Contextual Valence Assignment. Applied Artificial Intelligence 22(6), 558–601 (2008)
10. Thet, T.T., Na, J.-C., Khoo, C.S.G.: Aspect-Based Sentiment Analysis of Movie Reviews on Discussion Boards. Journal of Information Science 36(6), 823–848 (2010)
11. Wilson, T., Wiebe, J., Hoffmann, P.: Recognizing contextual polarity: An exploration of features for phrase-level sentiment analysis. Computational Linguistic 35(3), 399–433 (2009)

12. Wilson, T., Wiebe, J., Hoffmann, P.: Recognizing contextual polarity in phrase-level sentiment analysis. In: Human Language Technology and Empirical Methods in Natural Language Processing, pp. 347–354. ACL (2005)
13. Wilson, T., Wiebe, J., Hwa, R.: Recognizing Strong and Weak Opinion Clauses. Computational Intelligence 22(2), 73–99 (2006)
14. Zafarani, R., Liu, H.: Social Computing Data Repository at ASU Arizona State University, School of Computing, Informatics and Decision Systems Engineering (2009)

A System for Using National Bibliographies in Rights Information Infrastructures

Nuno Freire and Andreas Juffinger

The European Library, National Library of the Netherlands,
Willem-Alexanderhof 5, 2509 LK, The Hague, Netherlands
nfreire@gmail.com, andreas.juffinger@kb.nl

Abstract. In the process of digitising a book, a library needs to clear the rights associated with it. Rights clearance is largely a manual, time-consuming process which possibly costs more than the actual digitisation. To analyse the rights situation, a range of information is required, which is distributed across several national databases hosted in national libraries, publishers and collective rights organisations. National bibliographies are a key data source in these processes, as they are the only source to identify all the publications of a specific intellectual work in a country. However, national bibliographies are not designed and built to support rights clearance purposes. The information in bibliographic records results from cataloguing with users and library management in mind, and links between different publications of a single intellectual work are not available. This paper presents a system implemented in The European Library to integrate national bibliographies into the ARROW (Accessible Registries of Rights Information and Orphan Works) rights information infrastructure. The system makes it possible to identify all different publications with a common underlying intellectual work. This system forms the main source of bibliographic metadata due to the fact that The European Library is an aggregator of all Europe's national library catalogues.

Keywords: information extraction, work entity matching, national bibliographies, library catalogues, copyright.

1 Introduction

Libraries, publishers and collective rights organisations are discussing possible ways of maximising access to digital content in Europe without harming the rights of authors and copyright owners. It is commonly believed that libraries hold materials of public interest, which should be made digitally available for a broader public. Legal issues make it necessary to determine the exact copyright status of those works before a library can digitise a manifestation of that work. One of the major challenges in rights clearance is the significant fragmentation of rights information across multiple infrastructures, some of which are not remotely accessible. This makes the rights clearance process very demanding and expensive for libraries. Large scale digitisation projects can digitise thousands of books per week, and therefore there is a significant need to develop faster ways to clear the copyright status of the books.

C. Xing, F. Crestani, and A. Rauber (Eds.): ICADL 2011, LNCS 7008, pp. 88–97, 2011.

In the ARROW project[1] (Accessible Registries of Rights Information and Orphan Works), a single framework is being established to combine and access rights information. It proposes to create a seamless service across a distributed network of national databases containing information that will assist in determining the rights status of works. Once established, this infrastructure will provide valuable tools for libraries and other organisations to identify and contact rights holders in seeking rights clearance for the use of content.

National bibliographies are one of the key data sources in the rights clearance process. Their purpose is to list every publication in a country, under the auspices of a national library or other government agency. Depending on the country, all publishers in a country will need to send a copy of every published work to the national legal deposit, or in other countries, a national organisation will need to collect all publications. Given that the publisher domain is very heterogeneous and that thousands of publishers might exist in a country, national bibliographies are effectively the single point of reference with which to identify all the publications of an intellectual work by country. However, national bibliography catalogues are created for library management, preservation and, of course, for library users. Due to this fact they are not immediately usable for the purpose of rights clearance. The necessary information resides encoded, structured and unstructured, in the national bibliographies. For rights clearance, it is necessary to process the bibliographic data to make relevant information explicit, cleaned and normalised before it can be used.

This paper presents the system implemented in The European Library[2] for using national bibliographies in rights information infrastructures in order to identify all different publications with the same underlying intellectual work. This system supports, and is integrated into, the rights management framework of project ARROW.

This paper describes in Section 2 the rights clearance process of project ARROW, and gives an introduction to the data quality challenges of national bibliographies. Section 3 presents related work on the matching of works in library data. The system is presented in Section 4 and it is evaluated in Section 5. Section 6 presents the conclusions and briefly analyses the particular data problems that could not be overcome, and which will be addressed in future work.

2 The Problem

The main motivation behind ARROW is to support mass digitisation projects by finding automated ways to clear the rights of the books to be digitised. This rights clearance process is time consuming, since a library has to go through the following steps for each book:

- To identify the underlying work incorporated in the book to be digitised;
- To find out if the underlying work is in the public domain or in copyright, and whether it is an orphan work or out-of-print;
- To clearly describe the use that is requested for the book, such as digitisation for preservation, electronic document delivery, etc.;

[1] `http://www.arrow-net.eu/`
[2] `http://www.theeuropeanlibrary.org`

- To identify the rights holder(s) or their agent, such as a collecting society;
- To seek the appropriate permission, if necessary.

This process depends on the availability of existing bibliographic and rights data. There are already established information sources for printed material, in national bibliographies, books in print and the databases of rights organisations. Currently, these sources are not interoperable because of differences in data collection policies and data schemas. Bibliographic databases rarely include metadata about rights ownership and usage policies. Such information is usually held in a wide array of formats by publishers, collecting societies and authors.

ARROW addresses the interoperability of rights information along this process. It supports the identification of a work, the clarification of its rights status and the identification of the rights holders. The remainder of this section will present the complete ARROW workflow, and discuss the particular role of national bibliographies and related challenges.

2.1 The ARROW Rights Clearance Workflow

Figure 1 shows an overview of the basic workflow of ARROW. It starts from a library as a potential user that wishes to digitise a book, and shows the process that the ARROW system must support to provide a response containing the requested rights information. The process depends on data from several sources:

- National bibliographies' data aggregated in The European Library (TEL);
- Author data from the Virtual Authority File (VIAF);
- Publishing data from Books In Print databases (BIP);
- Rights holders data from Reprographic Rights Organisations (RRO);
- The International Standard Text Code (ISTC), a numbering system developed to enable the unique identification of textual works.

The initial steps of the workflow depend on The European Library's system for three tasks:

- To identify the exact record of the book that the library wants to digitise;
- To identify other records of books which share the same work;
- To gather data about the contributors of the work.

After all publication and contributor data is identified at The European Library, the ARROW workflow proceeds with the determination of the copyright status, the identification of publications still in commerce, the identification of rights holder(s), and finally seeks the appropriate permission from the rights holders.

2.2 National Bibliographies at the European Library

National bibliographies are typically created and maintained by national libraries. Whenever a book is published in a country, it is recorded in the corresponding national library catalogue from where the national bibliography is derived. However, each new publication of a work is recorded independently of all others, and has its own record.

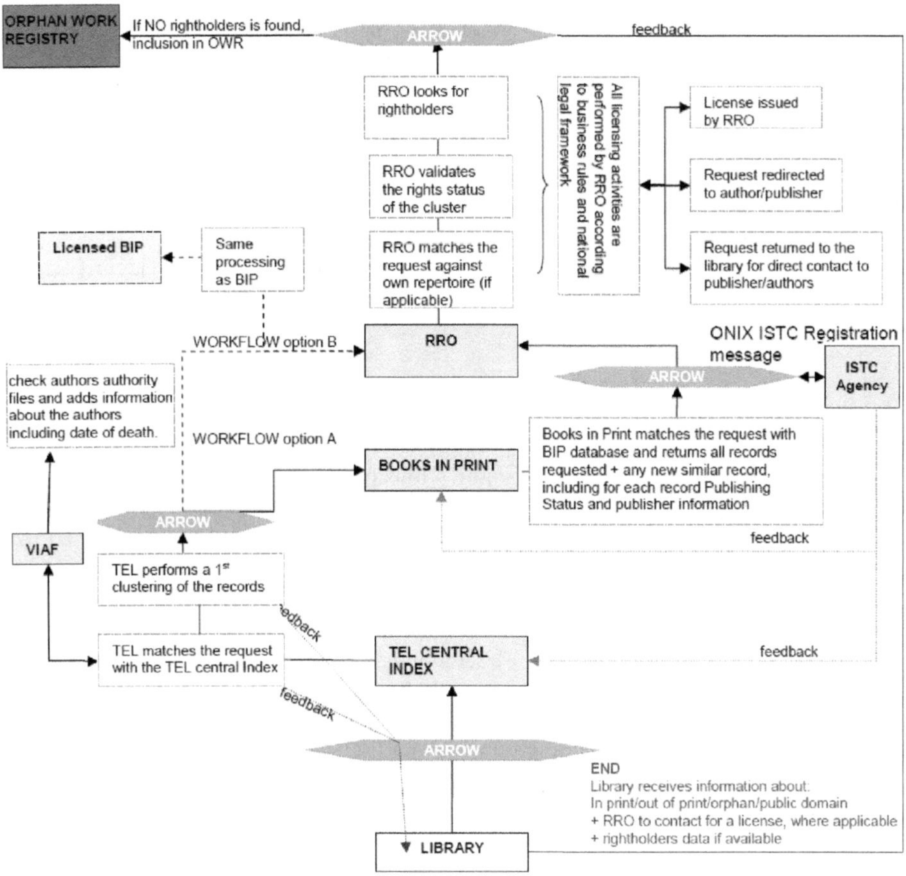

Fig. 1. The ARROW Rights Clearance Workflow

National bibliographies are created mainly having in mind the users of bibliographic data and the management of library holdings. So, although the required information exists within national bibliographies, it is not available in a structured form.

Besides the lack of the necessary data structure, national bibliographies also face data quality problems. In spite of many efforts that libraries undertake regarding the standardisation of cataloguing practices and bibliographic data formats, a profusion of heterogeneous data is still found in library catalogues.

Cataloguing rules still leave room for different interpretations, and the information that libraries record in catalogues is often too complex to be encoded in a structured form for machine processing. Librarians often have to resort to general notes fields to record particular information, or they have to work with the limitations of the information systems that are not always up-to-date with the standards, or which do not fully enforce and validate cataloguing practices. Therefore, the same information may be represented quite differently from library to library, and even within one library. In addition, these data sources are also subjected to general data quality problems, such as typing errors, misspellings, synonyms, homonyms, abbreviations, etc.

Since each publication of a work is recorded by libraries separately, the work may be represented in different ways. Starting from a particular publication of a work, the task of retrieving all other publications cannot be achieved easily by searching for equal values on titles and contributors [1]. Comparison and matching of work data has to be carried out taking into consideration the heterogeneity of data. Without it, the task of rights clearance would be performed on incomplete information, which could lead to incorrect results with legal consequences for libraries, or which could be unfair for rights holders.

Although national bibliographies seem promising data sources for identifying all the publications of a work, this process must be done with some uncertainty due to the characteristics of the data.

3 Related Work

Previous attempts to match different publications of a work have been carried out, but with different requirements than those of rights clearance.

The The International Federation of Library Associations and Institutions (IFLA) Study Group on the Functional Requirements for Bibliographic Records published its report 'Functional Requirements for Bibliographic Records (FRBR)' [3]. This report describes a model that identifies and clearly defines the entities of interest to users of bibliographic records, the attributes of each entity, and the types of relationships that operate between entities. The intent was to produce a new conceptual model that would serve as the basis for relating specific attributes and relationships to the various tasks that users perform with bibliographic data. One of the entities defined is the work*: a distinct intellectual or artistic creation* [3].

The publication of the FRBR report had the most impact on the evolution of traditional cataloguing rules. However, it also gave rise to a line of work on the development of techniques to transform the legacy data into an FRBR implementation model, a process often referred to as 'FRBRisation'[4]. Studies have shown that much of the information needed to FRBRise catalogues is already present in library catalogues [5], including the information about the work. In fact, the work entity is probably the entity defined in FRBR that attracted the most interest in FRBRisation research.

Several efforts have been undertaken to develop algorithms for the FRBRisation of bibliographic data, with the OCLC FRBR Work-Set Algorithm[3] being the most important reference. This algorithm is focussed on the clustering of data looking at the FRBR group 1 (Work, Expression, Manifestation and Item) but with special concern on works. Following this work, other efforts endeavoured to take FRBRisation a little further by allowing clustering of all entities of the FRBR model. An example is the BibSys FRBR conversion tool [4], which essentially builds a FRBR structure from bibliographic records and tries to cluster all entities of that structure by comparing their keys (the combination of properties found in a FRBR entity). In spite of all these efforts, authors further explain that "algorithms for eliciting FRBR structure" will only work as well as the bibliographic data on which they are based [4], and in national bibliographies data heterogeneity and data quality problems are clearly present.

---

[3] http://www.oclc.org/research/activities/past/orprojects/frbralgorithm/default.htm

The system described in this paper has its bases on a previous study conducted to evaluate the suitability of national bibliographies for identifying all the publications of a work to support rights clearance processes [2]. The conclusion of this study was that with the applied techniques, the identification of all publications is not sufficiently accurate for a fully automatic rights clearance process. To be able to identify more than 95% of the publications of a work, it was found that the algorithms also wrongly identified publications of different works. Therefore, human verification may be required, and the system described in this paper takes these results into consideration for its design.

4 The ARROW System at the European Library

This section presents the main design aspects of the system at The European Library. It starts with the description of the most relevant requirements from ARROW, and follows with the presentation of software architecture, some aspects of its data model, and concludes with some implementation and deployment details.

4.1 Functional Requirements

The main use case presented by the ARROW workflow is the *Work Clustering of Manifestations*. It has as input, a record from the national bibliography, chosen in the previous step of the workflow. The system should identify all other manifestations that potentially share the work, in part or totally.

Even though the clustering is to be done only with records of the national bibliography, elements which must be considered are data heterogeneity and missing data. Based on the results of a previous study [2], the manifestations should be clustered hierarchically in one *Primary Cluster* and several *Secondary Clusters*. The clusters should be formed as follows:

- *Primary Cluster* – This cluster should be formed by all manifestations that contain the same, or very similar, title, the same contributors, and the same language(s) of text.
- *Secondary Clusters* – These clusters should provide an organised view on similar manifestations, which may actually share the same work with the input manifestation, but were not included in the *Primary Cluster* due to data heterogeneity. These clusters should identify manifestations with the following similarities:
    - Same first contributor – because additional contributors are very often not catalogued in a structured way.
    - Similar titles by the same contributors – because titles are sometimes catalogued in different ways, particularly subtitles.
    - Manifestations catalogued without language of text – if the language is missing, it may be the same language as the input manifestation.
    - Manifestations catalogued without country of publication – the missing value may indicate that the manifestation was not published in the target country, but it may also be missing by error.

This organisation of the manifestations in different clusters aims to allow the ARROW workflow to proceed automatically in some cases, or reduce the manual effort in the verification of the results.

More comprehensive data about the contributors may be of value for other tasks of the ARROW workflow. For this reason, data about the contributors in each cluster should be complemented with data available in VIAF, such as variant forms of the names, dates of birth and death, and nationality.

Each cluster should be described by work level metadata, so that it can be used in the workflow for interoperability with publisher's bibliographic data and ISTC.

4.2 System Architecture

The system comprises six main software components, as shown in Figure 2. Together, these components provide the system with a repository for metadata, functionality for information extraction, similarity searching and clustering, and interoperability with external systems.

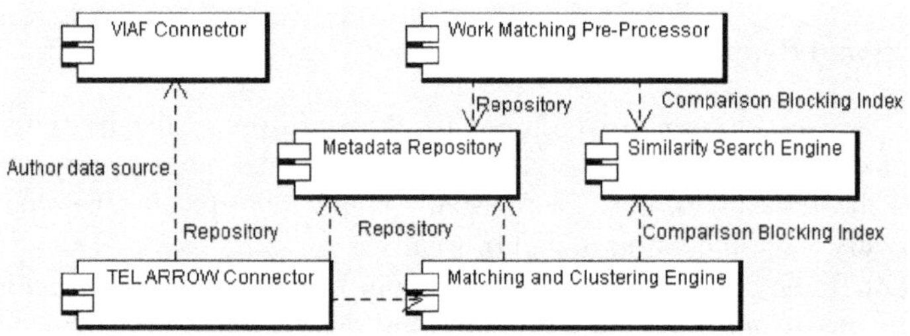

Fig. 2. The system components diagram

The *Work Matching Pre-processor* component is the main processor of bibliographic data from the national bibliographies. It addresses heterogeneity and quality issues in the data, so that the rest of the system may function on consistent data. It applies data cleaning, information extraction and data transformation techniques, in order to extract and normalise work metadata from bibliographic records.

The *Similarity Search Engine* allows the processing of similarity queries on work metadata. Its main role is to improve the processing performance of the *Matching and Clustering Engine*, by restricting the processing to a subset of the national bibliography, containing only records with a minimal level of similarity with the input record. Its implementation is based on indexing of character n-grams[4] of titles and author names.

The *Metadata Repository* component provides storage for the bibliographic records in the catalogues of national libraries and their corresponding work metadata.

The *Matching and Clustering Engine* component is responsible for the main processing of the ARROW requests. It applies duplicate detection techniques that do not require data to be encoded in the same way, and does not need the existence of identifiers to detect duplicates. The work clustering of manifestations are a form of

[4] An n-gram is a subsequence of n characters from a given character sequence.

duplicate detection, applied with defined criteria based on work metadata. These criteria defined how primary and secondary clusters should be formed.

The *VIAF Connector* component provides the retrieval of authority records about contributors from VIAF. These are used by the *TEL-ARROW Connector* for enrichment of the work metadata provided to ARROW.

Communication and exchange of data with ARROW is carried out by the *TEL-ARROW Connector*. This component implements the web services interface of ARROW and the ARROW data formats.

4.3 Data Model

Several data formats are used at the national libraries, typically from the family of MARC formats. Although at this stage, only MARC21 is supported in the ARROW workflow, The European Library system was designed to be independent of the MARC format. In this way, support for other formats such as UNIMARC, or even national variations of MARC21, can be added to the system without major implications to the system. The independence from the source MARC formats is achieved through an internal data model for works.

Work metadata is extracted from the MARC records by the *Work Matching Preprocessor* component when the national bibliographies are ingested into the system. The work metadata model aims to fulfil all the data needs of the system itself, and the needs of the ARROW workflow.

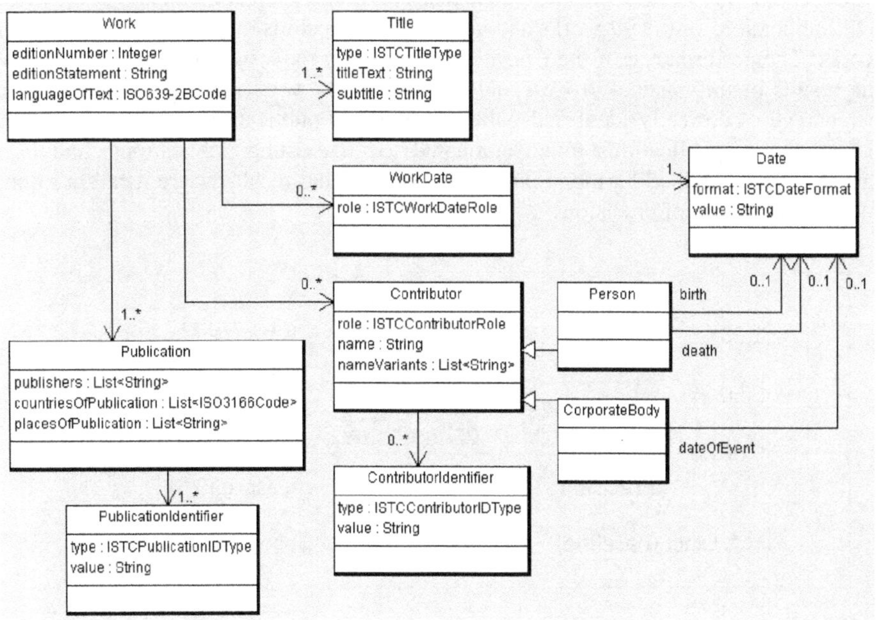

Fig. 3. The system internal work data model diagram

Two standard options for representing the work metadata were considered: the Functional Requirements for Bibliographic Records (FRBR) and the work metadata of ONIX ISTC[5]. The final decision was to use ISTC as a basis since the concept of work in ISTC is closer to the objectives of ARROW than that of FRBR, which is focussed on library users' needs, making it less suitable for rights clearance purposes.

The internal work data model (shown in Figure 3) uses ISTC entities and ONIX code lists extensively, but it contains adaptations for these particular applications. The structure is a subset of the ISTC, since libraries' MARC records do not contain all the data of ISTC. On the other hand, data fields that are not available in ISTC were added to support the match and cluster manifestation requests (for example, countries of publication, contributor name variants and life dates).

5 Evaluation

The system was evaluated on the pilot countries of ARROW: France, Germany, Spain and the United Kingdom. This section presents the evaluation of the *Work Clustering* results.

For the evaluation, 99 works were analysed. A semi automatic process identified 1,593 records of publications of these, or similar works. These records were inspected, and the publications of the same work where identified by a person, taking into consideration any information available in the bibliographic record, whether it was structured and machine readable, or whether it was in unstructured data fields which could only be understood by humans. The final evaluation collection identified 578 publications of the 99 works and another 1,015 records of publications of similar works. The performance of the *Cluster Manifestation* requests was compared against the results of this manual process and measurements were made for precision (the percentage of correctly identified publications in all publications found), recall (the percentage of publications found compared to all existing publications) and $F_{0,5}$-measure (the weighted harmonic mean of precision and recall, where recall has double the importance of precision).

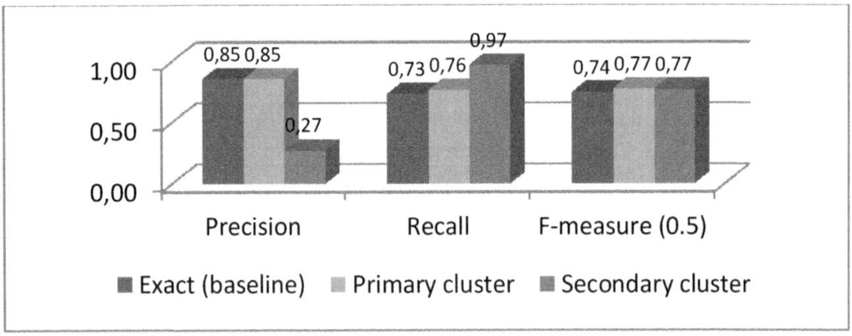

Fig. 4. Evaluation results of *Work Clustering*

[5] http://www.editeur.org/106/ONIX-ISTC-Registration-Format/

The measurements were made at three levels: at the *Primary Clusters*; at the *Secondary Clusters*; and, as a baseline, where the results were obtained without any similarity matching (exact comparison). The results are presented in Figure 4. At the *Primary Clusters* the system achieves a precision of 0.85 with a recall of 0.76. The *Secondary Clusters* allow the identification of nearly all publications with a recall of 0.97, however the precision lowers considerably to 0.27. The final 3% of the cases that would allow reaching 1.0 recall are cases where the records have a missing or invalid language of text, or have very unclean data in the title field.

6 Conclusion and Future Work

The system of The European Library provides ARROW with the capability to identify the common underlying work of a set of publications, which is a vital part of the rights clearance process. The evaluation of the system revealed that it reaches a recall of 0.97 and that some relevant publications of a work are still not identified. However, the analysis of the missed cases gives an insight into how the system could be improved: Regarding matching of titles, it could be supposed that a machine learned similarity metric for titles in bibliographic records could better discover the patterns in titles with superfluous data. When language information is missing, techniques for language detection could be applied, or it might also be reasonable to assume that the language of the publication is identical to others with the same title published in the same country. The presence of contributors in descriptive and natural text fields could be identified by applying named entity recognition.

Acknowledgements. This work was carried out through the ARROW project, which is funded under the eContentplus programme of the European Comission.

We would like to thank the Associazione Italiana Editori, the Bibliothèque nationale de France, and the University of Innsbruck for their support in the design and evaluation of the system.

References

[1] Freire, N., Borbinha, J., Calado, P.: Identification of FRBR Works within Bibliographic Databases: an Experiment with UNIMARC and Duplicate Detection Techniques. In: Goh, D.H.-L., Cao, T.H., Sølvberg, I.T., Rasmussen, E. (eds.) ICADL 2007. LNCS, vol. 4822, pp. 267–276. Springer, Heidelberg (2007)

[2] Freire, N., Juffinger, A.: Using national bibliographies for rights clearance. In: The Proceeding of the ACM/IEEE Joint Conference on Digital Libraries 2011 (to appear, 2011)

[3] IFLA study group on the Functional Requirements for Bibliographic Records, "Functional requirements for bibliographic records: final report". UBCIM publications, 19. K. G. Saur

[4] Aalberg, T., Žumer, M.: Looking for Entities in Bibliographic Records. In: Buchanan, G., Masoodian, M., Cunningham, S.J. (eds.) ICADL 2008. LNCS, vol. 5362, pp. 327–330. Springer, Heidelberg (2008)

[5] Hegna, K., Murtomaa, E.: Data Mining to find FRBR? International Cataloguing and Bibliographic Control 32, 52–55 (2003)

Exploiting Attribute Redundancy for Web Entity Data Extraction

Yanxu Zhu[1], Gang Yin[1], Xiang Li[1], Huaimin Wang[1,2],
Dianxi Shi[1], and Lin Yuan[3]

[1] College of Computer Science and Technology,
National University of Defense Technology, 410073 Changsha, Hunan, China
[2] National Laboratory for Parallel and Distributed Processing,
National University of Defense Technology, Changsha 410073, China
[3] College of Electronic Technology,
Information Engineering University, 450004 Zhengzhou, Henan, China
{zhu.yannick,jack.nudt,shockleyle,
dxshi1998,fkefss}@gmail.com, whm_w@163.com,

Abstract. Web entities are often associated with many attributes that describe them. It is essential to extract these attributes for Web entity data extraction. This paper proposes a novel approach using duplicated attribute value pairs. We start by constructing a initial seed set of attributes including names and enumerable values, and a training set of Web pages from target website; After that we locate the position of each attribute by matching attribute values within the pages of the site with values contained in the seed set; Thirdly we choose the position with the highest supportiveness as path for extraction, which we use to extract other attribute value pairs with the same template. Finally, we conduct an extensive experimental study with large real data set to demonstrate the effectiveness of our extraction approach.

Keywords: Web Entity Data Extraction, Attribute Redundancy.

1 Introduction

The World Wide Web contains a wealth of information about immense amounts of entities, such as electronic products (its attributes such as price, description and comments), open source software (attributes include home page, developers, programming language, operating system support, and so on), hotels, books, tickets, etc. These entities extracted from heterogeneous Websites can be integrated into some more valuable databases, which provide data source for many up-level Web applications (such as comparative shopping, vertical search). Current research shows that more than 40-50%[1] Web data are generated by templates. Templates provide structure clues for entity information extraction, extraction rules could be generated automatically by mining repeated patterns of templates[2][10][11], but Websites are heterogeneous, extraction rules for one Website is not suitable for another. It is inevitable to design extraction algorithms and customized training sample, which is time-consuming. So entity data extraction from heterogeneous Websites has been a

C. Xing, F. Crestani, and A. Rauber (Eds.): ICADL 2011, LNCS 7008, pp. 98–107, 2011.

hot topic in the field of information extraction. Finding a template from multiple input pages for data extraction was first studied in [3][4]. The algorithm is exponential in the input string length as it has to explore all possibilities. A set of heuristics is introduced to lower the complexity by limiting the space to explore and to backtrack. In [5], a more efficient method is given based on sophisticated tag path analysis, but it is still not applicable for immense amounts of pages.

Because of templates, many identical or similar data with different templates can generate different Web pages, which causes the phenomenon of data redundancy especially among Web sites of the same kind. Redundant data leads to many researches about duplicate elimination for improving efficiency of extraction [6][12][13]; on the other hand redundant data becomes implied association among heterogeneous Websites. Which could be used to mining different templates[7] for extraction. This has become a new direction of entity data extraction for heterogeneous Websites.

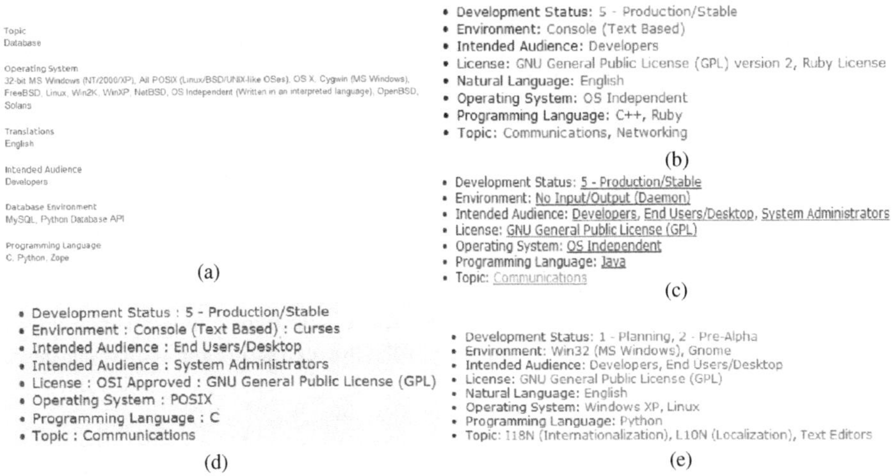

Fig. 1. Five different websites have similar attributes. (a)-(e) are from sourceforge.net, rubyforge.org, forge.ow2.org, alioth.debian.org and developer.berlios.de respectively. They are all from Open source software area.

Redundancy of Web data includes two main types: entity-level and attribute-level redundancy. The former refers to that different Websites contain identical entities, whose attributes are textually similar, such as [8] analyses this type of redundancy with some research about academic paper and hotel entities shared among different Websites; The latter means that different Websites doesn't contain identical entities but textually similar enumerated attributes. Figure 1 gives an example of attribute-level redundancy. (a) to (e) in Figure 1 show detail pages of project entities from five Websites in the area of open source software (OSS for short). They have 4 identical attributes (Intended Audience, Operating System, Programming Language and Topic) and one similar attribute (Environment and Database Environment are similar). This paper propose a novel approach to extraction attributes of entities from heterogeneous Websites using a seed attribute set shared by these Websites and some pages from target Website as the training set. The approach firstly locates the position of each attribute in pages of target Website by the training set. Then the approach extracts attribute value of each entity from Website with the position as extraction rules.

2 Preliminary

The approach in this paper is built on the basis of the three basic properties, these three properties are characteristics of the Web site based on template:

(1)The various website pages from same area contain the same type of entities. (2)The attributes of entities are textually similar. (3) Pages within a same web site share a similar structure follow the same template.

Property (2) means that Web sites have a lot of duplicate content. Just like Figure 1 shows, OSS forges share some similar attributes like Intended Audience, Operating System, Programming Language and Topic etc. Property (3) indicates that an attribute in multiple pages of a same site appears in the same position (such as tag path of DOM tree[9]), so once we locate the position of one attribute in a Website, we can extract value of this attribute using this position as extraction rule.

Our approach starts by constructing a seed set A of attributes. Each element in A consists of two parts: name of attribute and its default values which could be enumerated. The seed set A is constructed by human editors using exploratory search[14][15] pages of Websites (3.1 introduces several methods for preparing seed set A). Note that each element in A contains attribute values for an entity from a single entity page. A new site being extracted will typically contain pages for entities that already have attributes in A. Thus, we scan the pages in target Website to find values that match attribute values in the seed set A. These matching attribute values within the pages are used to learn extraction rules for extracting the remaining pages of the new site. The seed set is augmented with the newly extracted values not included and the process is repeated for other new sites.

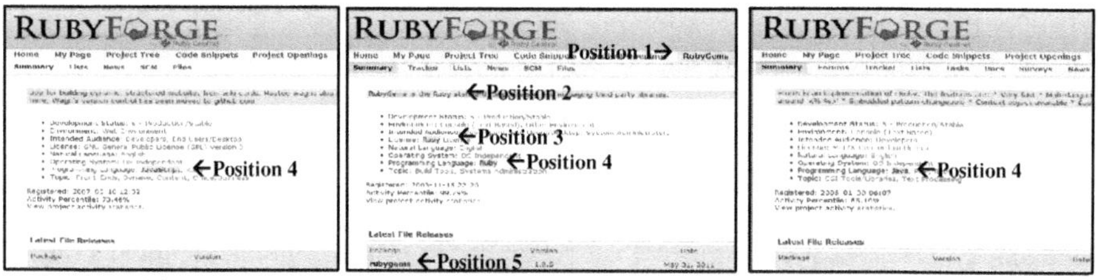

(a) (b) (c)

```
<li> Programming Language: <a href="/softwaremap/trove_list.php?form_cat=280">JavaScript</a>
.......
<li> Programming Language: <a href="/softwaremap/trove_list.php?form_cat=306">Ruby</a></li>
.......
<li> Programming Language: <a href="/softwaremap/trove_list.php?form_cat=198">Java</a>
```

(d)

Fig. 2. A Case study of our approach. (a)(b)(c) are 3 training pages from Rubyforge.org.(d) is pieces HTML code of three pages. Position i (i=1,2,3,4,5) means the position where data match the values in seed set.

Problem Statement. Consider a seed set $A=\{A_1,A_2,\ldots,A_n\}$ of attributes, $A_i=<AN_i, AV_i>(i=1,2,\ldots,n)$, $AV_i=\{AV_{i1},AV_{i2},\ldots,AV_{ij}\}$(j is the number of default values of each attribute A_i), AN and AV mean attribute name and attribute value respectively. Given a set of template-based Websites $W=\{W_1,W_2,\ldots,W_m\}$ belonging to a specific area (e.g., products, OSS), extract values of each attribute A_i of each entity from each page of web site $W_i \in W(i=1,2,\ldots,m)$.

In order to describe our approach clearly, firstly we give a case study below. Section 3 will give some key definitions and propose the detail methodology of our approach.

Case Study. Consider the seed set A={<Programming Language, {Ruby, JavaScript, Java}>}, given 3 training pages P1, P2 and P3 of Website rubyforge.org. We scan the pages to find values that match attribute values in A. Figure 2 shows the result of matching. The attribute value Ruby is matched at 5 positions in P2, the attribute value JavaScript and Java are matched at one position in P2 and P3 respectively. It is clear that the position 4 is the most supportive, because it has three pages to support itself while others have only one. Then we check the HTML source code of the three pages, the position 4 can be described with tag path <html><table><tr><td><ul><li><a> of DOM tree. At last we use the tag path to extract values within other pages of the Website.

3 Methodology

Figure 3 interprets our approach. It has three key steps corresponding to Figure 3 (a) to (c). We describe each of these steps in greater detail in the following three sub-sections.

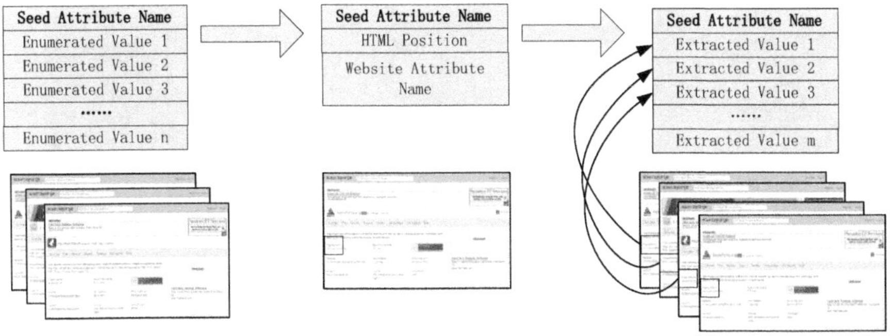

(a) Seed attributes and training pages from target website (b) Attribute position location in target website (c) Attribute value extraction

Fig. 3. The procedures of our approach. (a) shows the input of the approach including seed attribute set and pages from target website. (b) locates HTML position of seed attribute value , and gets its alias of target website. (c) uses HTML position and website attribute name to extract attribute values.

3.1 Seed Attributes Preparation

Unlike paper [8], We put forward our method to construct seed set of attributes. At present, seed sets are mainly constructed based on some domain taxonomies, like Probase, Freebase, WikiTaxonomy etc[1]. Taxonomies consist of concepts (e.g. open sources software), instances (e.g., Linux), attributes (e.g. Programming Language) and values (e.g., Linux's Programming Language is C), and relationships (e.g., OSS, as a concept, is closely related to Free/Libre and OSS). Actually taxonomy based methods[16] treat attribute and value as a kind of concept and instance of taxonomy, for example Ruby, JaveScript and Java are three instances of programming language. In order to get all the attributes of an entity like OSS, all the concepts related to OSS must be given. Figure 5 shows the query result of one taxonomy, which does not include other important attributes like Programming Language. Also taxonomies are usually general and huge, it a time-consuming job to query them. This paper uses a simple method based on the observation: most of Websites holding Web entities provide services for exploratory search, all the attributes of those entities are displayed with lists of hyperlinks[2].Figure 4 shows a snapshot of the exploratory search page of sourceforge.net. It is easily to identify the exploratory search pages by Crawling. It firstly finds the exploratory search page, and extracts all the attributes by mining Web lists with method in our earlier work[2].

Fig. 4. Seed Attributes. Categories, Platform, Dev Status, Programming Language, License are attribute names of OSS, other words below these names are their enumerated values

---

[1] http://research.microsoft.com/en-us/projects/probase/statistics.aspx

[2] http://rubyforge.org/softwaremap/trove_list.php
http://forge.ow2.org/softwaremap/trove_list.php
http://developer.berlios.de/softwaremap/trove_list.php
https://alioth.debian.org/softwaremap/trove_list.php

	Class	Attribute
1	open source software	logo
2	open source software	license
3	open source software	website
4	open source software	screenshot
5	open source software	developer
6	open source software	advantage
7	open source software	heart
8	open source software	name

Fig. 5. Snapshot of query to a taxonomy. The query gets all the attributes belonging to the concept OSS.

3.2 Attribute Position Location

In this step, the most difficult challenge is how to match the values in a seed set of attributes with those within the target pages. It relies heavily on the fact that attribute values across pages for the same real-world entity are textually similar. However, in practice, entity attribute values can vary between sites due to data entry errors, abbreviations, heterogeneous representations, etc. Now, suppose we use the Jaccard coefficient [17] to measure the similarity between attribute values, where each value is treated as a bag of words (using space as delimiter), and for two sets S1 and S2, their Jaccard similarity is defined as $JC(S1, S2) = |S1 \cap S2| \, |S1 \cup S2|$.

DEFINITION 1. (Matching) Given a pair of values u and v for attribute a, if the similarity JC(u, v) is greater than a given threshold T, then u and v are matched.

DEFINITION 2. (Position) If values u and v are matched, v is the value within one page from target Website, the path from the root node to the node containing v is called the position of v, which is also called tag path.

DEFINITION 3. (Supportiveness) If values of one attribute A are matched with k pages at position i, then the supportiveness of A at position i is k/m (m is number of training pages).

```
Algorithm 1 AttributeLocation
  input: Seed set of attributes A, training pages P
  output: Position of each attribute of A
    for each attribute A_i of A
      for each value AV_ij of A_i
        for each page p_k of P
          if AV_ij and P_k are matching at position PO_l
            record PO_l and count the number of matching
      calculate the supportiveness of PO_l
      find the maximum
      return PO_max
```

Algorithm 1 shows the detail of locating attribute position. Due to the use of three nested loops, the complexity is O(Km), K is the number of all the values in the seed set, m is the number of training pages. K usually is finite and small, in table 2 K is no more than one thousand.

3.3 Attribute Extraction

After we get position of each attribute, we can extract its values. The biggest challenge of this step is caused by embedded lists. Lots of Websites use list structure to presents attributes as list items, just like Figure 1. The positions of each item in the same list are identical. Figure 6 shows that the position of 8 attributes is <html><table>…<ul><li><a>. It means that we can extract lots of attributes besides target attribute. To resolve this problem, we propose a method with relative position based on an observation: attribute name and value are adjacent, there are no other words between name value pairs. The method locates attribute name accurately, then extract the value at the nearest right sibling leaf node of the DOM Tree.

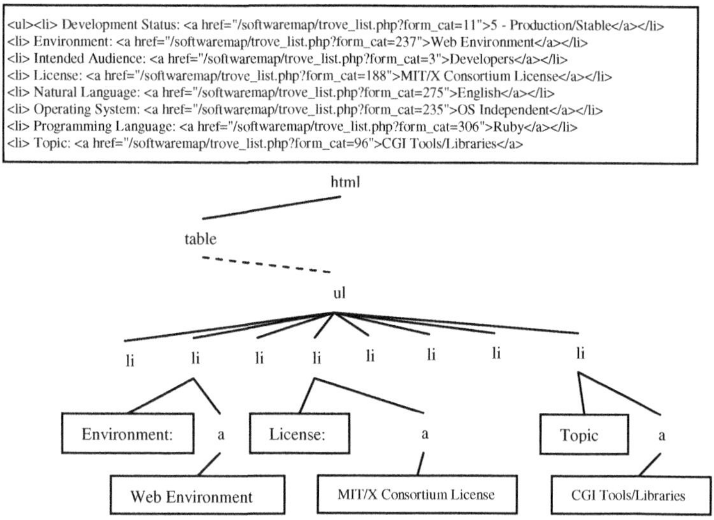

Fig. 6. Pieces of HTML code of a detail page from rubyforge.org. The upper part is the code, and the lower is the DOM Tree of this code. Squares represent leaf nodes which will be displayed on the page, and others are tag nodes which are invisible on the page.

DEFINITION 4. (Relative position) Given a tag path TP_i of a attribute A_i and the alias of A_i, we can locate the accurate position of the alias in a new page, then the path from it to the nearest right sibling leaf node is called relative position.

For example, the alias of attribute User Interface in rubyforge.org is Environment, the relative position of Web Environment is <Envionment><li><a>. Table 5 gives alias of each seed attribute for each target Website.

```
Algorithm 1 AttributeExtraction
  input: Position of each attribute A, the alias of A,
test pages P
  output: Each value of A in P
    for each attribute Ai of A
      for each page pk of P
        find the relative position RP of Ai in Pk
        return the value of RP
```

4 Experiment

This section reports our experiment on real-life web datasets which demonstrate the effectiveness of our approach.

Dataset. Table 1 lists 15 Websites from OSS area along with the number of projects and their rank in the Internet. We crawl detail pages of each project from these Websites. Figure 7 shows the snapshot of our data set containing 11 attributes and url of detail pages. The seed set of attributes are obtained from exploratory search page of sourceforge.net. Table 2 shows each attribute and the number of default values.

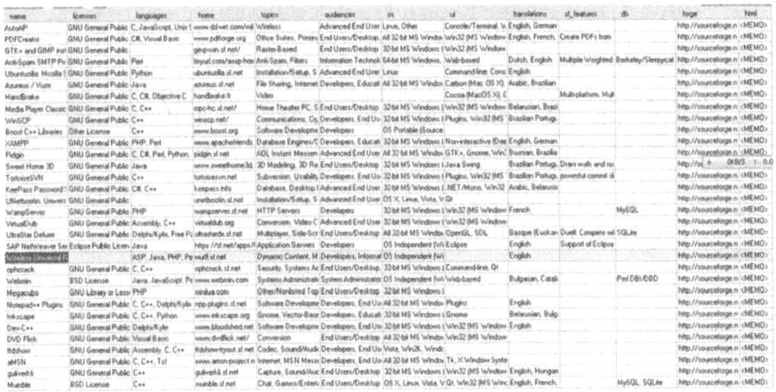

Fig. 7. Snapshot of Data set crawling from sourceforge.net. We crawl the Websites using extraction rules assigned manually. Each Website in Table 1 has a dataset, part of which is used for training, and the other for test.

Table 1. Data set description. We choose 15 website(forge) from OSS area. All the data are checked on 15/5/2011. The rank of websites are acquired from Alexa.com.

	Website	Projects	Rank
1	Alioth	916	3317
2	BerliOS	5,448	17,032
3	Rubyforge.org	9,222	12,939
4	freshmeat.net	43,947	13,389
5	CodePlex	23,224	2,026
6	forge.OW2.org	191	169,679
7	Gitorious	18,384	26,797
8	GNU Savannah	3,350	4,358
9	Google Code	296,139	2345
10	KnowledgeForge.net	264	3,310,775
11	Launchpad	23,326	6,704
12	Project Kenai	17,357	61,795
13	SourceForge	236,319	147
14	TuxFamily	2,270	27,798
15	Tigris.org	691	13,482

Experiment process and metrics. Because the data set are very huge, We choose 1000 records belonging to 1000 pages as ground truth randomly. We use our approach to extract 8 attributes in Tables 2 from the 1000 pages, compare the result with 1000 records, and calculate the precision and recall. Precision means how many extraction results are matched with ground truth correctly. Recall means how many extraction results are contained in ground truth.

Table 2. Detail description of seed set of attributes. Each column at row 1 is the name of attribute, each column at row 2 is the number of enumerated values.

Category	Platform	Development status	Programming language	License	Intended Audience	User Interface	Translations
262	4	7	100	78	4	10	7

Table 3. Precision and recall of extraction for all attributes

Attribute	Category	Platform	Development status	Programming language	License	Intended Audience	User Interface	Translations
Precision	90%	100%	100%	100%	100%	100%	100%	96%
Recall	40%	40%	33%	47%	67%	33%	33%	33%

Table 4. Mapping of attribute names of the seed set with the aliases used in each Websites. The header of the table consists of the name of each attribute, below which each cell represents the alias.

	Category	Platform	Development status	Programming language	License	Intended Audience	Environment	Translations
Alioth	Topic	Operating system	Development status	Programming language	License	Intended Audience	Environment	Natural Language
BerliOS	Topic	Operating system	Development status	Programming language	License	Intended Audience	Environment	Natural Language
Rubyforge.org	Topic	Operating system	Development status	Programming language	License	Intended Audience	Environment	Natural Language
freshmeat.net	None	Operating Systems	none	Implementation	Licenses	none	none	None
CodePlex	None	none	none	none	none	none	none	None
forge.OW2.org	Topic	Operating Systems	Development Status	Programming language	License	Intended Audience	Environment	Natural Language
Gitorious	None	none	none	none	none	none	none	None
GNU Savannah	None	none	none	none	none	none	none	None
Google Code	None	none	none	none	none	none	none	None
KnowledgeForge	None	none	none	none	Project licenses	none	none	None
Launchpad	None	none	none	Programming Languages	Licenses	none	none	None
Project Kenai	None	none	none	none	Source License	none	none	None
SourceForge	Categories Or topic	Platform or Operating Systems	Develop status	Programming language	License	Intended Audience	User Interface	Translations
TuxFamily	None	none	none	none	none	none	none	None
Tigris.org	种类	none	none	none	许可证	none	none	None

Table 3 shows the result of our experiment. The precision of all the attributes except Category and Translations are 100%. The reason for lower precision of Category is its diversity. Each Website may have different taxonomy of Category, not like other attributes whose values are flat. Another reason is that values of Category appear in natural language with a relatively higher-frequency, which leads to higher probability of position uncertainty.

Table 4 shows the alias of each attribute in each Websites. We can find that no all Websites contains all attributes in Table 2, which is the main reason that causes low recall. But if the Websites contain the attributes, the recall of approach will be more than 90%.

5 Conclusion

This paper proposes a novel methodology to extract attributes of entities from heterogeneous Websites. The approach constructs a seed set of attributes to be extracted, locate the position of each attribute in target Website, extract value of attribute using relative position. The paper gets some preliminary experiment result. It demonstrates the effectiveness of our extraction approach. More advanced work is under research.

References

1. Gibson, Punera, K., Tomkins, A.: The volume and evolution of web page templates. In: WWW, pp. 830–839. ACM Press, New York (2005)
2. Zhu, Y., Yin, G., Wang, H., Shi, D., Li, X., Yuan, L.: An Indent Shape based Approach for Web Lists Mining. In: Wang, F.L. (ed.) WISM 2011, Part II. LNCS, vol. 6988, pp. 113–121. Springer, Heidelberg (2011)
3. Agichtein, E.: Confidence Estimation Methods for Partially Supervised Relation Extraction. In: The 6th SIAM International Conference on Data Mining, ACM Press, New York (2006)
4. Agrawal, R., Bayardo, R.J., Srikant, R.: Athena: Mining-Based Interactive Management of Text Databases. In: Zaniolo, C., Grust, T., Scholl, M.H., Lockemann, P.C. (eds.) EDBT 2000. LNCS, vol. 1777, pp. 365–379. Springer, Heidelberg (2000)
5. Arasu, A., Garcia-Molina, H.: Extracting Structured Data from Web Pages. In: The 2003 ACM SIGMOD International Conference on Management of Data, pp. 337–348. ACM Press, New York (2003)
6. Elmagarmid, A., Ipeirotis, P., Verykios, V.: Duplicate record detection: A survey. IEEE Trans. Knowl. Data Eng. 19(1), 1–16 (2007)
7. Papotti, P., Crescenzi, V., Merialdo, P., Bronzi, M., Blanco, L.: Redundancy-driven web data extraction and integration. In: WebDB (2010)
8. Gulhane, P., Rastogi, R., Sengamedu, S., Tengli, A.: Exploiting content redundancy for web information extraction. PVLDB 3(1), 578–587 (2010)
9. Miao, G., et al.: Extracting data records from the web using tag path clusterting. In: WWW, pp. 981–990. ACM Press, New York (2009)
10. Jindal, N., Liu, B.: A Generalized Tree Matching Algorithm Considering Nested Lists for Web Data Extraction. In: The 10th SIAM, pp. 930–941 (2010)
11. Chang, C.-H., Lui, S.: IEPAD: Information Extraction Based on Pattern Discovery. In: The 10th International World Wide Web Conference, pp. 681–688 (2001)
12. Sivakumar, P., Parvathi, R.M.S.: An Efficient Approach of Noise Removal from Web Page for Effectual Web Content Mining. European Journal of Scientific Research 50(3), 340–351 (2011)
13. Liu, W., Meng, X., Yang, J., Xiao, J.: Duplicate Identification in Deep Web Data Integration. In: Chen, L., Tang, C., Yang, J., Gao, Y. (eds.) WAIM 2010. LNCS, vol. 6184, pp. 5–17. Springer, Heidelberg (2010)
14. Marchionini, G.: Exploratory search: from finding to understanding. Communications of the ACM 49(4), 46 (2006)
15. Huang, J., Wang, H., et al.: Link-based Hidden Attribute Discovery for Objects on Web. In: 14th International Conference on Extending Database Technology, pp. 473–484. ACM Press, New York (2011)
16. Wang, J., Shao, B., et al.: Understanding Tables on the Web. Technique report. Microsoft Research Asia (2011)
17. Manning, C., Raghavan, P., Schutze, H.: Introduction to Information Retrieval. Cambridge University Press, Cambridge (2008)

Understanding Playability and Motivational Needs in Human Computation Games*

Dion Hoe-Lian Goh and Chei Sian Lee

Wee Kim Wee School of Communication and Information,
Nanyang Technological University,
31 Nanyang Link,
Singapore 637718
{ashlgoh,leecs}@ntu.edu.sg

Abstract. Human computation games (HCGs) refer to applications that use games to harness human intelligence to perform various computational tasks. We examine how different types of HCGs affect players' perceptions and their motivational appeal, as these influence good HCG design. We focus on image tagging HCGs, where users play games to generate keywords for images. Three versions were created: collaborative HCG which required players to cooperate, competitive HCG where players worked against each other, and a control non-game manual tagging application. The applications were evaluated to uncover participants' playability perceptions, and the influence of motivational needs on usage intention. Results suggest that participants reported liking the collaborative and competitive HCGs over the control application. Further, using the trichotomy of needs theory, we found that an individual's need for achievement and power influenced intention to use the various applications.

Keywords: Human Computation Games, Playability, Image Tagging, Motivations, Needs.

1 Introduction

With the increase in the popularity of social computing, innovative applications have appeared. The human computation game (HCG) is one example that offers entertainment, but is designed to accomplish tasks or solve problems by harnessing collective human intelligence. While users are entertained by playing a game, they are also performing computations as a byproduct of gameplay [1]. The use of games is unsurprising given the significance of the burgeoning gaming industry worldwide.

HCGs have been employed in a number of areas such as image tagging [2] and location-based annotation authoring [3]. The ESP Game [2] is one of the earlier HCG examples in which two randomly paired players are tasked to create matching keywords to presented images. Points are earned based on specificity of the keywords, and coupled with a countdown timer, these elements add excitement and

* This work was supported by the Singapore National Research Foundation Interactive Digital Media R&D Program, under research Grant NRF NRF2008IDM-IDM004-012.

C. Xing, F. Crestani, and A. Rauber (Eds.): ICADL 2011, LNCS 7008, pp. 108–117, 2011.

hence motivation for players. Further, the resulting keywords can be used as tags for the images, and therefore harnessed by image retrieval algorithms.

HCGs are typically collaborative or competitive. Collaborative variants require users to work together to fulfill the games' objectives and the ESP Game is one example. Competitive games require users to work against each other, and the argument for competition is that it adds an additional dimension of challenge essential in sustaining interest [4]. For example, in KissKissBan [5], a blocker is introduced. The blocker generates keywords that other players could potentially use to describe an image. Players that enter keywords in the blocked list lose points while the blocker correspondingly gains points.

Additionally, important to the success of HCGs is an understanding of their appeal to different users, and why users are motivated to engage in them. As HCGs have different features, they potentially offer different benefits to players that could influence sustained usage [6]. Researchers have argued that studying motivational differences among users would lead a better understanding of user needs, usage patterns of games, and of course, game design that caters to different profiles of users [7]. However, such work has been confined largely to those catering to pure entertainment such as virtual worlds and arcade games [8].

Being an emerging area that yields promise in terms of usage and utility, there is currently active research in HCGs. However, to our knowledge, there is little research conducted in understanding how different game variations affect players' perceptions and how they appeal to players with different motivational needs. Thus, the goal of the present research is to compare two HCG variants against a non-game application to determine how users perceive them in terms of playability and motivational appeal.

2 Related Work

HCGs blend gaming and human problem solving, capitalizing on people's desire to be entertained and the fact that humans are better at solving certain problems than computers including image recognition and language understanding [9]. Two examples mentioned previously include the ESP Game [2] and KissKissBan [5]. Another example is Herd It, a Facebook collaborative HCG game for tagging music [10]. Players listen to a music clip and are quizzed on their opinions of the piece through multiple-choice questions. Points are awarded based on the percentage of the other players that agree with a player's choice. The selected answers to the quizzes may be used as tags to their respective songs. Indagator [3] is a competitive variation of a location-based content sharing HCG. Part of the game mechanics requires players to plant damage-inflicting traps and obstacles at various locations. Players who stumble upon these objects lose points while the obstacle-setter gains points.

In terms of evaluating games, playability is an important factor in influencing users' experiences and playing decisions. Attitudes towards mobile games was studied by [11] and factors such as ease of use, attractiveness, and enjoyment were strong predictors of users' intention to play them. Other attributes mentioned in the literature include players' engagement, sense of social presence and physiological arousal [12]. In the context of HCGs, researchers have suggested that incorporating elements of fun into user interfaces could create positive experience for the users

[3, 9]. Other suggestions to characterize fun include the ability to create a sense of competence for each player, providing challenging tasks, and creating a pleasant and interesting sensory experience [10].

Further, people play games for a variety of reasons and much research has been conducted on their motivations for doing so. Bartle [13] proposed four types of players in multiplayer computer games: Killers, Achievers, Socializers and Explorers. By understanding player types and their motivations, Bartle argued that game developers could provide features to appeal to all players types in order to be commercially successful. Although influential, [7] contended that Bartle's typology was not empirically tested and therefore sought to extend this work by deriving player motivations through empirical data. Using survey responses from players of various Massively-Multiplayer Online Role-Playing Games (MMORPGs), three motivational components were extracted: achievement, socializing, and immersion.

The review of the literature in motivations suggests that most motivational studies have tended to focus on MMPORGs, with comparatively fewer examining games in general. As yet, there are no studies that have been conducted on the emerging genre of HCGs. Consequently, our present research studies motivations for playing HCGs. Here, we do so through the lens of the trichotomy of needs theory popularized by McClelland [14]. Essentially, this theory argues that behavior is motivated by three needs. People high in need for achievement (NAch) have an intrinsic desire to perform well against a standard of excellence. Those with a need for affiliation (NAff) have a strong desire to associate with others, to establish relationships and feel a sense of belonging to a larger community. Finally, people with a high need for power (NPow) are attracted to environments or situations in which they influence activities or thoughts of others usually to accomplish one or more objectives. Empirical evidence for the theory is supported by a large body of research in diverse domains from organizational settings to the individual (e.g. [15]). Because games provide environments for building relationships, achieving goals, and leading others towards accomplishing those goals [3, 16], we expect that the three motivational needs would have an influence in people's desire to play HCGs.

In sum, key contributions of our present research are that we investigate user perceptions and motivations for usage across three different representative applications: two main genres of HCGs, the collaborative and competitive games, and a non-game, manual tagging application.

3 Evaluation of the Applications

3.1 Applications Developed for the Study

To accomplish our study, three image tagging applications were developed. First, the collaborative HCG (see Figure 1) had a similar design to the ESP Game where pairs of players had to enter the same tag to score points. In Figure 1, the top of the interface shows the time remaining and current score. The image to be tagged is displayed in the middle while players enter tags in the text field above the image. The number of tags entered by the partner is also shown. Beside the image, off-limit tags ("man", "flag", "black") are displayed. The list of tags generated by the user for the image is displayed below it.

Next, the competitive HCG also required paired players to enter tags to a presented image. However, the first player in the pair to enter a tag that described the image was awarded points while the other player received none. Players were then presented with the next image, and so on until the time limit of five minutes was reached, similar to the collaborative game. To raise the level of challenge, off-limit tags are not displayed and players who entered such tags had points deducted, similar to [5]. The interface for the competitive HCG is similar to the collaborative version in Figure 1 except that the off-limit tags are not shown.

Third, the manual tagging application served as a baseline for comparing our results. It presented images and users entered between one to five tags before the next image was shown. Off-limit tags could not be entered, and the user had five minutes to tag as many images as possible. While typical manual image tagging applications do not have count-down timers, we decided to include it as a form of feedback.

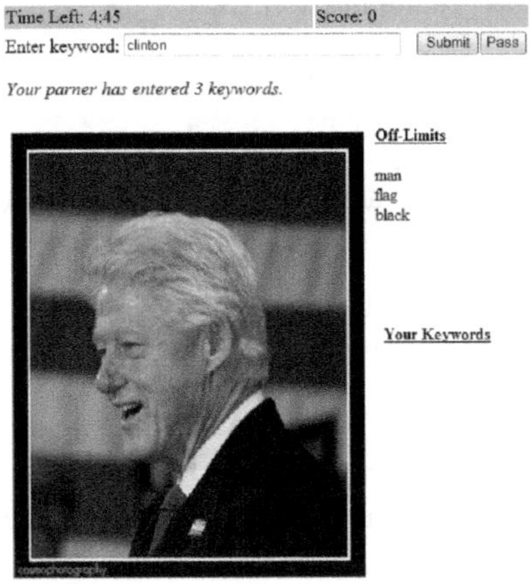

Fig. 1. Collaborative HCG developed for the study

3.2 Participants

A total of 103 undergraduate and graduate students were recruited from a local university. There were 58 males and 45 were females, with ages ranging from 19 to 37, and an average age of 26.8 years. The majority (85%) had a background in computer science, information technology or related disciplines, while the remainder were from disciplines such as arts and business. All participants reported understanding the concept of tagging with 68% reporting regular contribution of tags to content in various Web sites on a regular basis. However, most participants (86%) were unfamiliar with HCGs and had not played such games before. Nevertheless, the majority of the participants (about 64%) were frequent players of online games.

3.3 Procedure

Participants were randomly divided into three groups with each playing a specific game: 33 were assigned to the control application, 34 to the collaborative HCG and

36 to the competitive HCG. For the two HCG groups, participants were further randomly paired to play the games. The study was conducted in a lab across separate sessions. Although they were located in the same lab, participants did not know who their partners were, and communication among participants was prohibited.

For each session, the study commenced with a researcher providing a briefing on image tagging. For the collaborative and competitive game groups, the purpose of HCGs and their potentially useful role in supporting image tagging was explained. Next, participants were briefed on the usage of their respective applications. They were also instructed to generate tags that were descriptive of the images, being as specific as possible. Participants were then asked to try their assigned application as a means of familiarization. Finally, the study began and the participants played their assigned applications for one round, after which they completed a questionnaire that captured demographic data (reported earlier), perceptions of the application they played, and their motivations. Participants were also requested to provide qualitative comments about their assigned application.

3.4 Evaluation Measures

A questionnaire was designed to elicit participants' perceptions of playability of the applications they used in the following aspects. Questions were rated on a scale of 1 (strongly disagree) to 5 (strongly agree):

- **Challenge.** A game should be challenging enough to capture and sustain a player's interest, providing an enjoyable and interactive experience. However, it should not be too challenging such that the player feels frustrated and eventually gives up playing, nor should it be too easy such the player knows that success is inevitable and it becomes boring [17].
- **Immersion.** To maintain interest, a game must be able to capture the player's complete and undivided attention so that all of his/her skills are needed to deal with the challenges associated with the game. Research has shown that because of the deep engagement needed, players have a high emotional investment in the game, and this encourages sustained usage [18].
- **Control.** Players should perceive that they are able to translate their intentions into in-game behavior, and that their actions and strategies undertaken will impact the outcome of the game. In doing so, the game becomes highly replayable as players are motivated to improve their skills and explore different strategies to influence the gaming environment [16].

Additionally, to measure participants' need for affiliation, achievement and power, the Manifest Needs Questionnaire (MNQ; [19] was used. In the MNQ, each of these motivational needs contained five items, measured on a five-point scale indicating levels of agreement with a particular statement (e.g. "I seek an active role in the leadership of a group"). The MNQ has been widely used in a number of studies in various domains (e.g. [15]).

Finally, the questionnaire also asked participants about their intention to use their assigned application. The purpose was to quantitatively determine factors would influence such intentions [7, 11].

4 Results

4.1 Playability Perceptions

Table 1 shows the means and standard deviations of participants' playability perceptions of the applications. One-way ANOVAs indicated that there were significant differences in Intention to Use, $F(2, 100) = 13.24$, $p < 0.001$; Challenge, $F(2, 100) = 12.62$, $p < .001$; and Immersion, $F(2, 100) = 10.27$, $p < 0.001$. The difference in Control was statistically nonsignificant, $F(2, 100) = .75$, $p = .475$. Post-hoc comparisons using Tukey's test revealed the following:

- **Intention to Use.** Participants reported being significantly more likely to use the collaborative and competitive HCGs than the manual application. However, differences between the two HCGs were nonsignificant.
- **Challenge.** Participants reported that the collaborative and competitive HCGs were more challenging than the manual application, but no differences were observed between the two games.
- **Immersion.** Like Challenge, participants felt that both the collaborative and competitive HCGs could better capture their attention than the manual application. Difference in perceptions between the two games was nonsignificant.
- **Control.** As there were no statistical differences between the three applications, it could be inferred that participants felt that they could control the outcome of the applications to a similar degree.

Table 1. Means and standard deviations for participants' perceptions

Variable	Game Type					
	Manual (N = 33)		Collaborative (N = 34)		Competitive (N = 36)	
	Mean	SD	Mean	SD	Mean	SD
Intention to Use*	3.10	.97	3.82	.63	3.98	.61
Challenge*	2.81	.66	3.43	.61	3.53	.63
Immersion*	3.03	.79	3.62	.57	3.68	.59
Control	3.40	.76	3.57	.61	3.57	.52

* Statistically significant differences between the three applications at $p < .05$.

Qualitative feedback provides support for the above results. For the collaborative HCG, participants reported that a major one reason for why they enjoyed playing the game was the need to work together to score points ("*liked the team work idea*"). For the competitive HCG, the need to outdo one's partner to score points was the major attraction ("*if I can type faster, I can win a lot of points*"). In contrast, when asked about what their impressions about the manual tagging application, many participants reported there was "*nothing*" they liked about it, and also that it was "*boring to use*".

4.2 Motivational Needs

To investigate the influence of motivations and playability perceptions on participants' intention to use the application, three hierarchical regression analyses were conducted, each with two steps (see Table 2). In hierarchical regression, the main question at each step is whether the additional variables are significant, given those already in the model. The first block comprised the playability perception variables of Challenge, Immersion and Control, which have traditionally been regarded as important determinants of intention to use. In the second block, we entered the three motivations (NAff, NAch, NPow) to ascertain if these added anything to the prediction model over application perceptions.

Table 2. Hierarchical regression results for the three applications

	Dependent Variable (Intention to Use)					
	Manual (N = 33)		Collaborative (N = 34)		Competitive (N = 36)	
Independent Variables	Block 1 B	Block 2 β	Block 1 β	Block 2 β	Block 1 β	Block 2 β
Step 1						
Challenge	.14		.25		.27	
Immersion	.03		.37*		.23	
Control	.25		.03		.26	
Step 2						
Challenge		.07		.11		.32*
Immersion		.04		.27		-.08
Control		.28		.03		.26
NAff		-.13		.17		-.01
NAch		.22		.64**		.40*
NPow		-.20		-.48**		-.38*
R^2	.11	.18	.28	.58	.40	.63
ΔR^2		.07		.30*		.30**

* p < .05; ** p < .01

For the manual tagging application, both Block 1 and Block 2 show that neither the playability perception variables nor motivations were significant in predicting intention to use. In addition, the incremental improvement in R-square over Block 1 was about 7%, but the additional variance explained was nonsignificant.

Next, Block 1 for the collaborative HCG shows that only Immersion has a significant positive influence on intention to use. However, its influence disappears when the motivational needs are introduced in Block 2. Here, participants with high NAch ($\beta = .64$, $p < .01$) were more likely to be satisfied with the collaborative HCG. Interestingly, this relationship was inverse for NPow ($\beta = -.48$, $p < .01$), where those with high NPow were likely to be less satisfied with the application. The addition of the motivation variables yielded a significant 30% improvement in R-square over Block 1. This suggests the importance of NAch and NPow in determining intention to use as opposed to playability perceptions.

Finally, for the competitive HCG, Block 1 shows that the playability perception variables alone were not significant in predicting intention to use. With the introduction of motivational needs, Block 2 indicates that Challenge ($\beta = .32$, $p < .05$) and those with high NAch ($\beta = .40$, $p < .05$) were more likely to play the game, as were those with low NPow ($\beta = -.38$, $p < .05$).

5 Discussion

Unsurprisingly, our results showed that participants liked the game-based (collaborative and competitive) approach to tagging images as opposed to manual tagging, concurring with research (e.g. [9]) demonstrating that games indeed serve as motivators for harnessing human intelligence. More specifically, both game genres were perceived to be more challenging than the manual tagging application. Like Challenge, Immersion was rated significantly higher for both games. However, Control did not differ significantly which suggest that that the games are playable by users with different levels of gaming expertise. These findings augur well for image tagging HCGs since a challenging game that requires deep engagement fosters sustained usage [16]. Consequently, Intention to Use was also rated higher for both games than the manual tagging application.

Interestingly, none of the playability or motivational variables was significantly associated with intention to use the manual tagging application. This suggests the need to investigate other factors not accounted for in our study in future research. Further, our hierarchical regression analyses of the two HCGs suggest that an individual's motivations appear to be good predictors of game usage, perhaps even more so than perceptions of playabilty, in line with prior work (e.g. [8]. Specifically, Block 2 of our regression analyses shows that only for the competitive HCG, one playability variable, Challenge, was positively associated with intention to use. Here, the need to work against one's partner while thinking of appropriate image tags led to this finding. In contrast, two of the motivational needs (NAch and NPow) were significantly associated with intention to use the two HCGs.

Next, our finding that NAff was not a significant predictor of usage was somewhat unexpected, especially so for the collaborative HCG that required players to work together. A possible explanation is that in contrast to MMPORGs that have shown a strong association with NAff [7], the collaborative HCG was casual in genre, in which player relationships were ephemeral. The design of the game meant that collaboration was only needed for the duration of a session and thus, the need to establish relationships was not possible or needed. This suggests that the association between motivations and intention to use is not straightforward, and therefore calls for more research that includes different game genres.

Further, NAch was positively associated with usage intention of both collaborative and competitive HCGs. Despite having different rules, both games rewarded players based on how well they achieved their respective objectives. Comments from participants support this observation. One participant found the collaborative HCG appealing because he had to "*find out what the other person was thinking about*". In contrast, the manual tagging application lacked a scoring system and it did not seem to appeal to participants who had high achievement motivations. Instead, comments on reasons for liking the application were a mix of factors suggesting that motivations do not play a role, including "*like to work alone*".

Finally, it was interesting to note that NPow had an inverse relationship with intention to use both HCGs. While both games provided participants the opportunity to demonstrate their prowess in image tagging, the short-term nature of each gaming session did not give them the opportunity to influence or lead others. Further, the game mechanics and the anonymity of the partners/opponents meant that leadership or influence of others was not needed for success. Some evidence for this can be seen in our participants' qualitative comments. One high NPow participant noted that the competitive HCG "*is only for testing oneself*". Our results suggest the need to incorporate additional features that could appeal to this segment of participants.

6 Conclusion

The following implications may be derived from our findings. One, games have the potential to harness human computation, and both the collaborative and competitive variants have similar appeal. Two, our work has highlighted the need to extend beyond playability to include motivations in the study of HCGs. In addition, our work suggests a number of design implications for HCGs:

- Feedback mechanisms are a vital component to be incorporated. A major appeal of the HCGs is the scoring system that allowed participants (especially those with high NAch) to gauge their performance.
- Rewards for good performance should be considered. This has the potential to spur players to outdo others and promote sustained use. One such possibility is the leaderboard that lists high performing players.
- Features for community-building should be designed. Developing community-oriented features around casual HCGs could help counter their short-term nature. Examples include discussion boards, chat rooms and social media sites.
- Game that harness human computation should consider demonstrating how the generated computations are useful. Better yet, major contributors should be acknowledged. In doing so, users who see themselves as influencers or leaders (high NPow) are able to see that their work has impact on the community.

While our present work has yielded insights, there are limitations to be noted. We evaluated one particular type of each game genre in a specific domain of image tagging. For better generalizability, investigations using different game designs and different domains of human computation should be undertaken. Next, our findings were obtained through a single lab experiment. Conducting a longitudinal study involving repeated use of the applications over time would be beneficial in validating our results. Our participants were primarily undergraduate and graduate students. Replication of this study in other contexts (e.g. more diverse age groups) would be useful to better understand perceptions of HCGs. For future work, there is a need to design and evaluate a wider range of HCG genres such as action or role-playing. Further, a more in-depth study of personality and other individual differences such as familiarity and personal experience, and their relationship to the perception and use of HCGs would also be instructive.

References

1. Goh, D.H., Ang, R.P., Chua, A.Y.K., Lee, C.S.: Evaluating Game Genres for Tagging Images. In: Proceedings of the 6th Nordic Conference on Human-Computer Interaction, pp. 659–662. ACM Press, New York (2010)

2. von Ahn, L., Dabbish, L.: Labeling Images with a Computer Game. In: Proceedings of the 2004 Annual SIGCHI Conference on Human Factors in Computing Systems, pp. 319–326. ACM Press, New York (2004)
3. Lee, C.S., Goh, D.H., Chua, A.Y.K., Ang, R.P.: Indagator: Investigating Perceived Gratifications of an Application that Blends Mobile Content Sharing with Gameplay. Journal of the American Society for Information Science and Technology 61, 1244–1257 (2010)
4. Vorderer, P., Hartmann, T., Klimmt, C.: Explaining the Enjoyment of Playing Video Games: The Role of Competition. In: Proceedings of the Second International Conference on Computer Games, pp. 1–9. ACM Press, New York (2003)
5. Ho, C.J., Chang, T.H., Lee, J.C., Hsu, J.Y.J., Chen, K.T.: KissKissBan: A Competitive Human Computation Game for Image Annotation. In: Proceedings of the 2009 ACM SIGKDD Workshop on Human Computation, pp. 11–14. ACM Press, New York (2009)
6. Lee, C.S.: Managing Perceived Communication Failures with Affordances of ICTs. Computers in Human Behavior 26, 572–580 (2010)
7. Yee, N.: Motivations for Play in Online Games. Cyberpsychology & Behavior 9, 772–775 (2006)
8. Ryan, R.M., Rigby, C.S., Przybylski, A.K.: The Motivational Pull of Video Games: A Self-Determination Theory Approach. Motivation and Emotion 30, 347–364 (2006)
9. von Ahn, L., Dabbish, L.: Designing Games with a Purpose. Communications of the ACM 51, 58–67 (2008)
10. Barrington, L., Turnbull, D., O'Malley, D., Lanckriet, G.: User-Centered Design of a Social Game to Tag Music. In: Proceedings of the 2009 ACM SIGKDD Workshop on Human Computation, pp. 7–10. ACM Press, New York (2009)
11. Ha, I., Yoon, Y., Choi, M.: Determinants of Adoption of Mobile Games under Mobile Broadband Wireless Access Environment. Information & Management 44, 276–286 (2007)
12. O'Brien, H.L., Toms, E.G.: What is User Engagement? A Conceptual Framework for Defining User Engagement with Technology. Journal of the American Society of Information Science and Technology 59, 938–955 (2008)
13. Bartle, R.A.: Designing Virtual Worlds. New Riders, Berkeley (2004)
14. McClelland, D.: Human Motivation. Cambridge University Press, Cambridge (1985)
15. Shockley, K.M., Allen, T.: Investigating the Missing Link in Flexible Work Arrangement Utilization: An Individual Difference Perspective. Journal of Vocational Behavior 76, 131–142 (2010)
16. Sweetser, P., Wyeth, P.: GameFlow: A Model for Evaluating Player Enjoyment in Games. Computers in Entertainment 3 (2005), `http://doi.acm.org /10.1145/1077246.1077253`
17. Goh, D.H., Ang, R.P., Tan, H.C.: Strategies for Designing Effective Psychotherapeutic Gaming Interventions for Children and Adolescents. Computers in Human Behavior 24, 2217–2235 (2008)
18. Brockmyer, J.H., Fox, C.M., Curtiss, K.A., McBroom, E., Burkhart, K.M., Pidruzny, J.N.: The Development of the Game Engagement Questionnaire: A Measure of Engagement in Video Game-Playing. Journal of Experimental Social Psychology 45, 624–634 (2009)
19. Steers, R.M., Braunstein, D.N.: A Behaviorally Based Measure of Manifest Needs in Work Settings. Journal of Vocational Behavior 9, 251–266 (1976)

A Metadata Framework for Cloud-Based Digital Archives Using METS with PREMIS*

Jan Askhoj, Shigeo Sugimoto, and Mitsuharu Nagamori

Graduate School of Library, Information and Media Studies,
University of Tsukuba, 1-2 Kasuga, Tsukuba, Ibaraki 305-8550, Japan
Janaskhoej@gmail.com, {sugimoto,nagamori}@slis.tsukuba.ac.jp

Abstract. An increasing number of organizations are using cloud computing to create and store digital records. To ensure safe storage and long-term preservation, standards for metadata are needed. This paper proposes a metadata application profile for cloud archives.

We use guidelines from the Singapore Framework for Dublin Core Application Profiles to define the functional requirements, domain model and description set profile that form the basis of the proposed application profile. In our profile, we use METS as a transmission and package format, extending it with metadata from the PREMIS data dictionary and Dublin Core Metadata Element Set.

Using the proposed application profile, we create an example METS information package using predefined criteria. We find that the profile meets the functional requirements, and that it simplifies metadata provision for business systems, compared to systems that do not allow pre-registration.

Keywords: Archives, Records Management, Data Transfer, Cloud Computing.

1 Introduction

In recent years an increasing number of documents and records are being created and stored in the cloud. However, partly transferring responsibility for content to a third party, over whose systems customers have very little control, makes it difficult to ensure long-term preservation and access to records. There are a number of risks. A Cloud Service Provider may change its API, it may change its payment model, making it prohibitively expensive to maintain stored records, or in the worst-case scenario, it may go bankrupt, resulting in a total loss of records. This problem is made worse by a lack of interoperability between systems. Cloud computing is a relatively new technology and systems; protocols and APIs vary between providers. [1] [2]

We have previously worked on a common reference model to solve the problems of archiving in the cloud by dividing services into layers and defining functionality offered by the different entities of a cloud archive. However, a functional model does not solve the problems of records transfer and storage. There is also a need for a data model to create standardized information packages that can meets the requirements of archive systems and ensure reliable, long-term preservation.

* This research is supported in part by JSPS Grant-in-Aid-for-Scientific Research.

C. Xing, F. Crestani, and A. Rauber (Eds.): ICADL 2011, LNCS 7008, pp. 118–127, 2011.

2 Background and Characteristics of Cloud Archiving

Based on the definition above, it becomes clear that there are a number of differences between archiving in a cloud computing environment and traditional systems, where data is stored in a local data center.

Firstly, there are different Service Models for cloud computing available to organizations. These can be categorized in a layered model. From top to bottom, we have Software as a Service (SaaS), Platform as a Service (PaaS) and Infrastructure as a Service (IaaS) [3]. Here, each layer builds on services offered by the layer below, and in turn offers services to the layer above. An organization wanting to use cloud services to create an archiving systems must decide what service model is suitable and a data model must be able to work across such services.

Secondly, because of the abstracted nature of cloud computing, costumers have little control over the storage of the bits that make up individual records. Lock-in to an externally provided API can make record and platform migration difficult [4].

Finally, the networked and scalable nature of cloud computing means that sharing of computing resources between organizations becomes possible. Organizations do not need detailed knowledge of the technical infrastructure of cloud services, as long as they know 1) the API in use and 2) the types and formats of data being offered. Such knowledge of APIs and data formats can be utilized to simplify information transfer and the creation of preservation metadata.

3 Previous Work and Related Research

During our previous research, we have found that the widely used Reference Model for an Open Archival Information System OAIS [5] is not directly applicable in a cloud environment [6]. To help solve this problem, we developed a Layered Model

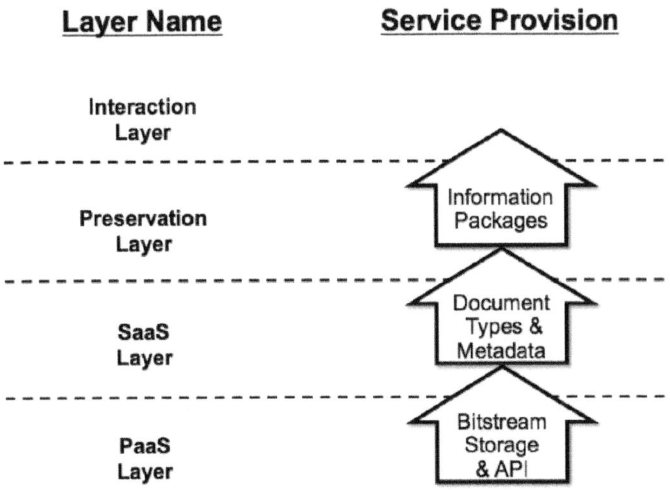

Fig. 1. Layered Model showing the services provided by each layer

combining the OAIS Model functional entities and information types with a layered model used for describing cloud Services.

The model consists of the following four layers (from bottom to top).

1. **PaaS Layer:** This layer provides a trusted repository for storage and a computing platform. All objects in the PaaS Layer are simple strings of bits that make no sense as information outside the context of a system that can read and represent them.
2. **SaaS Layer:** Business systems creating Digital Objects for archiving. The systems in the SaaS Layer represent bit-strings from the lower layer as documents and metadata that can be used by systems and users.
3. **Preservation Layer:** Ensures that archive systems can access Information Objects by offering: Registration, harvesting, conversion, representation info, and packaging.
4. **Interaction Layer:** This layer contains the Archive systems and other entities that access and use Information Packages for archiving purposes.

4 Problem Definition

Although we have developed an abstract model, no complete application profile (rule set and description for the design of metadata) exists for businesses and archives for the transfer and preservation of records and metadata in the cloud. There are, however, a number of application profiles for preservation of electronic content, such as web pages or electronic documents. An example of such a profile is the one used by the British Library's web archive [7]. However, these profiles are only partly usable, because they do not address the specific problems of archiving in the cloud.

Whereas our previous work defines the functionality, information flow and services of a cloud archive, an application profile is needed to describe what metadata is to be used, how it is exchanged, and which entity is responsible for producing what. Furthermore, since different kinds of metadata (structural, descriptive and administrative) may be added at different times by different systems, the metadata records must be flexible and expandable. Traditional archive systems, as specified in the OAIS model, rely on Producers to submit Submission Information Packages (SIP) in a specific format. Producing these is a burden to the producing organizations. We want to provide a solution that ensures as high a level of automation as possible to simplify the submission process.

5 Approach

We have based our design model on the Singapore Framework for Dublin Core Application Profiles. This is a framework useful for creating and documenting DC metadata application profiles that ensure interoperability [8]. The Singapore Framework specifies the components that make up an Application Profile. These are: Functional requirements, Domain model, Description Set Profile (DSP), Usage guidelines (optional), and Encoding Syntax Guidelines (optional). For the purposes of this paper, we have chosen to focus on the mandatory parts of the profile.

6 Application Profile Design

6.1 Functional Requirements for an Application Profile

The ultimate goal of our research is to build a complete archiving solution that can be deployed using cloud technology, and ensures that records can be securely stored and preserved. To that end, we want to create a system where producers can easily save content and metadata to a cloud based storage solution, where it will be automatically converted into generic Information Packages that can be accessed and used by electronic archive systems, ensuring long-term preservation.

To provide guidelines for data transfer and package creation we want to create an application profile that provides guidelines for package formats, metadata schema definitions and the inclusion of additional metadata schemas.

We have also defined a number of areas that are out of scope for the profile. These include metadata in schemas or formats that have not been pre-registered and metadata from schemas that do not have proper documentation.

6.2 Defining a Domain Model

In figure 1 we presented the conceptual model that serves as the starting point of our research. We have built on this by adding basic functional entities, actions and relations from the functional requirements and our previous research. The result is a Domain Model with 5 entities, as can be seen in figure 2.

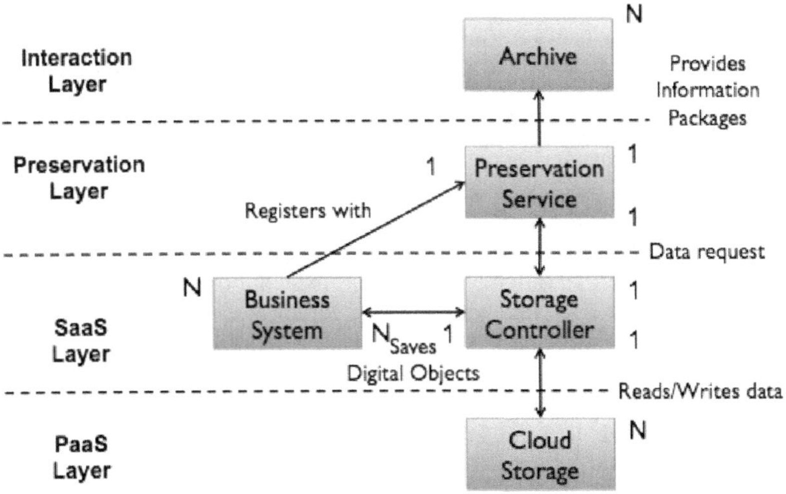

Fig. 2. Domain model specifying functional entities, actions and relations

In the model, Business Systems register preservation related information (system information, metadata schemas etc.) with the Preservation Service. This preregistered data is static and can be reused for multiple objects, allowing the Preservation Service to automatically generate preservation metadata. It also serves to abstract business systems; by using generic metadata based on business system type. The Preservation Service sends back a confirmation with information used to access Cloud Storage. When Digital Objects (DO) are declared records in a Business System, they are sent to the storage controller and saved in Cloud Storage. At that time, the Storage Controller creates a permanent generic URI for the DO and metadata in Cloud Storage. A save confirmation is sent back to the Business System. When the save is complete, the Business System sends a notification to the Preservation Service containing the permanent URI of the stored objects, along a predefined amount of metadata for preservation purposes (specifically, metadata that has not been pre-registered and is not provided by the Preservation Service). The Preservation Service then creates a generic Information Package by carrying out the following steps: 1: Collating and normalizing saved and received metadata. 2: Adding preservation information (based on registered information, event information, previously registered Business System information and external Data Sources), 3: Creating an Information Package using the resulting metadata. 4: Adding a package description.

6.3 Description Set Profile

A Description Set Profile (DSP) is designed to create constraints on metadata. A DSP constrains the resources to be described, their properties and how values are referenced. According to the DC website, a DSP uses a number of existing metadata vocabularies and applies syntactic and formatting restraints on these.

6.3.1 Container and Schema Selection Using METS

To create information packages and transfer data, a container format for metadata transmission is needed. We have chosen to use the METS Metadata Encoding and Transmission Standard [9], a widely used and actively maintained standard, expressed in XML. We have chosen METS for a number of reasons: It allows the inclusion of other metadata schemas, it can express structurally complex objects and several solutions using METS already exist.

To represent metadata for preservation, we have chosen the PREMIS metadata dictionary [10]. PREMIS defines core preservation metadata (semantic units) needed to support long-term preservation [11]. Whereas there are other schemas for archiving metadata, such as the UK National Archives, Requirements for Electronic Records Management Systems Metadata Standard [12], PREMIS benefits from having a documented data dictionary in XML that has been made to work with METS.

Business Systems can use any number of schemas, as long as they are clearly defined and registered with the Preservation Service. We have chosen the Dublin Core Metadata Element Set version 1.1 to represent Descriptive Metadata [13].

6.3.2 Metadata Element Selection

METS consists of a number of different sections. Of these, only the Structural Map is required. We have chosen to use all sections except two that lie outside our use-case: Structural Links (used when archiving web-pages) and Behavior metadata (used to describe the behavior of executable objects). We have divided our metadata into the following sections, following the METS guidelines: 1. A Root element common for all data objects, providing a unique identifier assigned by the Storage Controller along with XML and Namespace definitions. 2. A Header element describing the background of the METS element, such as the created/modified date and name/URI information about the entities responsible, and finally the status of the package. 3. Descriptive metadata from the originating Business System, represented by Dublin Core. 4. Administrative metadata, consisting of cloud-related and other technical metadata coupled with provenance information. The Administrative metadata section is also where we include PREMIS metadata. 5. File inventory, listing the files stored in the cloud that comprise the virtual METS package. 6. Structural map, showing the hierarchical structure (if any) of the objects listed in the File inventory. For all sections, we are using a combination of a subset of the existing METS metadata elements and elements we have created ourselves.

For PREMIS, we have chosen all metadata for Object, both mandatory and optional, excluding container elements. We use the Dublin Core Metadata Element Set to represent Business System metadata.

6.3.3 Defining Metadata Constraints

As the final part of the Description Set Profile (DSP), we have defined constraints for all terms in the METS profile. We use the following constraint types: Mandatory (y/n), Repeatable (y/n), Controlled Vocabulary/Free Text/Container.

6.4 Decisions for Implementing PREMIS with METS

When designing a METS profile, especially one that incorporates external schemas, a number of decisions must be taken to ensure optimal usability and interoperability [14]. We have used the Library of Congress "Guidelines for using PREMIS with METS for exchange" as a guideline for implementation [15]. The guidelines help in explaining a number of decisions that need to be taken by an organization when integrating the two schemas. For example, which METS sections should be used for PREMIS metadata elements. In our case, the most challenging questions to answer have been the following:

1. Choosing between overlapping elements from different schemas. A number of metadata elements in METS and PREMIS have similar functionality. For example, METS specifies *mime type* as an optional attribute of the File section, whereas PREMIS uses the more granular *format*. We have identified 8 elements from both schemas that have similar purpose and have chosen between these on a case-by-case basis, trying to reach a balance between specificity and simplicity.
2. How to deal with locally controlled vocabularies. It is likely that business systems may use their own metadata vocabularies. However, the only place where such metadata may be expresses is in the descriptive metadata section of a METS package. Such metadata may be an important source of information and should be preserved. For documentation and representation purposes, all external metadata schemas should be documented and registered, preferably via the use of namespaces.
3. How to represent structural relationships between complex Data Objects. In our profile, we use the Structural map section of METS for this purpose. For simple (non-construct) DOs, a METS package consists of 3 objects: the DO itself, a

metadata file and the METS package file itself. For construct DOs, both the DO and its individual object are treated as a complete DO, and a METS package is created for each one. For example, an Email with one attachment will result in three METS packages: One for the email body (a html or similar object), one for the attachment (in this example a word file), and one for the construct object (email with attachment). The structure of construct objects are expressed as a hierarchy, where root objects are level 0, subordinate objects are level 1 and so on.

6.4.1 Metadata Schema Representation

Before moving on to the evaluation, we would like to present an example of metadata as presented in our profile. As mentioned previously, we have focused on the mandatory parts of the Singapore Framework, i.e., Functional requirements, Domain model, and Description Set Profile (DSP). Usage guidelines and Encoding Syntax Guidelines are not presented here. Figure 3 shows an example from our schema.

Element	METS Section	External Schema Field	TYPE	MANDA-TORY	REPEAT	ORIGIN	Explanation	Example 1 - PDF file	Rationale for Origin
messageDigestAlgorithm	Administrative metadata 2 (amdSec) - PREMIS Metadata	1.5.2.1	Controlled Vocabulary	M	N	B(reg)	Specific algorithm used to construct message digest for the DO.	MD5	Part of PREMIS Data Dictionary

Fig. 3. An example representation of a metadata element

The first field from the left gives the name of the element, "message DigestAlgorithm. Next is the METS section used, here the Administrative Metadata section (amdSec) of METS, specifically the PREMIS metadata. Next is the ID of the field in an external schema, the PREMIS Id. Next is the type of field, such as controlled vocabulary, container, etc. The next two fields are the restraints Mandatory and Repeatable. Following this is the origin of the element showing where in the cloud system the element has been generated. Here the origin is B(reg) meaning pre-registered metadata from the originating business system. Following is an explanation (here taken directly from PREMIS) and an example of the element. Finally is the rationale/reason why the field is included in the schema.

7 Using the Application Profile with the Layered Model

One of the goals of our research is been to build a bridge between the abstractable, service based nature of cloud computing and the world of archives with its need for systems and metadata to support long-term preservation. Using PREMIS on its own, archives are able to give comprehensive descriptions of resources for preservation purposes, however PREMIS does not provide guidance on how element creation may be distributed between entities, especially when used with other standards.

Our application profile is built on a layered model that is equally applicable when it comes to metadata. Using the model we have divided the application profile metadata into layers: The PaaS layer is responsible for low-level storage metadata. This is managed by storage controller and is not part of the finished Information

Package. In the SaaS layer is the business system descriptive metadata, non-registered metadata produced by business systems and storage metadata from the Storage Controller (the two latter become part of PREMIS in the Preservation Layer). In the Preservation Layer is pre-registered and auto-generated PREMIS metadata along with the metadata used in the creation of METS packages. Finally, in the Interaction Layer we find complete Archiving Information Packages coupled with archive specific Package Information and Package Descriptions.

8 Application Profile Evaluation

8.1 Selection of Evaluation Method

During the design phase of our application profile, we have been fortunate enough to find a number of related papers dealing with application profile design. Unfortunately most of these papers contain very little information about profile evaluation.

When evaluating, the most important criteria is whether the profile can solve the problem defined in the Functional Requirements. However, whether this is the case or not can be hard to determine without first applying the application profile in the real world. By then, a lot of work may already have been put into solution, making it difficult and costly to implement changes.

The Singapore Framework lists two important success criteria: Longevity and Interoperability. Whereas these are important things to keep in mind when designing the profile (using open standards, creating good documentation etc.), both are hard to use as objective criteria before the profile has been implemented and used.

In this paper, we have used an alternative way of evaluating the proposed profile that does not require an implementation. One of the goals of our functional requirements is to make it easy for producers to save content and metadata to a cloud storage solution. To create an Information Package with sufficient preservation metadata, as specified in PREMIS, producers must provide a certain amount of metadata for the Preservation Service. Our profile is built on a model that makes it possible to share data via a common platform and allows business systems to pre-register information. This should reduce the amount of metadata that must be provided by business systems, thereby reducing cost. To test this hypothesis, we have analyzed a METS package manually created using the profile to determine the amount and sources of metadata.

8.2 Information Package Metadata Sources

In the OAIS model, preparation of SIPs for submission to an archive is the responsibility of the Producer. SIPs need to contain Content Information (The Digital Object to be archived with associated representation information) and some Preservation Description Information.

We have identified six potential sources of preservation metadata: Preregistered information about the producing business system, preregistered information about the Digital Objects being submitted, the Digital Objects themselves, explicit descriptive metadata describing each Digital Object, information about events occurring during the preservation process, information from external data sources.

8.3 Building an Example Information Package

Using the proposed application profile, we created a generic METS Package by hand, using the following criteria: The target DO is a PDF file with two related objects. The DO is encrypted and digitally signed. It is assigned one type of Rights information. During import,,information from one Event is registered. The business system uses all 15 elements from the Dublin Core Metadata Element Set all 64 mandatory and optional PREMIS elements (excluding Container elements) for the Object entity.

8.4 Findings

In our scenario, business systems provide 52% (55 elements out of a total of 106) of the metadata in an information package. The Preservation Service provides 42% (45 elements). The remaining 6% of the information is from external sources. By examining the metadata elements one by one, we found that 43% (46 elements) of the Information Package metadata could be pre-registered, either by the business system or preservation service. These were mainly static elements such as those describing system information and those defining metadata types (as opposed to values). Finally, we found that 17% (18 elements) could be auto-generated during the preservation process. An example of this is information dealing with file properties.

Based on these findings, we can conclude that our application profile can simplify metadata provision for business systems, compared to systems that do not allow pre-registration. Furthermore, there is a potential for the automation of metadata provision, further reducing the amount of metadata that must be explicitly provided.

We did a further examination of the metadata that must be provided by business systems and that cannot be automated. We found that many of the elements described complicated attributes of Digital Objects, such as structural relations, encryption or rights information. The more complex the Digital Objects to be preserved, the more metadata must be provided by business systems, increasing cost for producers.

9 Conclusion

In this paper, we have presented an application profile for a cloud-based digital archive system. The profile builds on previous research undertaken to define functionality, information flow and services of such a system. The finished application profile serves as a guide to producers and users of "preservation as a service", and is an important step towards building an actual cloud archiving solution.

We have used the Singapore Framework for Dublin Core Application Profiles as a working guide to designing the different components of the profile. Although parts of the framework are still in draft, we found it a valuable source of information. Using METS as the container format has helped further simplify profile design. Although we have had to make our own decisions in a number of cases, we have been able to successfully extend our METS packages with PREMIS and Dublin Core.

Based on our evaluation of an example METS package created by hand for evaluation purposes, we believe that the proposed application profile will make it possible to create Information Packages that can be used in a cloud archive, as described in our Domain Model. It combines metadata from a number of sources to create a complete set of preservation metadata. We have found that our profile can make it easier for business systems to provide metadata by pre-registering data.

Finally, it should be noted that our example METS package only provides one scenario for evaluation. Depending on the type of DO and its properties, the amount of metadata provided by different entities can be expected to vary. Once a working implementation of our profile has been built, there is scope for re-evaluation using real Digital Objects with a broader range of criteria.

References

1. Jain, L., Bhardwaj, S.: Enterprise Cloud Computing: Key Considerations for Adoption (2010)
2. Mell, P., Grance, T.: The NIST definition of cloud computing. National Institute of Standards and Technology 53(6) (2009)
3. Lenk, A., Klems, M., Nimis, J., Tai, S., Sandholm, T.: What's inside the Cloud? An architectural map of the Cloud landscape. In: ICSE Workshop on Software Engineering Challenges of Cloud Computing, CLOUD 2009, pp. 23–31 (2009)
4. Zend. Simple Cloud API Reference, `http://simplecloud.org/api` (accessed January 14, 2011)
5. CCSDS Secretariat. Reference Model for an Open Archival Information System (OAIS) (2002), `http://public.ccsds.org/publications/archive/650x0b1.PDF` (accessed November 2, 2011)
6. Askhoj, J., Nagamori, M., Sugimoto, S.: Archiving as a service: a model for the provision of shared archiving services using cloud computing. In: Proceedings of the 2011 iConference, pp. 151–158 (2011)
7. Enders, M.: A METS Based Information Package for Long Term Accessibility of Web Archives. In: Proceedings of the 7th International Conference on Preservation of Digital Objects, Vienna, Austria (August 2010), `http://www.ifs.tuwien.ac.at/dp/ipres2010/papers/enders-70.pdf` (accessed June 10, 2011)
8. Nilsson, M., Baker, T., Johnston, P.: The Singapore Framework for Dublin Core Application Profiles, DCMI Recommended Resource (2008)
9. Gartner, R.: METS: Metadata Encoding and Transmission Standard. JISC Techwatch Report TSW (February 05, 2002)
10. Gartner, R.: PREMIS - Implementation Strategies Update 2: Core Elements for Metadata to Support Digital Preservation. RLG Diginews 8(6), 2002–2007 (2004)
11. Caplan, P.: Understanding PREMIS. Library of Congress (2009)
12. UK Archives. Requirements for electronic records management systems 2: Metadata Standard (2002), `http://www.nationalarchives.gov.uk/documents/metadatafinal.pdf` (accessed June 11, 2011)
13. Hillmann, D.: Using dublin core. Dublin Core Metadata Initiative (2008)
14. Dappert, A.: The PREMIS Data Dictionary: Information you need to know for preserving digital documents (2008)
15. The Library of Congress. Guidelines for using PREMIS with METS for exchange (2008), `http://www.loc.gov/standards/premis/guidelines-premismets.pdf` (accessed August 2, 2011)

Coding FRBR-Structured Bibliographic Information in MARC

Trond Aalberg[1], Tanja Merčun[2], and Maja Žumer[2]

[1] Norwegian University of Science and Technology, Trondheim, Norway
[2] University of Ljubljana, Ljubljana, Slovenia

Abstract. The lack of support for the FRBR model in current bibliographic standards has been a major bottleneck for the implementation and use of this model in library databases. In this paper we present solutions for coding FRBR structures using MARC and show that it is possible to code even more complex FRBR structures within the current format. This solution promises a migration path for library systems without losing the compatibility with existing standards.

1 Introduction

The majority of bibliographic information that is managed by libraries today is based on a family of MARC formats that implement the generic record syntax defined in the *ISO 2709 Format for information exchange* [7]. These formats were originally intended to support exchange of bibliographic records, but in reality they also specify the model and requirements for the bibliographic information and have a significant impact on the way information is structured and stored in individual library systems. With the FRBR entity-relationship model proposed in *IFLA Functional Requirements for Bibliographic Records* [15], the library community has been introduced to a new conceptual framework for bibliographic data. The major benefit of the model is that it facilitates the implementation of a more efficient and advanced end-user application that would enable new ways of finding, exploring and discovering relevant content. Additional benefits are increased reuse value of bibliographic information and cross domain interoperability.

Despite the general acknowledgement of the model as a common framework for the next generation of library systems, implementations and large scale use of the model are missing. The alignment of MARC formats with the FRBR model seems to be the main challenge ahead as it is the lack of support for FRBR in current MARC formats and data that presents a major bottleneck for implementing FRBR in library systems.

In this paper we present two experiments related to FRBR expressed in MARC records. Our experiments show that it is possible to code FRBR in MARC, but there are requirements related to both: data quality and data structuring. A practice that ensures consistent cataloguing and inclusion of the required data would significantly improve the quality in terms of FRBR-based

C. Xing, F. Crestani, and A. Rauber (Eds.): ICADL 2011, LNCS 7008, pp. 128–137, 2011.

interpretation of the records, but more advanced features of the formats will have to be used in combination with certain changes in cataloguing practice to fully support the more complex FRBR structures.

2 MARC and FRBR

MARC format is a compact and rather simple data structure where fields, identified by three-character tags, are used to organize the data. Fields are either fixed length control fields or variable length data fields that have further subfields identified by a single character code. A data field may additionally have a set of single character indicators used to guide the interpretation of the field. The core record structure is defined by the ISO 2709 standard, but the meaning and usage of tags and codes is specific to each MARC format, with MARC 21 [8] and UNIMARC [6] as the two main internationally used standards. Current MARC formats are mainly intended to support simple search and display of bibliographic records. Concepts and structures in the record are based on bibliographic standards and practices that date back to the card catalogues.

FRBR is an entity-relationship model that defines four entities for different abstractions of intellectual and artistic products: *Item*, *Manifestation*, *Expression* and *Work*. Work is the most abstract and enables us to refer to intellectual creations at the conceptual level. An expression is the intellectual product that we recognize as unique content in the shape of specific text, sound or images. Manifestation and item entities are equivalent to the concepts of publication and copy. FRBR includes additional entities for persons and corporate bodies, as well as entities that appear as subject entries. The model defines the basic relationships that bind the entities together, as well as other common bibliographic relationships e.g. between different works.

Most fields in MARC 21 can be related to FRBR entities and relationships, although the match is far from perfect [9]. The interpretation of existing records has been explored in numerous projects and prototypes [5,11,2,12]. Problems inherent to FRBR-based interpretations were identified already a decade ago [4] and are related to a model that imposes requirements that the data do not meet. Techniques that compensate for such data have been utilized in some prototypes and tools [3,13], but major challenges must be solved before it is realistic to convert library catalogues to FRBR in large scale.

Any algorithm that attempts to apply the FRBR model on existing MARC records will have to deal with the following issues:

Find the set of entities described in the record. Records may be simple and describe only one work, expression and manifestation, often with one person or corporate body as the creator of the work. This is a default interpretation, but it will never be valid for all records in a collection. Some records describe more complex structure such as multiple embodied expressions, multiple persons, or related works.

Associate correct data with each entity. Fields and subfields can, in some cases, be directly mapped to an entity type, but there are also a number of fields that may have values related to different types. For complex structures with multiple instances of the same type we also need to map particular values to particular instances.

Create identifiers. Bibliographic entities are mainly identified by description and selected attributes can be used as key values. Entities may appear in different contexts (as content, as subject or as related entity) and a key challenge is to identify equivalent entities consistently across all records in a catalogue.

Find relationships. This is both a question of relationship type as well as the identification of the endpoints of relationships. Type and endpoints can in some cases be sufficiently expressed by the data, in other cases both type and endpoints are ambiguous.

Our experience in the interpretation of existing MARC records is based on a FRBRizer tool for experimental conversions of MARC records to FRBR [1]. For a given collection of MARC records we create rules for the above mentioned tasks, which are used to automatically create an XSL-based transformation that can be applied to a collection of MARC records coded in XML. The conversion produces a record for each distinct entity with typed links to related entities. Entities are identified by key attributes derived from the data fields. The tool supports both UNIMARC and MARC 21 and has been used in various experimental conversions [14] including the Norwegian BIBSYS database, The Slovenian national bibliography, test collections from the Swedish library agency BTJ and the Norwegian Deichmanske library. For each collection there is typically a need to define a specific set for rules, although there are some common rules that will apply to most formats within the same family. The tool can be used to explore and evaluate alternative FRBR-based interpretations of large collections, but it is its ability to process all parts of a record and create more complex structures that differentiates our tool from solutions others have developed. The FRBRizer tool was used in the following studies to perform experimental conversions based on alterantive rules for the processing. The rule-based approach with automatic creation of the conversion program allowed for an iterative and explorative process. Results were evaluated by transforming the result sets to HTML-based presentations or to the dot graph format for more visual inspections.

3 Improving Data Quality

The main objective of our first study was to explore how data quality affects the possibility to extract FRBR entities and relationships from bibliographic data. The study was performed in cooperation with the Deichmanske library in Norway as part of a project on new ways of presenting data in library catalogues and linking to external sources. The experiment was based on a test collection consisting of all records found in the library catalogue for 4 different authors: J.R.R Tolkien (242 records), William Shakespeare (577 records), Knut Hamsun (842 records) and Per Petterson (61 records). The first two are well know English writers with a multitude of translations, adaptations and related works, the latter two are Norwegian authors with works that have translations into many languages. The records are originally created by the Norwegian library service based on the Norwegian NORMARC format and national cataloguing rules. A first set of conversion rules was made for this set in cooperation with cataloguers from the library. Main problems we identified in the result were:

Lack of titles to identify works. Uniform or preferred titles are particularly needed for proper work identification. The default title for a work would otherwise be the title used on the publication which is, in many cases, different between expressions and manifestations of the same work. Our initial conversion created a large number of works that were logically equivalent, but had not been merged because the titles used for the works were not the same.

Irrelevant works. Relying mainly on the publication (manifestation) title, the result included many entities that should not be considered as works. Generic or collective titles (such as "Novels") did not represent distinct intellectual products, but the records did not have any information that could systematically be used to distinguish between titles that represented distinct works and titles that did not.

Uninterpretable added entries. Added entries are used for additional persons or titles related to the publication, which may include alternative titles, titles of the various parts, and even titles of other related works. As there was no way to distinguish between the meaning of these fields, most of this information could not be interpreted.

Relationships The lack of relator codes or inconsistent code values made it difficult to relate persons to the correct entity. These codes present values that indicate the role of the person (author, translator, etc.) and are the main source for relating added entry persons to either the work, expression or manifestation.

Different practices in the same collection We generally found it difficult to define rules, particularly more advanced ones, that would be valid for all records in the collection. This was caused by a cataloguing practice that has changed over time as well as by variations in the way different types of publications were being catalogued.

After the first round of processing and analysis we defined data elements that could be corrected to improve the conversion results. To reduce the number of records to a manageable set, we only continued with the records for the Norwegian authors Knut Hamsun and Per Petterson (a total of 903 records). The major changes that were performed by the cataloguing department of the library are listed in table 1. A new set of rules was made for the corrected records and a final conversion was done. A qualitative comparison of the results before and after the corrections are shown in table 2. The numbers are only for works that are contained in one or more expressions. Works that only have authors or works by these authors as subjects, are not included.

For Knut Hamsun we were able to reduce the number of works from 218 to 93. The corrections and the new rules removed a large number of equivalent works

Table 1. Correcting the data in the test collection

Corrections	# changes
Language codes	corrections in 5 records
Uniform titles (240$a)	additions and corrections in 109 records
Title statement (245$a)	corrections in 6 records
Indicators for collective titles (245)	added in in 113 records
Added entries (700/740)	144 new entries and used uniform title in analytic entries
Indicators for analytic entries (700/740)	added 239 indicators in 119 records
Correcting relator codes	48 new entries and many corrections

Table 2. Quality comparison of improvements

Test set	Works	Expressions	Manifestations
Hamsun original set	218	367	519
Hamsun corrected set	93	342	553
Petterson original set	13	40	48
Petterson corrected set	40	82	59

identified by different titles in the original data as well as a large number of false works (by the use of an indicator for 245). Some additional works were added because the indicator for 700 and 740 could now be used to locate contained works. For Per Petterson collection the result shows an increase from 13 to 40 works. The increase was caused by the possibility to locate contained works in compiled collections of short stories etc. In our experiment we focused on manifestations that embodied multiple expressions, expressions in different languages as well as subject entries (records presenting related works are not included in the result). All changes made conform to the Norwegian cataloguing rules and the NORMARC format.

4 Utilizing the Structure

The second study presents a more systematic experiment with two test collections created from a smaller number of retrieved records for a work Little women (60 records) and works by Stieg Larsson (40 records). In contrary to the first study, related works were also included in the test and one of our tasks was to set up a mechanism for establishing and defining links between these works. Consulting RDA (Resource Description and Access) and using some of the new MARC 21 fields, records were unified and enriched with missing data and relationships in a way that would produce as perfect records and consequently as perfect frbrization results as possible using the following methods:

Consistent use of descriptive identification. All records were examined and transformed to consistently identify primary entities and agents, as well as to describe the main set of attributes (such as form of work, language, or content type) needed for the creation of access points (descriptive identifiers). Whenever the manifestation embodied one expression of one work, field 240 (or 130) was used to add the uniform title of work and language of expression. In case a manifestation contained more expressions of the same or different works, all expressions and works were identified using analytical entries in 700 fields. Attributes connected to the expressions or works were recorded within other fields (for example 336, 380, 600, etc.) in combination with field linking.

Field linking. In records where more expressions or works were embodied in the same manifestation, attributes within the same field were often different for different primary entities. Also agents and relationships between entities could be linked to only one of the entities described in the record and not all of them. To make the distinction and relate an attribute, related primary entity, or an agent to the correct expression or work, we have used $8 field link to connect them to the appropriate primary entity.

Relator code. To express the agent's role and connect it to the appropriate primary entity, we used relator code $4 in all 100 and 700 fields. This not only gave us the possibility to distinguish between the roles of agents, but also enabled us to link agents to the appropriate primary entity level.

Related primary entities. Authorized access points were created to identify and draw work to work, expression to expression and manifestation to manifestation relationships. Subfield $i was added to 700/730 and 776 field to express the nature of the relationship (adaptation, sequel, translation, etc.) and a set of subfields was defined by the entity being related. Related works were identified using creator, title, and form of work, while expression to expression relationship was described using title, content type, language, and agent (such as translator, narrator) and, in some cases, version. We additionally used indicators in the 700/730 field to clearly distinguish between analytical entries for contained works/expressions, added entries for persons, and added entry access points for related works.

Modifications to the records are done manually as the data was either unavailable in existing records, insufficient for identification or written in a way that was not machine interpretable. To test and examine the enriched and frbrized records, graphs were generated to show all the entities in the collection of records as well as all the relationships between these entities. They have shown that correct formation of access points and relationships as well as the addition of relator codes, field linking and uniform titles within MARC records enables us to correctly identify all the entities within records as well as draw complex relationships between entities that have so far been discussed mainly only on paper.

Figures 1 and 2 show an example of three related records and a merged graph from the Stieg Larsson example. Record 1 presents an English translation of the first book in the Millenium trilogy, record 2 is a Spanish translation of the third book in the trilogy and record 3 is a motion picture adaptation based on the third book. This example presents only a small part of the whole graph, but

Record 1

```
001     15269628
008     080421s2008 nyu 000 0 eng
020     $a9780307269751
041 1   $aeng $hswe
100 1   $aLarsson, Stieg, $d1954-2004. $4aut
240 10  $aMillennium. $n1. $pMan som hatar
        kvinnor. $leng $81
245 14  $aThe girl with the dragon tattoo /
        $cby Stieg Larsson ; translated from
        the Swedish by Reg Keeland.
250     $a1st United States ed.
260     $aNew York : $bAlfred A. Knopf, $c2008
300     $a465 p. ; $c25 cm.
336     $atext
337     $abook
338     $aprint
380     $aNovel
700 10  $aKeeland, Reg, $d1943- $4trl $81
700 11  $isequel $aLarsson, Stieg,
        $d1954-2004. $tMillennium.
        $n2. $pFlickan som lekte med elden.
        $kNovel $4aut $81
```

Record 2

```
001     10283300
008     090715s2009 sp 000|f spa d
020     $a978-84-233-4161-0
041 1   $aspa $hswe
100 1   $aLarsson, Stieg, $d1954-2004. $4aut
240 10  $aMillennium. $n3. $pLuftslottet som
        sprängdes. $lspa $81
245 13  $aLa reina en el palacio de las
        corrientes de aire / $cStieg Larsson ;
        traducción de Martín Lexell y Juan
        José Ortega Román.
260     $aBarcelona : $bDestino, $c2009.
300     $a854 p. ; $c23 cm.
336     $atext
337     $abook
338     $aprint
380     $aNovel $81
700 10  $aOrtega Roman, Juan Jose. $4trl $81
700 10  $aLexell, Martin. $4trl $81
700 11  $isequel to $aLarsson, Stieg,
        $d1954-2004. $tMillennium. $n2.
        $pFlickan som lekte med elden. $kNovel
        $4aut $81
```

Record 3

```
001     671812168
008     101026p20112009xxu148 vlswe d
041 1   $aswe $jeng $hswe
130 0   $aLuftslottet som sprängdes $lswe
        $81 $82
245 14  $aThe girl who kicked the hornet's
        nest / $cdirector Daniel Alfredson.
250     $aWidescreen ed.
260     $a[United States] : $bMusic Box Films,
        $c2011
300     $a1 videodisc (148 min.) : $bsd.,
        col. ; $c4 3/4 in.
336     $amoving image
337     $avideo
338     $aDVD
380     $aMotion picture
700 10  $aAlfredson, Daniel, $d1959- $4drt $81
700 10  $aNyqvist, Michael, $d1960- $4aus $81
700 10  $aRapace, Noomi, $d1979- $4aus $81
700 11  $ibased on $aLarsson, Stieg,
        $d1954-2004. $tMillennium. $n3.
        $pLuftslottet som sprängdes. $kNovel
        $4aut $82
```

Nodes and relationships for record 2

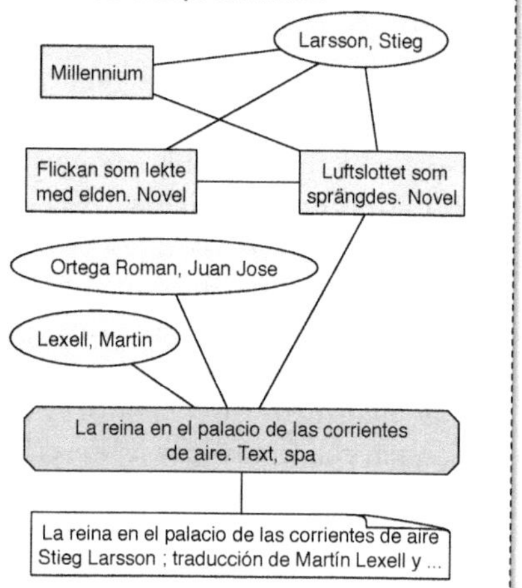

Fig. 1. Example records

even with just a few records we can see a networked structure that could be of great benefit for improving the functionality in end-user applications.

The complexity of the final result is illustrated in table 3. In total, 71 works, 99 expressions and 100 manifestations were generated from the two examples, however in the Little women record set more than half of the manifestations embodied two or more expressions as opposed to the Stieg Larsson example where there were only three such manifestations. The table also shows the rather large number of relationships that we have created to make links between related entities. Most of the relationships have been done on the work level, somewhat less on the expression level, and only a few links have been created on the manifestation level. Formulating relationships and correctly describing multiple

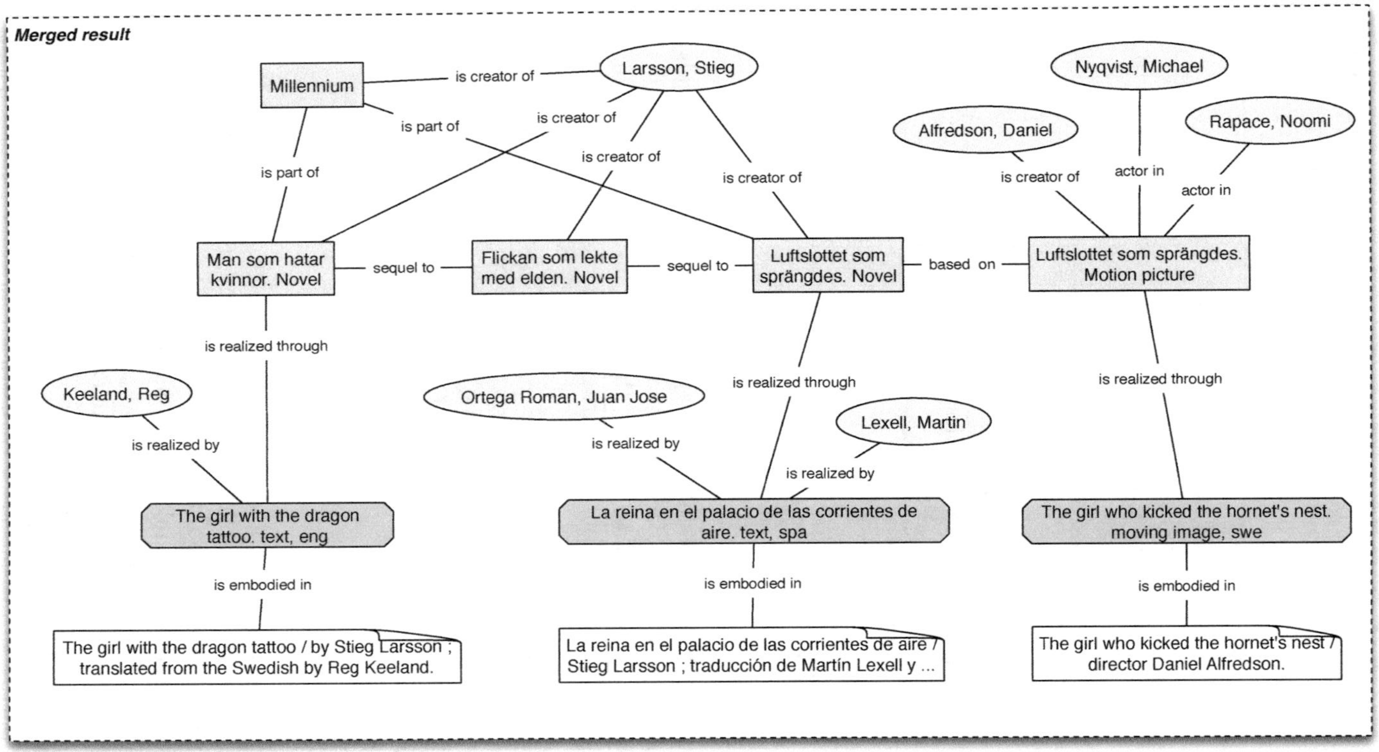

Fig. 2. The merged graph is created from the three example records. Note boxes are manifestations, rounded rectangles are expressions, rectangles are works and ellipses are persons. Relationships created in the processing of the records are bidirectional, but the figure only includes the label for one direction.

Table 3. Entities and relationships

		Little Women		Stieg Larsson	
Entities	Work		58		13
	Expression		69		30
	Manifestation		60		40
Relationships	Work to work		38		18
	is an adaptation of	17		8	
	is a successor to	2		4	
	is an imitation of	7		0	
	has part	9		3	
	has as subject	3		3	
	Expression to expression		11		5
	is an adaptation of	6		5	
	is an abridgment of	4		0	
	Manifestation to manifestation		2		2
	is a reproduction of	2		2	
	Person to work		76		23
	is creator of	73		20	
	is subject of	3		3	
	Person to expression		16		19
	Person to manifestation		1		2

works or expressions in manifestation have proven to be the most difficult tasks and some of the problems we faced have not really addressed so far (not even by the RDA). As described, we utilized the possibilities of field linking and indicators within MARC 21, but there are still some issues such as difficulties in forming expression to expression relationships. This experiment showed that a detailed structure combined with a very consistent descriptive identification is necessary in order to formulate good frbrized results. While RDA provides some of the mechanisms for this, it does not make a requirement for libraries to use them, which will prove to be problematic if we really want to implement the full FRBR idea in the near future.

5 Conclusion

Both experiments illustrate that some of the problems related to FRBR-based interpretations are caused by cataloguing practice rather than the MARC format itself. By unifying, adding the missing data, constructing analytical entries and access points, we were able to implement the main FRBR ideas and even more complex structures in MARC 21 format. While this approach has proved to be effective and the performance of the format was significantly improved, the frbrization process needed to be complemented by improved data quality and structure to achieve satisfactory results. The amount of work invested in not only transforming the legacy data, but also in creating new FRBR-based records using this method, however, suggests that while the method can translate the FRBR model into practice and therefore carries good semantic quality [10] the framework is maybe too inefficient to be adopted in practice and has therefore poor pragmatic quality. Demonstrating that existing MARC 21 format in combination with RDA rules can accommodate the complexity of FRBR is a promising start, but further investigations would be needed to discover a more effective method for bringing FRBR-based data into library catalogues, such as the use of identifiers for primary entities, transformation of MARC formats or a selection of some other format that would be better suited for recording FRBR structures.

References

1. Aalberg, T.: A process and tool for the conversion of MARC records to a normalized FRBR implementation. In: Sugimoto, S., Hunter, J., Rauber, A., Morishima, A. (eds.) ICADL 2006. LNCS, vol. 4312, pp. 283–292. Springer, Heidelberg (2006)
2. Aalberg, T., Žumer, M.: Looking for entities in bibliographic records. In: Buchanan, G., Masoodian, M., Cunningham, S.J. (eds.) ICADL 2008. LNCS, vol. 5362, pp. 327–330. Springer, Heidelberg (2008)
3. Freire, N., Borbinha, J.L., Calado, P.: Identification of FRBR works within bibliographic databases: An experiment with UNIMARC and duplicate detection techniques. In: Goh, D.H.-L., Cao, T.H., Sølvberg, I.T., Rasmussen, E. (eds.) ICADL 2007. LNCS, vol. 4822, pp. 267–276. Springer, Heidelberg (2007)
4. Hegna, K., Murtomaa, E.: Data mining to find: FRBR? In: IFLA 68th General Conference and Council, August 18-24. IFLA, Glasgow (2002)
5. Hickey, T., O'Neill, E.T.: FRBRizing OCLC's WorldCat. Cataloging & Classification Quarterly 39(3/4), 239–251 (2005)
6. International Federation of Library Associations and Institutions. UNIMARC Manual: Bibliographic Format. IFLA Universal Bibliographic Control and International MARC Core Programme, UBCIM (1994)
7. International Standard Organization. ISO 2709:2008. Information and documentation – Format for information exchange (2008)
8. Library of Congress. MARC 21 Format for Bibliographic Data. Network Development and MARC Standards Office (October 2010)
9. Library of Congress Network Development and MARC Standards Office. Functional Analysis of the MARC 21 Bibliographic and Holdings Formats (April 2006)
10. Lindland, O., Sindre, G., Solvberg, A.: Understanding quality in conceptual modeling. IEEE Software 11 (1994)
11. Manguinhas, H.M.A., Freire, N.M.A., Borbinha, J.L.B.: FRBRization of records in multiple catalogs. In: Proceedings of the 10th Annual Joint Conference on Digital Libraries, JCDL 2010, pp. 225–234. ACM, New York (2010)
12. Mcgrath, K., Bisko, L.: Identifying FRBR work-level data in bibliographic records for manifestations of moving images. The Code4Lib Journal 1(5) (December 2008)
13. OCLC. FRBR Work-Set Algorithm (August 2009), `http://www.oclc.org/research/activities/past/orprojects/frbralgorithm/default.htm`
14. Pisanski, J., Žumer, M., Aalberg, T.: Frbrisation: towards a bright new future for national bibliographies. In: World Library and Information Congress: 75th IFLA General Conference and Assembly, August 23-27. IFLA, Milano (2009)
15. The International Federation of Library Associations and Institutions. Functional requirements for bibliographic records. UBCIM Publications - New Series, vol. 19 (1998)

Metrics for Metadata Quality Assurance and Their Implications for Digital Libraries

Ya-Ning Chen[1,2], Chun-Ya Wen[3], Hui-Pin Chen[2], Yen-Hung Lin[4], and Hon-Chung Sum[3]

[1] Graduate Institute of Library and Information Studies, National Taiwan Normal University, Taipei, Taiwan
[2] Computing Center, Academia Sinica, Taipei, Taiwan
[3] Research Center for Information Technology Innovation, Academia Sinica, Taipei, Taiwan
[4] Institute of History and Philology, Academia Sinica, Taipei, Taiwan
{arthur,cancer,winefast,yenny,htshen}@gate.sinica.edu.tw

Abstract. This study aims at developing a set of common metrics for metadata quality at data element level. The proposed metrics are used to assure the quality for metadata that converted from heterogeneous metadata formats and sources into a Dublin Core based centralized digital repository. This study adopted the metrics provided by Bruce and Hillmann [2] as a basis to develop the proposed metrics as follows: conformance to expectations, provenance, accuracy, completeness and accessibility. A hybrid approach of automatic and manual had developed for assessing metadata quality in practice. Target subjects were two individual projects selected from the Taiwan E-Learning and Digital Archives Program as use case to illustrate the details and examine the feasibility of the proposed metrics. Finally, this study discussed the implications of proposed metadata quality metrics for digital libraries in the following perspectives: project management, metadata management, hidden quality problems and accessibility.

Keywords: metadata, quality assurance, evaluation, digital libraries.

1 Introduction

Generally, metadata is defined as "data about data", organizing resources for digital repositories. In the past decades many metadata formats have been developed and adopted by various communities, and much effort has been contributed to implementation and creation of metadata. As known, metadata plays an important role in resource description and discovery for digital libraries. Therefore, high quality metadata will facilitate access to resources stored in digital repositories, whereas poor quality metadata will cause resources invisible for users and remain unused.

In fact metadata quality is a complicated issue because a standard definition is still under developing. [2] pointed that metadata quality is difficult to define. [5] attempted to define that good quality metadata is compliant to community practice and standards. On the other hand, [10][11] regarded that quality metadata provides enough

C. Xing, F. Crestani, and A. Rauber (Eds.): ICADL 2011, LNCS 7008, pp. 138–147, 2011.

information or description of resource for users to find, review, and reuse resources. [4] offered a definition of that high quality metadata supports the functional requirements of the system it is designed to support. Furthermore, [4][14][16] arrived at a common agreement for metadata quality as "fitness for purpose or use". Based on the above discussion, one may draw a summary for that metadata quality involves fitness for purpose. It means that assessment for metadata quality are dependent on purpose, ranging from use analysis [9][16][17], to process [1][3], assurance [6], and cost effectiveness [12][13]. Without question, existing metadata metrics have developed for meeting various purposes. Nevertheless, most of them are focused on a domain-specific digital repository, like elearning objects [10], language archives [5], and government information [8].

Many digital library projects have been sponsored by the Taiwan E-Learning and Digital Archives (hereafter TELDAP) across diverse domains, such as libraries, museums, archives, herbariums and personal collections. All various domain-specific metadata formats and records are exported into the TELDAP Union Catalog (hereafter TELDAP UC) as a centralized digital repository for retrieval and dissemination. These formats including MARC, CDWA, EAD and Darwin Core are required to convert into Dublin Core (hereafter DC) for TELDAP UC. Therefore, this study aims at developing a set of common metrics to assure quality for metadata during conversion from local repositories to TELDAP UC. In other words, it becomes an issue to examine to some degree what quality metadata of individual projects has achieved for TELDAP UC. Furthermore, how a set of common metrics can fit to assure metadata quality for heterogeneous domain-based metadata records.

2 Methodology

This article attempts to develop metadata quality metrics at data element level during converting metadata from heterogeneous sources and formats into a DC-based centralized digital repository. It also provides suggestion for quality improvements. The proposed metadata quality metrics adopted [2] as a basis to develop a set of common metrics. Two target subjects were selected from TELDAP as use case to elaborate the proposed metadata metrics. One was from archives and the other was from biodiversity domain. The remainder of this article is structured as follows. First, related works based on literature review motivate the contribution of our proposed approach. Next, detail of metadata metrics and their development is illustrated. Then two use cases are employed to examine the feasibility of proposed metadata metrics in Section 5. In Section 6 the discussion will be presented. Finally, concluding remarks are drawn for future research in Section 7.

3 Related Work

In the early stage, [8][9] provided four assessment categories as metrics for metadata quality in the following: accuracy, completeness, profile and serviceability. By wider application of OAI-PHM to harvesting metadata from distributed servers, [16] explored the usage of DC elements as an approach for assessing metadata quality.

On the other hand, [5] developed algorithm to automatically evaluate metadata quality based on OLAC.

In 2004, [2] provided a set of quality metrics for metadata as follows: completeness, accuracy, provenance, conformance to expectations, logical consistency and coherence, timeliness, and accessibility. They also extended metadata quality metrics into three levels: semantic structure, syntax structure and data value. Then, [6] attempted to build a framework for metadata quality assurance, and offered several criteria in the following: purpose of metadata, cataloging rules, maintenance, interoperability, and validation. In addition to information quality framework, [15] furthering built up a mapping link to the framework developed by [2] for metadata quality metrics. According to the previous literature review, one may draw generalization for metadata quality metrics as follows:

- Although there's a lack of agreed definition on assessing the metadata quality in digital repositories, several studies such as [5][11] adopted metrics developed by [2] as framework to refine their quality metrics.
- It seems apparent that automatic measurement is a current trend for metadata quality metrics, such as [5][10][11][13][16]. However, few studies have adopted a hybrid approach of automatic and manual for metadata metrics, such as [7].
- Most studies [1][5][8][10][16] applied metadata metrics for a centralized digital repository by adoption of a specific metadata format for similar domain-based objects and their metadata. Furthermore, metadata quality metrics can be divided into different levels: data element and data value, but most studies such as [3][6][7][9][11][17] still emphasized metrics at data element level.
- However, it becomes an issue to investigate on how metrics assure the quality for heterogeneous domain-based metadata at data element level.

4 Metadata Metrics for Quality Assurance

Many digital projects across libraries, museums, archives and herbaria in Taiwan are sponsored by TELDAP. The TELDAP UC has built up as a centralized digital repository to collect and distribute metadata and their digital objects. In order to aggregate various metadata from heterogeneous sources, DC is selected as a metadata crosswalk mechanism for TELDAP UC. A variety of specific domain format-based metadata and their digital objects are converted into TELDAP UC, such as MARC, CDWA, EAD and Darwin Core. Currently, TELDAP UC consists of 3.26 million DC metadata records, 3.35 million images and 6.66 million digital objects in April 2011. However, it becomes an urgent issue to assure metadata quality of various heterogeneous formats and sources to facilitate resource discovery, representation and future reuse.

This study adopted the principles provided by [2] as a basis to design metadata metrics for quality assurance at data element level. The metadata metrics were used to measure the quality of metadata which have existed in or are going to export into TELDAP UC. Overall, five principles were elaborated as follows: conformance to expectations, provenance (including logical consistency and coherence), accuracy, completeness and accessibility (including timeliness). Consequently, a radar chart consisting of the above five principles will generate as a measuring result for each individual projects.

4.1 Conformance to Expectations Metrics

The conformance to expectations measures the degree whether the individual projects achieved in exporting the predicted amount of metadata records into TELDAP UC, in terms of project management. Essentially, metadata is important in facilitating resource identification and discovery for users. One of TELDAP objectives is to offer digitized objects with appropriate metadata for resource description. The metrics aim to examine the degree to which how many metadata records are provided by individual projects. The calculation formula is simple as below:

Amount of exported metadata records ≒ Amount of predicted metadata records

4.2 Provenance Metrics

The provenance measures the authority of metadata. The authority has several meanings for TELDAP UC. In terms of record's origin, metadata has to clearly show who the primary responsible body is for creating records. This perspective is related to measure the reputation of institutions or projects who maintains metadata for digital objects. Metadata also plays an important role in facilitating the commercial use of digital objects in light of promotion and diffusion. How to indicate rights information who owns digital objects will build a connection between owners and commercial vendors. On the other hand, digital objects need more information to be unambiguously identified from each other, such as title, identifier, and format. It is also important to examine whether creation of metadata records is according to certain content rules and controlled vocabularies. In summary, three items for the provenance metrics are listed in the following:

- Are Title, Identifier, Format, Subject, Rights and Publisher elements of DC provided in metadata records?
- Do metadata records use any controlled vocabulary, like LCSH, AAT and LC Name Authority file, as data value for Subject, Coverage, Creator, and Contributor?
- Do metadata professionals adopt any content rules similar to AACR, CCO and ISAD(G) to manage metadata at data element and/or value level?

4.3 Completeness Metrics

The metrics are to measure the degree of completeness how many metadata elements are used to describe the objects. Some metadata formats define all of their elements as optional such as DC and LOM, and some define several elements as mandatory such as MARC 21, CDWA and EAD. [11] argued that not all data elements are relevant for all resources, and not all metadata elements are equally relevant to all contexts. Usually, the more elements are adopted to describe objects, the more contextual information can deliver users to know what objects are. Besides six elements are defined as mandatory to offer basic information as discussed in provenance metrics, projects are encouraged providing more DC elements and their values in TELDAP UC to build up a richer context of objects for users. Therefore, the more DC elements are used in TELDAP UC, the higher score will get.

4.4 Accuracy Metrics

The accuracy is to measure the degree which the metadata values are correct, that is, how well they describe the object. The metrics attempt to measure whether the values are put into the appropriate metadata elements, rather than whether the values are correct or not. It seems very easy for developing software to achieve automatic check against accuracy if software can base rules on official crosswalk documents, such as crosswalk document offered by the Getty Research Institute for mapping between different metadata formats. In fact, this problem is difficult for automatic operations owing to the context of aggregating objects and metadata formats. The aggregating objects that metadata described are either original items or digitized counterparts are the first issue needed to clarify. According to the DC 1:1 principle, original and their digital objects are regarded as different objects for metadata. It is imperative to draw a clear line between original and counterpart objects for examining and re-contextualizing their corresponding elements and values. On the other hand, one may find that some metadata formats describe more than two distinctive objects. For instance, EAD is a tag-based metadata format to describe compounded granular or level archives such as fonds, series, files and items. When converting metadata from EAD to DC, only elements of targeted level archives will be converted and others leave intact. Therefore, accuracy metrics count on manual check.

4.5 Accessibility Metrics

The accessibility metrics are used to measure the degree of availability which metadata records and their digital objects of source repositories are accessible. Most studies defined the accessibility for metadata records only. This study extended the accessibility to include both metadata and their digital objects. The reason lies in that TELDAP UC is purposed to be a portal site in guiding users back to source repositories for retrieving more comprehensive and timely descriptions of metadata elements and higher resolution of digital objects. Therefore, the timeliness metrics were integrated into accessibility metrics in this study.

According to TELDAP UC's context, criteria for accessibility were designed into two categories: linkages to source metadata records and digital objects. In terms of accessibility of metadata records, three items were proposed to measure the quality ranging from existence, to validity and uniqueness. First of all, individual projects have to offer URLs leading users from TELDAP UC to access more detailed metadata descriptions of source repositories. Next, software will check whether URLs are valid. Software will also examine whether each URL has its own metadata record. In other words, our metrics will investigate whether different metadata records in TELDAP UC point to the same URL or metadata record of source repositories. If one URL is shared by more than 2 records, it means that the same object is described by more than two and is not conformant with DC's 1:1 principle. It could also confuse users' perceptions. On the other hand, each metadata record has its own digital object. The similar metrics of metadata linkages were also revised to examine the existence, validity and uniqueness for URLs of digital objects. The metrics of linkages to digital objects will assess whether one metadata record has its own objects, as well as how many objects are valid and unique. To sum up, six items are created for accessibility as below:

- What is the percentage of URLs available to access to source metadata records?
- What is the percentage of URLs valid to access to source metadata records?
- What is the percentage of URLs unique to access to source metadata records?
- What is the percentage of URLs available to access to source digital objects?
- What is the percentage of URLs valid to access to source digital objects?
- What is the percentage of URLs unique to access to source digital objects?

5 Use Case

5.1 Case 1: A Digital Herbarium Project for Specimens

This use case is a digital herbarium project focused on building up a digital repository for specimens. The radar chart shown as Fig. 1 reveals interesting points for overall measurement as follows:

Automatic Evaluation Results. In terms of conformance to expectations metrics, this project sharply failed in exporting the expected amount of metadata records into TELDAP UC, and only 13.90% attained. On the other hand, this project got 85.40% and 100% for the completeness and accessibility metrics respectively. Only Type and Format elements were totally not used by this project. After check, we found that this project was without appropriate planning and investment on metadata creation. However, further improvements for this project were offered in the follow points:

- A default value "physical object" was suggested for Type element during export.
- To extract dimensional data suitable for Format element from Description element, such as height, width and length.

Human Evaluation Results. In terms of provenance, it revealed that this project adopted content rules to control consistency on resource description and syntax, especially for Identifier, Date and Coverage elements. As for the accuracy, it indicated that this project didn't follow the DC 1:1 principle and definitions of some elements as follows:

- In Identifier element, this project offered the metadata record no. rather than an identifier for original object of specimen.
- The Language element should be empty owing to specimens are similar to objects of visual arts. Providing the URL of digital repository was not suitable for Source.

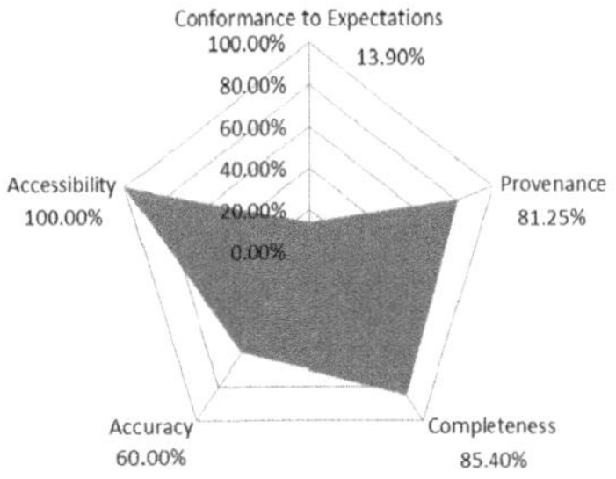

Fig. 1. A Radar Chart of Metadata Quality Metrics for Use Case 1

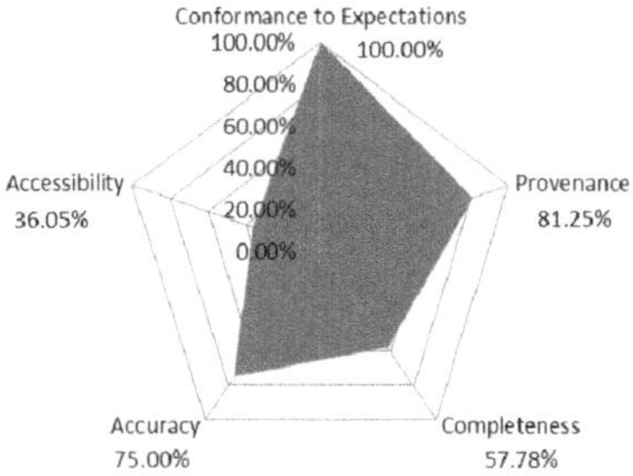

Fig. 2. A Radar Chart of Metadata Quality Metrics for Use Case 25.2

5.2 Case 2: An Archival Project for Diplomatic Records

This use case is a digital archival project focused on building up a digital repository for diplomatic records. The radar chart shown as Fig. 2 uncovers interesting points for overall measurement as follows:

Automatic Evaluation Results. In terms of conformance to expectations and accessibility metrics, this project was wholly successful in offering the expected amount of metadata, but failed in accessibility. In terms of completeness, several DC elements were totally not provided, including Creator, Contributor, Format, Publisher, Source, Relation and Coverage, and score was 57.78%. Apparently this project delivered so little background information that neither tells users a story that embedded in objects nor links back to source repository for more information. After serious check, the fact is that this project failed in converting these metadata elements. However, it suggested that this project has to convert metadata again for offering basic background to users of TELDAP UC.

Human Evaluation Results. The target objects which convert from source repository to TELDAP UC were original archival items. This project offered more information for other DC elements, but some are not appropriate. Information of digital objects such as authorship and rights were put at Type. This project also converted information of fonds, series and files into Description of item-level archival metadata for TELDAP UC, such as scope, and biography or history information of fonds and series. Therefore, the measurement offered suggestion for this project as follows:

- Information related to digital objects as well as fonds, series and files was deleted.
- Names of fonds, series and files can be concatenated together into Relation in order to indicate the contextual relationship of archival hierarchical provenance.

6 Discussion

6.1 Keep Project Management and Metadata Management in Accordance with Digital Lifecycle

Although original purpose of the proposed metadata metrics is to assure quality of metadata at data element level, it is found that metrics are also suitable for project management. The proposed metadata metrics can be used at all stages of lifecycle of project management. At the project planning stage, the proposed metrics can be used to assess what project's expecting outputs and results will attain and what resources and costs are required to perform the planning tasks. It means that metrics are useful for decision making on whether the project is worthy to invest or not, based on the measurement between goals and investments. When project moves into implementation stage, the proposed metrics can be a basis to audit whether on-going tasks are compliant to the predictive goals and requirements. Further, project staffs can be inspired to detect the errors and find out the reasons for improvements. Finally, when project steps forward into the closing stage, the proposed metrics can be employed as evaluation tool to measure whether project achieve the expecting outputs and results. Therefore, the proposed metrics is not only applied at final stage of projects, but also every stage for project management and a basis for decision making. On the other hand, the proposed metadata are also corresponding with every stage of lifecycle of digital objects for metadata management, ranging from planning, to policy for developing and auditing metadata, and access for users.

6.2 Explore Hidden Problems Based on Human Evaluation

Some proposed metadata metrics are complementary to each other, such as completeness, provenance and accuracy, but a much deeper investigation by human evaluation is necessary to explore the potential issues for provenance and accuracy. Take use case 1 as example, this project got 85% for completeness, but some elements such as Source, Language and Identifier were not used correctly. As for the use case 2, authorship information of digitized objects was put into Type, and also lacked contextual information such as "Fonds-Series-Files" information at Relation to indicate the hierarchical provenance between the archival items and their higher hierarchical archives. Furthermore, another issue is related to readability for display. In use case 2, more than 10 historical persons involved with diplomatic records were put at Subject element and each person was listed at one line. Such display is not so friendly for users to browse metadata. In terms of consistency and coherence in provenance, this study suggested that combine all historical persons altogether with a single label instead of multiple ones at Subject.

Indeed problems related to provenance and accuracy metrics are beyond capability of automatic checking software, and need human evaluation to find them out. According to the above discussion, one may find that a hybrid approach of automatic check with human evaluation can achieve better metadata quality assurance, because hidden problems can be uncovered by manual check rather than automatic.

6.3 Take a Deeper Investigation on Accessibility

In addition to metadata and their objects, the DC 1:1 principle is also applied to accessibility in this study. It means that the proposed accessibility metrics are used to audit the valid connections between metadata records of TELDAP UC and their source metadata and objects stored in local repositories. Although automatic checking software can detect whether URLs are valid and duplicate, however, it can't judge whether the connections lead to correct match between TELDAP UC and local repositories. Human evaluation can achieve, however, that is too labor-intensive and time-consuming. Therefore, it becomes an issue to examine the feasibility how to implement the ideas such as Web 2.0 social media to invite users to feedback incorrect matching.

7 Conclusion

This study is successful in developing a set of common metadata quality metrics for TELDAP UC. The proposed metrics have applied around 100 heterogeneous domain-specific and format-based individual projects to prove their feasibility for TELDAP UC. Initially the proposed metrics for metadata quality are used by individual projects, but they can be leveraged in a retrospective way to examine the overall metadata quality of TELDAP UC. This study also illustrates how to customize a framework of metadata quality metrics into a practical approach. Further this study sheds light on project management, metadata management, hidden problems and accessibility for digital libraries. However, future work will develop a wholly automatic approach for enhancing metadata quality metrics at data element level, and expand quality assurance into data value level.

Acknowledgments. The work described in this paper was supported partially by Taiwan E-learning and Digital Archives Program (TELDAP) sponsored by the National Science Council of Taiwan under NSC Grants: NSC100-2631-H-001-014, NSC 100-2631-H-001-020 & NSC 100-2631-H-001-012.

References

1. Barton, J., Currier, S., Hey, J.M.N.: Building Quality Assurance into Metadata Creation: An Analysis Based on the Learning Objects and e-Prints Communities of Practices. In: Proceedings of 2003 Dublin Core Conference: Supporting Communities of Discourse and Practice - Metadata Research and Applications, pp. 39–48 (2003), http://dcpapers.dublincore.org/ojs/pubs/article/view/732/728
2. Bruce, T.R., Hillmann, D.I.: The Continuum of Metadata Quality: Defining, Expressing, Exploiting. In: Hillmann, D.I., Wesbrooks, E.L. (eds.) Metadata in Practice, pp. 238–256. ALA, Chicago (2004)
3. Currier, S., Barton, J., O'Beirne, R., Ryan, B.: Quality Assurance for Digital Learning Object Repositories: Issues for the Metadata Creation Process. Research in Learning Technology 12(1), 5–20 (2004), http://eprints.rclis.org/bitstream/10760/5237/1/03barton.pdf
4. Guy, M., Powell, A., Day, M.: Improving the Quality of Metadata in Eprint Archives. Ariadne 38 (2004), http://www.ariadne.ac.uk/issue38/guy/intro.html
5. Hughes, B.: Metadata Quality Evaluation: Experience from the Open Language Archives Community. In: Chen, Z., Chen, H., Miao, Q., Fu, Y., Fox, E., Lim, E.-p. (eds.) ICADL 2004. LNCS, vol. 3334, pp. 320–329. Springer, Heidelberg (2004)
6. Kelly, B., Closier, A., Hiom, D.: Gateway Standardization: A Quality Assurance Framework for Metadata. Library Trends 53(4), 637–650 (2005)

7. Ma, S., Lu, C., Lin, X., Galloway, M.: Evaluating the Metadata Quality for the IPL. Proceedings of the American Society for Information Science and Technology 46(1), 1–17 (2009)
8. Moen, W.E., Stewart, E.L., McClure, C.R.: The Role of Content Analysis in Evaluating Metadata for the U.S. Government Information Locator Service: Results from an Exploratory Study (1997), http://www.unt.edu/wmoen/publications/GILSMDContentAnalysis.htm
9. Moen, W.E., Stewart, E.L., McClure, C.R.: Assessing Metadata Quality: Findings and Methodological Considerations from an Evaluation of the U.S. Government Information Locator Service. In: Proceedings of Proceeding ADL 1998 Proceedings of the Advances in Digital Libraries Conference. IEEE Computer Society, Washington, DC (1998), http://doi.ieeecomputersociety.org/10.1109/ADL.1998.670425
10. Ochoa, X., Duval, E.: Towards Automatic Evaluation of Learning Object Metadata Quality. In: Roddick, J., Benjamins, V.R., Si-said Cherfi, S., Chiang, R., Claramunt, C., Elmasri, R.A., Grandi, F., Han, H., Hepp, M., Lytras, M.D., Mišić, V.B., Poels, G., Song, I.-Y., Trujillo, J., Vangenot, C. (eds.) ER Workshops 2006. LNCS, vol. 4231, pp. 372–381. Springer, Heidelberg (2006)
11. Ochoa, X., Duval, E.: Towards Automatic Evaluation of Metadata Quality in Digital Repositories (2009), http://ariadne.cti.espol.edu.ec/M4M/files/TowardsAutomaticQuality.pdf
12. Stvilia, B., Gasser, L.: Value-based Metadata Quality Assessment. Library & Information Science Research 30(1), 67–74 (2008)
13. Stvilia, B., Gasser, L., Twidale, M.B., Shreeves, S.L., Cole, T.W.: Metadata Quality for Federated Collections. In: Proceedings of the 9th Conference on Information Quality, pp. 111–125. M.E. Sharpe, New York (2004)
14. Stvilia, B., Gasser, L., Twidale, M.B., Smith, L.C.: A Framework for Information Quality Assessment. Journal of the American Society for Information Science and Technology 58(12), 1720–1733 (2007)
15. Shreeves, S.L., Knutson, E.M., Stvilia, B., Palmer, C.L., Twidale, M.B., Cole, T.W.: Is Quality Metadata Sharable Metadata: The Implications of Local Metadata Practices for Federated Collections. In: Proceedings of ACRL 20th National Conference, pp. 223–237 (2005)
16. Ward, J.: A Quantitative Analysis of Unqualified Dublin Core Metadata Element Set Usage within Data Providers Registered with the Open Archives Initiative. In: Proceedings of the 3rd ACM/IEEE-CS Joint Conference on Digital Libraries, pp. 315–317. IEEE Computer Society, Washington, DC (2003)
17. Wilson, A.J.: Toward Releasing the Metadata Bottleneck: A Baseline Evaluation of Contributor-supplied Metadata. Library Resources & Technical Service 51(1), 16–28 (2007)

Research and Practice of Electronic Resources Preservation in Tsinghua University Library

Ting Zeng[1], Li Dong[1], Chao Li[2], and Gang Chen[3]

[1] Tsinghua University Library,
Tsinghua University,
100084 Beijing, P.R. China
[2] Research Institute of Information Technology,
Tsinghua University,
100084 Beijing, P.R. China
[3] Department of Electrical Engineering,
Tsinghua University,
100084 Beijing, P.R. China
{zengting,dongli}@lib.tsinghua.edu.cn,
lichao00@tsinghua.org.cn, gangchen08@gmail.com

Abstract. Our university library decide to build up a local electronic resources preservation and management platform, which can both provide the preservation of subscribed electronic resources, and provide the access to those resources. In this paper, we introduce some of our research and practice for the task, including the framework design, AIP and metadata framework design and practice. In addition to deciding on the more traditional descriptive and administrative metadata, particular care needs to be given to the choice of structural and preservation metadata, as well as to integrating the various metadata components. We have selected METS structural, PREMIS preservation and QDC descriptive metadata for the Tsinghua University Library's digital preservation system and used METS as the container format for integrating the various metadata components, and developed a metadata framework to aid in the preservation of digital resources.

Keywords: Electronic resources preservation, AIP, metadata framework, METS, PREMIS, QDC.

1 Introduction

Tsinghua University library, as the responsible institution of CALIS electronic resources purchasing group, has established a good relationship with a lot of database providers. In order to provide long-term access to some important electronic resource databases, our university decide to build up a local electronic resources preservation and management platform, which can both provide the preservation of subscribed electronic resources, and provides the access to those resources. In this paper, we introduce some of our research and practice for the task, including preservation system framework design, AIP and metadata framework design and practice.

C. Xing, F. Crestani, and A. Rauber (Eds.): ICADL 2011, LNCS 7008, pp. 148–157, 2011.

2 A Preservation Framework

Our aim is to construct a digital preservation system that can:

(1) Accommodate different collections and their corresponding data storage systems,
(2) Preserve digital objects of not only international formats but also native/local special formats,
(3) Allow common technical staff (who is not digital preservation professionals) in digital libraries to manage and preserve their digital holdings,
(4) Offer convenience for institutions to look for proper professional digital preservation services, and save the costs of hiring professionals and purchasing professional facilities.
(5) Besides, the system should be a complete solution that is affordable (limited budget for digital preservation) and flexible (can integrate growing range of methods/tools /services, and accommodate new storage systems or new formats).

Compared with the related work [1], AOMS-II owns the following features:

(1) Practical preservation service discovery and description mechanism: AOMS is based on web service rather than semantic web service.
(2) Comprehensive risk notification: AOMS-II considers and manages the risks of not only file formats/versions obsolescence, but also storage media, hardware, software, OS, etc.
(3) Aid in scheduling on preservation actions: Consequently, the corresponding risks may be interlaced, and AOMS-II helps collection managers schedule the preservation actions effectively and efficiently.
(4) Extensibility to multiple different data storage systems and native/local special formats

For these objectives, AOMS-II should demonstrate the architecture to leverage the efforts of the different preservation initiatives by integrating the growing range of methods, tools and services being developed into a flexible, extensible Web-based framework (Fig. 1).

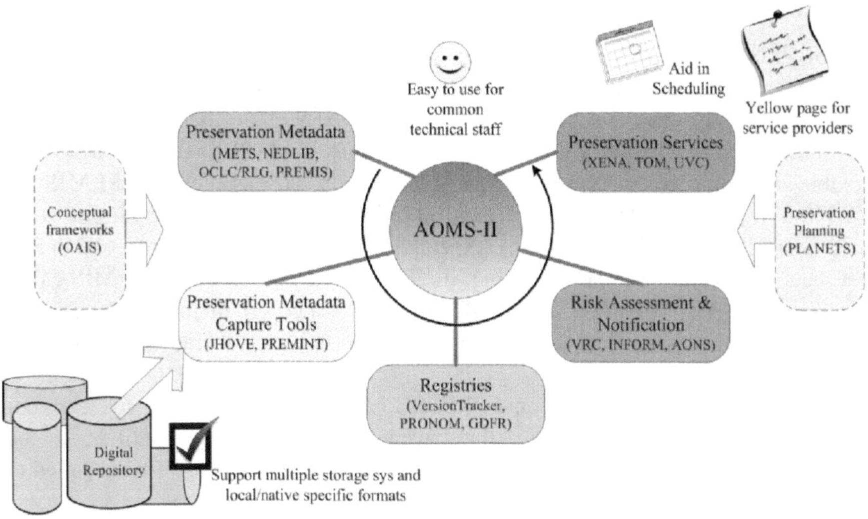

Fig. 1. A flexible and extensible Web-based framework

3 AIP and Metadata Framework Design and Practice

3.1 AIP and Metadata Framework Design

We have selected METS structural, PREMIS preservation and QDC descriptive metadata for the Tsinghua University Library's digital preservation system and used METS as the container format for integrating the various metadata components, and developed AIP and metadata framework to aid in the preservation of digital resources.

(1) Using QDC and PREMIS with METS

The Metadata Encoding and Transmission Standard (METS) is "a standard for encoding descriptive, administrative, and structural metadata regarding objects within a digital library". Actually, the semantics of only structural metadata are defined in METS, including the (possibly hierarchical) structure of digital objects, the names and locations of the files comprising those objects, and a small number of their technical properties. METS also provides placeholders for descriptive metadata and four types of administrative metadata: technical, digital provenance, source and rights metadata. A METS document functions as a container for all the relevant metadata for a digital object [2].

METS provides a robust and flexible way to define digital objects. Qualified Dublin Core (QDC) provides ways to describe objects, and builds on the widely used Dublin Core. Finally, the Preservation Metadata Implementation Strategy (PREMIS) provides ways of describing objects and processes that are essential for digital preservation. All of these three metadata are content-type independent, which makes it possible to define shared usage guidelines for the various content-types held in the archival store, so we have selected METS structural, PREMIS preservation and QDC descriptive metadata for the Tsinghua University Library's digital preservation system.

(2) Issues in Using PREMIS with METS

Use of PREMIS in the METS container format has become an accepted method for recording provenance and other metadata that supports preservation. Though many organizations are now implementing PREMIS in METS, there has been little consolidation of PREMIS-METS practices. Because both PREMIS and METS are flexible and extensible, so there are some key issues to consider in using PREMIS with METS. We have decided to make local decisions according to our use cases and document PREMIS-METS decisions (especially about 13 key issues proposed in Sally Vermaaten' study, "A Checklist and a Case for Documenting PREMIS-METS Decisions in a METS Profile,") in the METS profile as a beneficial internal practice and an efficient way of sharing and comparing metadata strategies, thereby facilitating best practices [3].

One of the main principles behind PREMIS is that you need to be very clear about what you are describing. PREMIS defines five kinds of things (called entities) you can talk about: Intellectual Entities, Objects, Agents, Events and Rights. This is called the PREMIS data model [4].

PREMIS do not actually define any metadata pertaining to Intellectual Entities because there are plenty of descriptive metadata standards to choose from (MODS,

QDC, DC, etc). There are four types of metadata pertaining to Objects, Events, Rights, Agents defined in PREMIS.

A METS file consists of different sections. Of those sections, the administrative metadata section (amdSec) is the one that is most relevant for the use of PREMIS, in particular its technical (techMD), digital provenance (digiProvMD), and rights (rightsMD) subsections. We have decided to break the PREMIS metadata pertaining to the four types of PREMIS entity into METS administrative metadata subsections. The PREMIS Object entity contains mostly technical metadata (e.g., information about size, format, fixity), and so can fit in the METS technical metadata section. The PREMIS Event entity is about actions taken on the object, so is appropriately included under the METS digital provenance section, and PREMIS Rights fits under the METS rights section. The Agent entity is associated with either Rights or Events, so should be in the same section as its related entity [2].

For redundancies between PREMIS and METS, we decide to record those elements redundantly because we need to support both preservation and access of digital resources. For format specific technical metadata, we decide to record them in PREMIS <objectCharacteristicsExtension> so as to keep all technical preservation-related metadata together in PREMIS, which could decrease the need for expressing element values redundantly. For structural metadata, METS structural map is used because it's highly flexible and more expressive. For linking between PREMIS and METS elements, because the METS document might be exchanged and parsed, and the metadata will be extracted and stored outside the METS document, it is important that links remain stable and unambiguous.

(3) Data Model

AIPs are archival information packages as defined in OAIS [5]. AIP is a package describing one archival object in preservation system. In order to deal with updates efficiently, we have decided to separate structural information about the relationship of the files in a representation from the descriptive information and from submission provenance information. We have designed a generic data model fitting for all content-types. There are three kinds of objects in this data model: Intellectual Entity, Representation, and Submission. Intellectual Entities are conceptual, and might be called "bibliographic entities." A representation object is the set of all file objects needed to render an Intellectual Entity. A submission object describes one submission event. Their relationships are shown in Fig. 2. The generic data model, therefore, contains three kinds of metadata AIPs representing different kinds of objects.

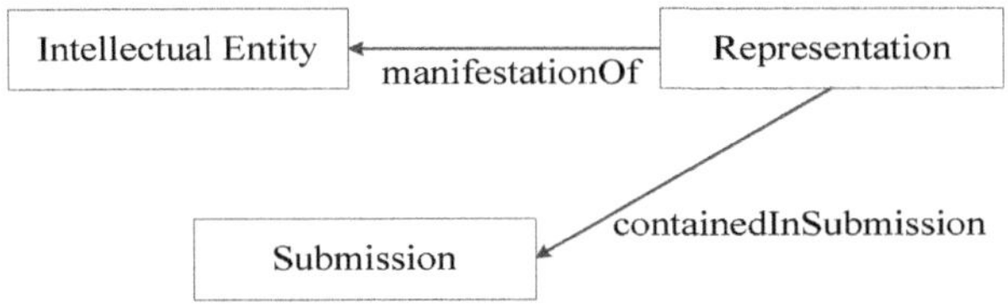

Fig. 2. Generic Data Model

Because eJournal is the first content-type for our digital preservation system, we have designed the eJournal data model by inheriting and extending the generic data model further. In order to deal with updates and support access function (browse, navigation, etc.) efficiently, we have decided to split logically separate metadata subsets that are expected to be updated independently (journal, issue, article) into separate AIPs. There are five kinds of objects in this data model: Journal, Issue, Article, Representation, and Submission [6]. Their relationships are shown in Fig. 3.

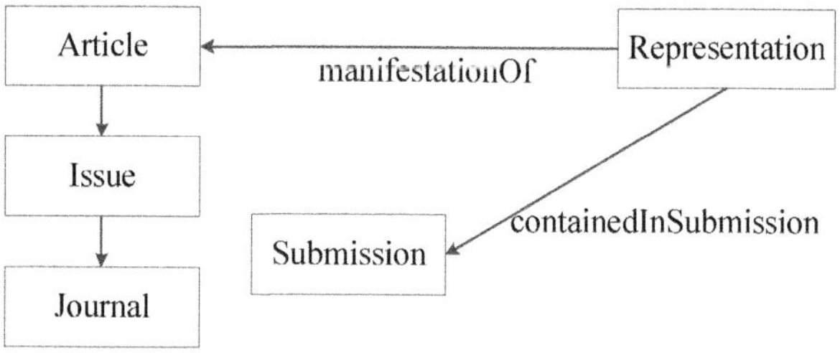

Fig. 3. eJournal Data Model

(4) Preservation Metadata Framework

We have selected METS structural, PREMIS preservation and QDC descriptive metadata to construct our metadata framework. The metadata framework is module based as shown in Fig. 4. The framework includes a descriptiveMD module (descriptive metadata), a RightMD (rights metadata) module, a TechMD module (technical metadata), a structuralMD (structural metadata) module and a digiProvMD module (digital provenance metadata). The descriptiveMD module is an indispensable part of our preservation metadata scheme because digital resources in our preservation system will be inaccessible without descriptive metadata; structuralMD is used for expressing the hierarchical structure of the digital object; RightMD is used for aggregating information about rights and permissions that are directly relevant to preservation and access of objects; digiProvMD is used for aggregating information about actions that affect objects and an accurate and trustworthy record of events is critical for maintaining

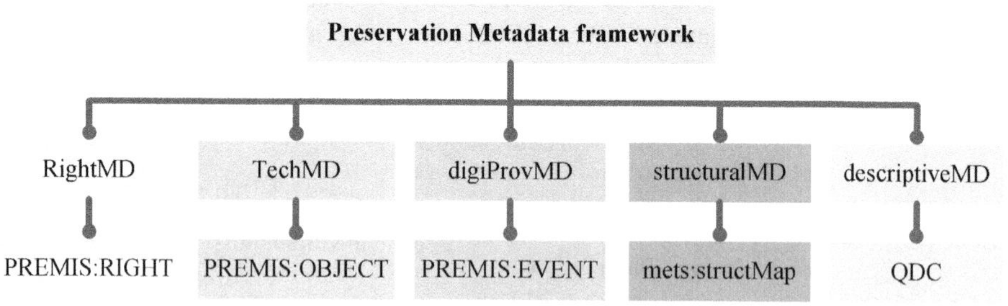

Fig. 4. Preservation Metadata framework

the digital provenance of an object, which in turn is important in demonstrating the authenticity of the object [7]; and TechMD records the technical features of digital objects, so it is the most important part of our preservation metadata scheme (The preservation action function of the metadata framework mainly lies in this module).

3.2 AIP and Metadata Framework Practice

We have applied the AIP and metadata framework design to the preservation of subscribed IEEE electronic resources, and provide the access to those resources too.

(1) AIPs for IEEE eJournals

IEEE Data will be sent on a weekly basis, being offered by publication type (IEEE Periodicals, IEEE Conferences, etc.). Under each publication type, there will be a directory indicating the week number. Data is organized by the week number. Under the week # directory there will be zip files. Each zip file pertains to a publication title. These Zip files are submission information packets (SIP) as defined in OAIS. Each submission may contain several SIPs that need to be unpacked and virus checked before they can be processed further.

An unpacked SIP contains content files (PDF data and extended objects), descriptive metadata for articles, issues and journals (xml metadata). Since a SIP may contain one or several volumes, issues and articles for one journal, the content needs to be split up into identified packages with well-defined structural relationships. Because the publisher supplied structural relationships between article, issues and journal objects

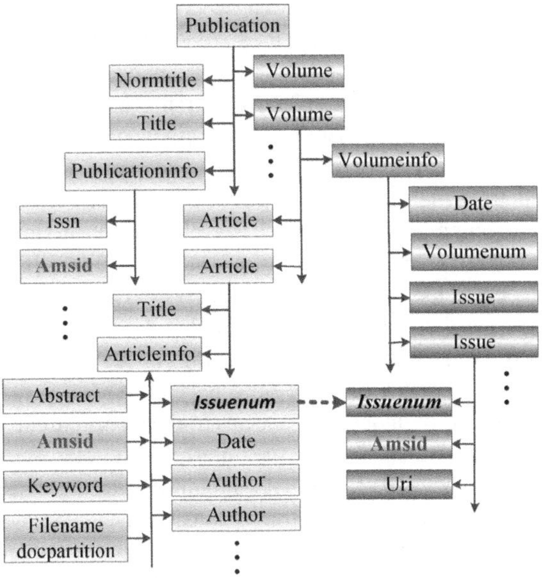

Fig. 5. The structure of xml metadata (IEEE)

have been captured through explicit metadata, so these relationships are extracted and represented in a uniform way, as specified in our METS, PREMIS and QDC application profiles.

Publisher supplied metadata have been represented using in-house formats (IEEE Publications Metadata Description and XML Documentation). We extract metadata either from the publisher supplied metadata or directly from the content. The latter is typically the case for technical metadata. The structure of xml metadata is shown in Fig. 5. The extracted metadata is then normalized according to our METS, PREMIS and QDC application profiles.

According to the eJournal data model shown in Fig. 3, we need to produce five kinds of metadata AIPs representing different kinds of objects: journals, issues, articles, representations, and submissions. Each one is realized as a separate METS file. The first three objects are purely logical objects intended to hold relevant, mostly descriptive, metadata at that level. The latter two are different.

(2) Producing AIPs for different kinds of objects [6]

Our workflow is the following.

First, producing the AIP for a submission object. A submission object describes one submission event, including all the zipped SIP files and a record of all activities performed during ingest. We store the original data as it was provided by the publisher in the archival store linked to from its submission. The submission object references SIPs in its <mets:fileSec> and each SIP has a <mets:amdSec> of its own to hold its provenance metadata. The provenance metadata for submission is mainly about some events free of side-effects, such as "accession", "uncompressUnpack", "metadataExtraction" and "virusCheck". These events are recorded at the submission level in order to avoid redundancy, since they are identical for all files of a submission. The provenance metadata for SIPs is also about some side-effect free events, such as "integrityCheck", "formatIdentification", "validation" and " propertyExtraction", etc.

Second, producing AIPs representing three kinds of objects: journal, issue and article. These objects are purely logical objects intended to hold relevant, mostly descriptive, metadata at that level. Descriptive metadata is expressed using the QDC metadata extension schema to METS and is embedded in a single <mets:dmdSec> section of the METS file. A separate Tsinghua University Library's QDC profile describes the elements in use and their meaning. By analyzing xml metadata (shown in Fig. 5) of every SIP, we can get descriptive metadata pertaining to these three kinds of objects: journal, issue and article, then writing to the <mets:dmdSec> of their respective METS files. The set of these objects represents a hierarchical data model with well defined links from underlying entities to the direct parent. For example, each new article links to the issue, each new issue links to the journal. We store relationships between those AIPs in the AIPs themselves in addition to the database. This ensures that the archival store is a closed system that is consistent and complete within itself. As the issue or journal AIPs do not contain information about the order of the articles, and as articles may be ingested out of sequence, the position of the article within an issue must be stored within the article's descriptive metadata.

Then, producing AIPs for representations. Each representation of an article is stored in a separate METS file. A representation links its actual content files together, records all events that have happened to its content files (such as uncompressing, migrating, extracting properties) and links to related versions of those files (such as the original compressed file, or the file that was the source for the migration or normalization). Representations are linked to their article and submission objects using the PREMIS extension schema.

(3) Issues encountered in the practice

There are mainly three issues encountered in the practice.

The first issue is related to the eJournal data model as shown in Fig. 3. PREMIS information should be connected to the object or objects to which it relates in the METS structural map. However, it is helpful to explicitly state in a profile what level of object PREMIS information describes [3]. A diagram such as Fig. 3 in Angela Dappert and Markus Enders' case study, "Using METS, PREMIS, and MODS for Archiving eJournals," may be valuable for describing how different metadata sections fit together in complex structures [6]. So we have borrowed much from their work and designed the eJournal data model as shown in Fig. 3 for IEEE eJournals. But according to the structure of xml metadata (IEEE) as shown in Fig. 5, the practical logical structure of IEEE eJournals consists of four levels, that is, publication – volume – issue – article. So we need to decide whether to extend the eJournal data model and add volume as an intermediate layer (purely logical object) between journal and issue or not. If we add such an intermediate layer, we need to produce the metadata AIP representing the volume object too. After careful consideration, we decide not to add such an intermediate layer at present. Although the issue is part of an annual volume, this has not been given its own AIP because volume does not play a significant role in the delivery of access to the journal and we want to keep the eJournal data model consistent for all eJournals too.

The second issue is related to descriptive metadata of logical objects. There are lots of descriptive metadata fields defined for publication, volume and article in the IEEE Publications Metadata Description and XML Documentation. Our initial decision was to preserve as much descriptive metadata as possible for incoming collections to support access function. In anticipation of optimizing the population of as-yet-to-be-built discovery and delivery systems, the ingest workflow attempted to capture and structure descriptive metadata in as standard a way as possible across diverse collections, while losing none of the nuances. But this approach requires a substantial analytical effort to create detailed and appropriate AIPs for logical objects. We decide later that the main aim of our preservation repository is for preservation, not for access. That is, it will provide some generic access functions and supplement the native discovery and delivery systems, such as IEL platform, but not to replace them. So we simplify the ingest process by extracting and normalizing only those descriptive metadata deeming useful and necessary according to our QDC application profile. Of course, we're able to analyze the original data in the archival store again and get more information easily later if it becomes necessary in the future.

The third issue is related to preservation metadata generation and storage. AOMS-II uses a semi-automated way to gather the required preservation metadata [8]. Some of the metadata are captured by automated tools, and some of the metadata are input by digital collection managers. The preservation metadata capture tool is responsible for identification of a digital object's format, the corresponding software to create/render it, the OS that can host the software and the information about the storage hardware storing the bitstream of the object. In our preliminary tests, we found that the objects in IEEE eJournals collection are well formed, all the objects are usually in one or two formats and stored in one storage system and on one type of hardware. So, if we scan and identify the preservation metadata one by one for a collection, thousands or millions of the results will be similar. To reduce the size of the overall AIPs (information is just stored once where possible) and save the storage space, and to make the identification process more efficient, AOMS-II also provides an interface in which users can input the basic information of one collection (such as collection ID, collection name, and collection catalog, etc.) and submit several sample digital objects for identification. If users choose to identify by the unit of collection, then the corresponding preservation metadata is for one collection, not an individual object. We use such mechanism for generation and storage of format, creatingApplication and environment related preservation metadata. This is very convenient for the independent control of preservation metadata and makes it easy to update such metadata for lots of objects at a time without changing a large number of AIPs.

(4) Using metadata for preservation risk management

Effective digital preservation depends on a set of preservation services that work together to ensure that digital objects can be preserved for the long-term. These services need digital preservation metadata, in particular, descriptions of the properties that digital objects may have and descriptions of the requirements that guide digital preservation services [9]. We have used metadata for preservation risk management at present.

We have designed a semi-automatic system for managing multiple digital preservation risks of digital resources in our preservation repository[8]. AOMS-II help collection managers monitor the obsolescence status related to multiple digital preservation risks and assist discovering optimum preservation service(s) to get rid of the digital preservation risks. AOMS-II considers and manages the risks of not only file formats/versions obsolescence, but also storage media, hardware, software, OS, etc. The preservation metadata in our preservation repository stores information describing the digital collections from multiple preservation aspects, which is one of the important bases for preservation risk detection. Another important base for preservation risk detection is the registries. The registries in AOMS-II not only include the international published registries such as PRONOM and GDFR, etc. but also include self-built registries for native special formats/software. The preservation risk detection component in AOMS-II periodically compares preservation metadata about software and formats with corresponding registries that store information about the latest available authoring, rendering or viewing software and recommended formats. Moreover, the risk detection component periodically checks the preservation metadata about storage media, hardware, OS etc. for issues like life expectancy due, outmode, and potential incompatibility risks. As there are no such kind of registries now can support these kinds of detections, AOMS-II maintains its own local registries. For the limitation of time and budget, AOMS-II did not maintain "super" local registries for all the OS, storage media and hardware in world. Sufficiency-is-good principle is followed in the construction and maintenance processes of the local registries. In the current version of AOMS-II, we just built the registries based on an IT investigation of Library in Tsinghua University and use an internet crawler to get the corresponding information from a famous IT market website in China – Zhong Guan Cun Online.

4 Conclusion

Currently, our digital resources preservation and service system included about 300 million records (more than 4TB full text files) of IEEE electronic resource with licence. Through the practice of long-term preservation of IEEE digital resources, we have accumulated a lot of technical experience and in the future, we can apply our experience to more digital resource system required long-term preservation.

References

1. Li, C., Ma, N., Xing, C., Jiang, A.: An Integrated Approach for Smart Digital Preservation System Based on Web Service. In: Buchanan, G., Masoodian, M., Cunningham, S.J. (eds.) ICADL 2008. LNCS, vol. 5362, pp. 347–350. Springer, Heidelberg (2008)
2. Guenther, R.: Battle of the Buzzwords: Flexibility vs. Interoperability When Implementing PREMIS in METS. D-Lib Magazine (May/June 2008), doi:10.1045/july2008-guenther
3. Vermaaten, S.: A Checklist and a Case for Documenting PREMIS-METS Decisions in a METS Profile. D-Lib Magazine (September/October 2010), doi:10.1045/september2010-vermaaten
4. PREMIS Working Group, PREMIS Data Dictionary. Version 2.0 (March 2008), `http://www.loc.gov/standards/premis/v2/premis-report-2-0.pdf`
5. CCSDS, Reference Model for an Open Archival Information System (OAIS). CCSDS 650.0-B-1, Blue Book (the full ISO standard) (January 2002), `http://public.ccsds.org/publications/archive/650x0b1.pdf`
6. Dappert, A., Enders, M.: Using METS, PREMIS and MODS for Archiving eJournals. D-Lib Magazine (September/October 2008), doi:10.1045/september2008-dappert
7. Caplan, P.: Understanding PREMIS (2009), `http://www.loc.gov/standards/premis/understanding-premis.pdf`
8. Li, C., Xing, C.X., Dong, L., Huang, M.B.: A Semi-Automatic System for Managing Multiple Digital Preservation Risks of Digital Libraries in China. In: Proc. 9th ACM/IEEE-CS Joint Conf. on Digital Libraries, p. 425 (2009), doi:10.1145/1555400.1555495
9. Dappert, A., Farquhar, A.: Implementing Metadata that Guides Digital Preservation Services. In: Proceedings of the 6th International Conference on Preservation of Digital Objects, iPres 2009 (2009), `http://www.planets-project.eu/docs/papers/Dappert_MetadataAndPreservationServices_iPres2009.pdf`

User Tagging for Digital Archives: The Case of Commercial Keywords from the Grand Secretariat

Shu-Jiun Chen

Research Center for Information Technology Innovation, Academia Sinica, Taiwan,
128 Academia Road, Section 2, Nankang, Taipei, Taiwan
sophy@sinica.edu.tw

Abstract. To make digital archives more accessible to industries, this study used tags given by cultural creative industries to items in digital archives to analyze the differences between the terms used by commercial users and scholars from the archive agency. We analyzed the self-created commercial tags (60%) and the tags adopted from academic terms (40%). The results showed that terms provided by the archive agency were still more likely to be based on domain knowledge. In comparison, the superordinate terms are more likely to be needed by the commercial users. This study suggests that the research findings of six types of semantic relationship and eight types of linked properties could be further transformed into metadata best practices for the digital archive agency, thus bridging the divide between the commercial tags and academic terms.

Keywords: user tagging, digital archives, cultural & creative industries.

1 Introduction

User tagging is a kind of folksonomy. The tags given by users searching for information often include alternative terms, such as social tagging, social bookmarking, collaborative tagging, etc. [1]. User tagging is designed to organize, access and utilize information effectively, similar to other library and information systems. However, compared to traditional collection-indexing languages used by professionals from cultural heritage institutions, user tagging serves as a user-centered approach to enhancing the searching and sharing of cultural heritage information [2]. Bar-Ilan et al. recognized that the emerging method is also a kind of "wisdom of the crowds" [3]. In recent years, relevant studies have probed into the complementary relationship between the folksonomy and the library's controlled vocabulary [4], [5], [6]. Previous empirical studies on the user tagging in the cultural heritage institutions, including libraries [7], [8] and museum communities [9], [10], are focused on the students and the general public as main users. But studies on the archive data are rare, let alone the investigation into the user need of the industry. In this study, user tagging would be applied through the perspective of the cultural creative industries in order to explore the differences between the terms used by commercial demanders of the digital archives and those used by the scholars of the archive agency.

C. Xing, F. Crestani, and A. Rauber (Eds.): ICADL 2011, LNCS 7008, pp. 158–167, 2011.

In Taiwan, the National Science Council established the National Digital Archives Program in 2002, with participants including more than one hundred libraries, museums, archives, academic institutions, ministries and agencies, and universities and colleges[11]. To make digital archives more accessible to industries, this study took the archives of the Grand Secretariat from the Institute of History and Philology at Academia Sinica as its research object. This database was originally stored in the Grand Secretariat Archives at the Forbidden City, and included many kinds of documents dating back to the Ming and Qing dynasties (1368-1911 A.D.), including more than 310,000 items, such as edicts, memorials, and all kinds of records; all of these are rare historical materials useful for studies on the political, social, economic, military, and legal system in Ming and Qing dynasties[12]. This study first observed and analyzed user tags given by the cultural creative industries to the database, and identified the extent of divide between the academic and commercial terms. Secondly, it analyzed the mode in which the industries utilize academic terms. Finally it analyzed the semantic relationship between the commercial and academic terms, as well as the types of linked properties involved.

2 Methods

2.1 Data Collection

This study used both quantitative and qualitative content analysis, with the tags given by cultural creative industries to digital archives as the research object, in order to explore the differences in terms used by commercial demanders of the digital archive and scholars from the archive agency. This study first took "The Database of The Grand Secretariat Archives" from the Institute of History and Philology at Academia Sinica as its research object, based on the eight digital collection items with commercial potential provided by the archive agency, and also asked the agency to provide about two hundred words of description and keywords for each collection item.

Secondly, we chose the cultural creative industries from four kinds of domain, including educational publishers, digital learning, life application and cultural product, digital entertainment and video channel, and visual arts and digital arts. A total of eleven industries were asked to review the images and descriptions for the eight collection items from The Grand Secretariat Archives, and provide two kinds of information, "possibility of value-added utilization" and "keyword" for each collection item from the commercial perspective. We acquired a total of 648 tags as analysis samples.

2.2 Data Analysis

This study used content analysis for classification and examination of the keywords given to each collection item by the participating industries, including quantitative and qualitative research techniques. In order to understand the level of differences in the academic and commercial users regarding collection keywords, this study first collected a total of 648 keywords for eight collection items provided by eleven industries, and then checked, classified, and coded each keyword along with the descriptions provided by the archive agency, to distinguish between (A) the keywords shown in the description from the archive agency, and (B) the keywords tagged only by the commercial users. Secondly, we excluded repeated keywords for each item, gathered the codes of these keywords, and calculated the quantities and proportions of the keywords of type (A) and (B).

For Type (A), in order to understand how the industries utilized terms from the archive agency directly for their tagging, this study took each item as an analysis unit, surveyed and summarized the relationships between tags and description texts. For Type (B), this study also regarded each item as an analysis unit, and analyzed and identified the semantic relationships and linked properties between each item and its keywords. As a result, the important semantic relationships proposed by Birger Hjørland would be adopted as the theoretical basis for this empirical research [13].

3 Results and Discussion

This study is based on eight items from The Grand Secretariat Archives provided by the archive agency. Each item includes three kinds of information: a digital image, the item title, and an item description. After the eleven industries reviewed the information and provided the tags for each item from the perspective of cultural creative application, we acquired a total of 648 tags as analysis samples. As shown in Table 1, overall, the industries think more than two-thirds of the digital collections are candidates for value-added utilization (51%), 17% of them are exceptional candidates for value-added utilization, and about one-third of them are less desirable for utilization (32%). According to the statistics in Table 1, most industries think the item G003 is "extremely possible" for value-added utilization; fewest industries think it is "impossible" for value-added utilization. After further comparing the quantities of tags, we found that among all the items within the scope of this study, these items were also most frequently tagged by the industries.

Table 1. Possibility of value-added utilization by industries

Collection Item Code	Quantity of industries that think it is extremely possible	Quantity of industries that think it is possible	Quantity of industries that think it is impossible
G001	1	5	5
G002	0	6	5
G003	5	4	2
G004	5	3	3
G005	2	7	2
G006	0	6	5
G007	1	7	3
G008	1	7	3
Subtotal (percentage)	15 (17%)	45 (51%)	28 (32%)
Total		88 (100%)	

3.1 Differences Analysis of Academic and Commercial Tagging

In this study, all the tags provided by the industries were further sorted into two types: terms adopted from academic terms for Type A, or commercial self-created terms for Type B. As Table 2 shows, whether or not repeated tags were excluded, new tags that belong to Type B made up the majority overall, with a total of 354 items (55%). When repeated tags were excluded, the percentage would increase to 62%, with a total of 329 items. In other words, about 40% of the description texts provided by the archive agency are used by industries for tagging, and the other 60% might not be able to immediately satisfy the commercial users' demands for digital collections. Therefore, are there any linked and semantic relationship between the 60% tags from the industries and the 40% tags provided by the archive agency? We will discuss this issue later in this study.

Furthermore, Table 2 shows the tagging proportion of the item G006 is obviously different from the average, for which Type A made up 72%, and Type B made up 28%. That means the industries mostly adopted the terms provided by the archive agency for their tagging, but seldom added new tags by themselves. We further observed this item and found that the length of description texts provided by the archive agency are 2.8 times longer than the average for the other seven items. Its image is the only one shown in black and white. As shown in Table 1, we found that G006 is one of the three items that most industries believed to be less possible for value-added utilization. Opposite to G006, Type A for the item G007 made up 28%, and Type B 72%. This means most industries would add new tags for this item, but seldom adopt terms provided by the archive agency. We further observed the content

Table 2. Proportion of academic and commercial tagging for each collection item

Collection Item Code	Quantity of tags in total		Quantity of tags in total after excluding those repeated	
	Type A* (%)	Type B** (%)	Type A* (%)	Type B** (%)
G001	32(46%)	38(54%)	23(40%)	34(60%)
G002	33(48%)	36(52%)	25 (41%)	33(59%)
G003	54(50%)	53(50%)	31(38%)	50(62%)
G004	45(46%)	53(54%)	31(39%)	49(61%)
G005	33(41%)	48(59%)	26(37%)	45(63%)
G006	54(72%)	21(28%)	32(62%)	20(38%)
G007	21(28%)	54(72%)	15(23%)	50(77%)
G008	22(30%)	51(70%)	18(27%)	48(73%)
Total	294(45%)	354(55%)	199(38%)	329(62%)
	648(100%)		528(100%)	

Notes:

1. Tags of Type A were taken from the collection titles or description texts provided by the archive agency
2. Tags of Type B were newly added to digital collections by the industries, never shown in the information provided by the archive agency

of the item G007 and found that the length of its description text is the shortest among all the eight items. It seems to show that (1) the more detailed the description provided by the archive agency is, the more industries would adopt agency terms in tagging, and (2) that rich information presented by the digital image itself might also affect the quantity of tags newly added by the industries, as well as determine whether further value-added utilization is possible.

3.2 Analysis of the Mode of Industries Utilizing Academic Terms

We further surveyed and analyzed the 294 terms from the archive agency that are adopted by the industries, and found that no matter whether the terms were from collection titles or description texts, they all resulted from four modes of tagging: "adopting all," "adopting conceptual segment," "adopting superordinate concept," and "combining partial concepts of terms." We will discuss these four modes below.

Adopting All. Industries adopted exact phrases of the collection titles or description texts in tagging. For example, the industries (noted as T0122, T0143) adopted the title of G001 "長江圖殘卷局部 (Partial Fragment of Yangtze River Painting)" for tagging.

Adopting Conceptual Segment. The industries did not adopt exact phrases of the collection titles or description texts in tagging, but first cut them into several conceptual segments. For example, the title for G0002 is "山西省墩堡圖殘本局部 (Partial Fragment of Shanxi Province Mound Fort Painting)," the industry (noted as T0243) adopted the title in tagging by cutting it into segments, such as "山西 (Shanxi)," "墩(mound)," "墩堡(mound fort)," and "山西省(Shanxi Province)."

Adopting Superordinate Concept. After the industries surveyed the description texts provided by the archive agency, they only adopted superordinate terms in tagging. For example, the description for G003 "cloth wrapper" is as below:

"此幅為清代所繪盛京應用織金坐龍▯ 緞地▯ 綢裏夾包袱式樣，原件尺寸：縱112公分，橫112公分。觀看時應順時針右轉45度，此樣精繪正龍紋與皇帝龍袍同式，皆正面向前呈坐姿，具有威嚴。左、右及上方各有代表富貴之牡丹花紋，象▯ 祥瑞之兆的雲紋點綴其間，顯得相當生動靈活，下角正中裝飾著波濤翻騰的浪花及點點飛▯ ，別有一番趣味。(The cloth wrapper of yellow satin with golden weaved pattern of seated dragon was made in Qing dynasty. Original size: length 112 cm, width 112 cm. Turn 45 degrees clockwise when viewing this dragon pattern, which is the same kind of pattern appearing on the emperor's dragon robes. They are all frontally seated with dignity. The peony pattern symbolizing good fortune appeared on the left, right and top, with cloud pattern symbolizing auspiciousness dotted among them, which looked quite lively.)"

The industry (noted as T0344) only adopted two terms from the description, "龍袍 (dragon robe)" and "祥瑞(auspiciousness)" for commercial tagging, and another industry (noted as T0342) took "皇帝龍袍(emperor's dragon robe)" and "牡丹花紋 (peony pattern)" for commercial tagging. The mode shows that the industries would take the terms provided by the archive agency as a reference for tagging, but only employ the terms with superordinate concepts, not academic and professional terms.

Combining Partial Concepts of Terms. The industries did not adopt exact phrases of the description texts provided by the archive agency, but took its partial concepts to combine into a new tag. For example, one paragraph of the description for G004 "包袱式樣 (cloth wrapper)" is shown below:

"……此幅為墨筆所繪皇史宬應用織金龍▯ 緞地▯ 綢裏夾包袱式樣...所繪為五爪雙遊龍，龍與蟒之分在爪，龍有五爪，乃皇帝、皇后、皇太子、親王、皇室成員所用紋飾；三爪，四爪為蟒，供一定品級之官員使用...(The decorative pattern of twin roaming dragons with five paws painted in ink on this cloth wrapper can only be used for royal family members, such as the emperor, empress, crown prince, and price)"

The industry (noted as T0422) borrowed two terms, "皇室成員 (royal family members)" and "紋飾 (decorative pattern)," and then combined them into one tag, "皇室成員紋飾 (decorative pattern for royal family members)." The mode shows that some new tags were the combination of terms borrowed from the archive agency and terms newly created by the industry itself. For example, one paragraph of the description for G001 "Partial Fragment of Yangtze River Painting" is shown below:

"……此幅長江圖殘卷...大約完成於明永樂六年至宣德五年（1408-1430A.D.),左自湖廣岳州府臨湘縣境內鴨欄磯起，右至興國州上山磯止...(This fragment of Yangtze River Painting was finished during sometime from the sixth year of Yongle period in Ming dynasty to the fifth year of Xuande period in Ming (1408-1430A.D.)"

The industry borrowed the term "明永樂 (Yongle period in Ming dynasty)" mentioned above, and combined it with the self-created term "古地圖 (ancient map)" into a new tag "明永樂古地圖 (ancient map in Yongle period in Ming dynasty)."

3.3 Analysis of the Relationship between Commercial Tagging and Academic Terms

As the analysis results show, in all the commercial tagging, 60% of the tags were added by the industries, and 40% were adopted from the terms provided by the archive agency. Therefore, in this study we try to analyze and discuss the semantic relationship and linked properties within the two types of terms for each item. In addition, we attempt to analyze the linked properties for tagging and items, in order to understand if there are any important properties that could be integrated with commercial tagging and academic terms for collection images.

Analysis of Semantic Relationship between Commercial Tagging and Academic Terms. In this study, after identifying the semantic relationships between commercial tagging and academic terms, the result shows six kinds of semantic relationships in Table 3 below, which are: genus-species relationship, instance relationship, equivalence relationships, part-whole relationships, positional relationship, and associative relationship.

Table 3. Types of semantic relationships between commercial tagging and academic terms

Type code	Type of semantic relationships	Implication	Case study	
			Academic term (Type A)	Commercial tagging (Type B)
S001	genus-species relationship	Relationship between the superordinate and the subordinate, indicating that an object belongs to a classification (e.g. A is a kind of B)	牡丹花紋(peony pattern) (TB0412)	中國紋飾(Chinese decoration) (TR0345)
S002	Instance relationship	Semantic relationship exists in the specific example of general concept and this concept (e.g. A is an example of B)	湯若望(Johann Adam Schall von Bell) (TB0612)	傳教士 (missionary)(TR0623)
S003	Equivalence relationship	The meaning of A is equal to B. A is equal to B.	清代(Qing dynasty) (TB0312)	清朝(Qing dynasty) (TR0323)
S004	Part-whole relationships	The relationship of whole and part (A is part of B). Part means a portion, a substance or a member that can constitute a whole	親王、皇后、皇太子、皇帝(prince, empress, crown prince, emperor)(TB0412)	皇族(royalty) (TR0423)
S005	Positional relationship	Using a concept to explain an object that is located under another designated concept, e.g., A is located in B	湖廣岳州府臨湘縣 (Linxiang County at Yuezhou fu in Huguang) (TB0112)	中國(China) (TR0112)
S006	Associative relationship	The relationship is identified psychologically, mostly uncertain relationship. In this study it includes derivation, creation, and association	龍(Dragon) (TB0823)	降龍十八掌、亢龍有悔、龍戰於野、潛龍勿用(18 Dragon-Subduing Palms, The Proud Dragon Repents, The Fighting Dragon In The Wilderness, The Hidden Dragon) (TR0823)

Analysis of Linked Properties between Commercial Tagging and Academic Terms and Collections. The goal of this part of the study is to understand the scope of the terms tagged by the industries for each item not originating from those provided by the archive agency. In other words, these terms represent the divide existing between the archive agency and the industries. If the relationships between commercial tagging and academic terms provided by the archive agency can be identified, the gap between academic (or professional) and commercial terms could be further narrowed. In this study, we only surveyed, analyzed and classified keywords appearing in the commercial users' tagging, a total of 354 terms that belong to eight types of linked properties between the tags and the collections: "time," "space," "type," "material," "physical characteristics," "subject," "people," and "commercial features." Please refer to Table 4 for details.

Table 4. Types of linked properties among commercial tagging and academic terms and collections

Type code	Type of lined properties	Implication	Case Study	
			Academic term (Type A)	**Commercial tagging (Type B)**
R001	Time	Production date of the collection item	明永樂六年(Sixth year of Yongle, Ming dynasty) (TB0112)	明朝(Ming dynasty)(TR0123)
R002	Space	Production place of the collection item	盛京(Shengjing) (TB0312)	中國(China) (TR0112)
R003	Type	Type of the collection item	乾隆己卯科廣西鄉試題名錄(Record for the Guangxi provincial exam in the year of Jimao of Qianlong period)(TB0512)	中國古書(Chinese ancient book) (TR0523)
R004	Material	Material used to produce the collection item	曆書(almanac)(TB0645)	紙(paper)(TR0623),
R005	Physical characteristics	The appearance, shape, color, design, pattern and decoration of the collection item	立姿正龍(Standing dragon)(TB0512)	龍形(Dragon shape) (TR0545)
R006	Subject	Affiliated disciplines or fields	天文(astronomy) (TB0612)	中國天文科技 (Chinese astronomy and technology)(TB0623)
R007	People	Individuals or groups that are identified by their career or the events they participated in	光緒(Emperor Guang-xu)(TB0823)	天子(Emperor) (TR0823)
R008	Commercial applications	From commercial point of view, the item can be used for value-added application in the future	包袱(Cloth wrapper) (TB0423)	四方巾(kerchief) (TR0423), 筆記本 (notebook)(TR0422), 絲巾(scarf)(TR0422), 屏風(screen)(TR0422)

4 Conclusions

This study examined commercial users of digital files and professionals in the archive agency from the perspective of cultural creative industries, and the differences in tagging between the two groups. In our research, we not only analyzed the divide between academic and commercial terms applied to The Grand Secretariat Archives, but also performed an empirical study on the modes in which the industries used academic terms, and the semantic relationships and linked properties between commercial terms and academic terms. In accordance with the results of the study, the conclusion is as below.

Descriptive terms provided by the archive agency can satisfy 40% of commercial tagging. In this study, after the industries surveyed the digital files provided by the archive agency, one-third of those files had the possibility of value-added utilization.

We found that about 40% of commercial tags were adopted from the terms provided by the archive agency, mainly taken from the collection titles or description texts, in four different modes: "adopting all," "adopting conceptual segment," "adopting superordinate concept," and "combining partial concepts of terms."

The divide between commercial tagging and academic terms could be linked with six types of semantic relationship. Through the self-created commercial tags (60%) and the tags adopted from academic terms (40%), we analyzed these two types of tagging and found six types of semantic relationships: genus-species relationship, instance relationship, equivalence relationships, part-whole relationships, positional relationship, and associative relationship. This study suggests these six types of semantic relationships be further transformed into a practical standard of metadata for the archive agency, thus bridging the divide between commercial tagging and academic terms.

Eight types of properties of commercial tagging are mainly taken from digital collections. The study found that self-created tags assigned by commercial users, can be classified into eight types of linked properties between tagging and items: "time," "space," "type," "material," "physical characteristics," "subject," "people," and "commercial features." The archive agency had already provided these terms in a general sense, but the study shows these terms are still more like those constructed with domain knowledge. In comparison, superordinate terms are more likely to be needed by the commercial users. This study suggests that the eight types of linked properties can be further transformed into metadata best practices for the digital archive agency, or can even be adopted for the development of methods of mapping between the two types of vocabularies, thus bridging the divide between commercial tagging and academic terms.

Acknowledgements. This work was supported in part by Taiwan E-learning and Digital Archives Programs (TELDAP) sponsored by the National Science Council of Taiwan under NSC Grants: NSC 99-2631-H-008-001, NSC 99-3113-P-001-003-PO.

References

1. Furner, J.: User Tagging of Library Resources: Toward a Framework for System Evaluation. In: Proc. of World Library and Information Congress: 73th IFLA General Conference and Council (2007), `http://archive.ifla.org/IV/ifla73/papers/157-Furner-en.pdf`
2. Furner, J.: Folksonomy. In: Bates, M. (ed.) Encyclopedia of Library and Information Sciences, 3rd edn. (2009), `http://polaris.gseis.ucla.edu/jfurner/papers/Folksonomy3.pdf`
3. Bar-Ilan, J., Zhitomirsky-Geffet, M., Miller, Y., Shoham, S.: The Effects of Background Information and Social Interaction on Image Tagging. Journal of the American Society for Information Science and Technology 61(5), 940–951 (2010)
4. Yi, K., Chan, L.M.: Linking Folksonomy to Library of Congress Subject Headings: an Exploratory Study. Journal of Documentation 65(6), 872–900 (2009)

5. Daly, E., Ballantyne, N.: Ensuring the Discoverability of Digital Images for Social Work Education: an Online "Tagging" Survey to Test Controlled Vocabularies. Weology 6(2) (2009), http://www.webology.org/2009/v6n2/a69.html
6. Smith, G.: Tagging: Emerging Trends. Bulletin of the American Society for Information Science and Technology 34(6), 14–17 (2008)
7. Eckert, K., Hanger, C., Niemann, C.: Tagging and Automation: Challenges and Opportunities for Academic Libraries. Library Hi Tech. 27(4), 557–569 (2009)
8. Lawson, K.: Mining Social Tagging Data for Enhanced Subject Access for Readers and Researchers. The Journal of Academic Librarianship 35(6), 574–582 (2009)
9. Trant, J.: Exploring the Potential for Social Tagging and Folksonomy in Art Museums: Proof of Concept. New Review of Hypermedia and Multimedia 12(1), 83–105 (2006)
10. Srinivasan, R., Boast, R., Becvar, K.M., Furner, J.: Blobgects: Digital Museum Catalogs and Diverse User Communities. Journal of the American Society for Information Science and Technology 60(4), 666–678 (2009)
11. Chen, S.J., Lee, D.T.: Taiwan's National Digital Archives Program: Past, Present and Future. In: National Central Library (ed.) Proc. of Conference on Cooperative Development and Sharing of Chinese Resources and Digital Archives of Chinese Library Resources, Taipei, Taiwan, pp. 113–140 (2010)
12. Museum of the Institute of History and Philology, Academia Sinica: The Grand Secretariat Archives (3) (2006), http://www.ihp.sinica.edu.tw/~museum/tw/artifacts_from_ethnic.php?class_plan=135
13. Hjørland, B.: Semantics and Knowledge Organization. Annual Review of Information Science and Technology 41, 367–405 (2007)

From Box to Bin – Semi-automatic Digitization of a Huge Collection of Ethnological Documents

Alf-Christian Schering[1], Ilvio Bruder[1], Susanne Jürgensmann[1], Holger Meyer[1], and Christoph Schmitt[2]

[1] Database Research Group, Dept. of Comp. Science, University of Rostock, Germany
[2] Dept. of European Ethnology & Wossidlo Archive, University of Rostock, Germany

Abstract. Especially in the field of cultural heritage preservation, one has to deal with information formerly collected on non-digital media such as paper, which is highly subject to deterioration and decay. The phrase "From Box to Bin" provokes, because it is connoted differently when not used in context of digitization. Most digitization projects follow a very straightforward workflow to convert documents into digital data. "From *Box* to *Bin*" literally depicts the workflow of our digitization project, based on a huge archive of ethnological documents, in which documents are filed in *box*es. On the one hand, our project is aimed at long time preservation of documents on film reels in metallic *bin*s stored under the earth's surface in a closed mine. On the other hand, documents are converted into *bin*ary data, which will be provided for research within a digital archive system. In this article, we describe our very specific digitization workflow and the experiences, we have had with it so far.

1 Introduction

For many years, contemporary digital libraries and archives have been focusing on preservation of cultural and scientific heritage from ancient history thru state-of-the-art knowledge. Most digitization projects adhere to a very straightforward workflow to convert these documents into digital data. [2] and [7] give a good overview about common digitization practices for cultural documents. A digitization process has organisational and technical aspects, which have to be considered systematically. A well-known reference model for a digital preservation workflow is the ISO OAIS model [4]. There are already systems supporting the digitization process using workflows, such as Goobi [3] and Zend [1], which mostly deal with flatly structured digital information and usually focus on simple 'document-by-document' processing. In contrast to this, our project *WossiDiA*, the *Digital Wossidlo Archive* [5], is targeted on an archive, which embraces one of the most heterogeneous and diverse collections of paper slips in European Ethnology. It's 2.2 million digital images of hand-written documents, representing knowledge gathered from Wossidlo's field research, are highly interconnected to each other and require a more individual document handling. Our project is funded by the DFG (German Research Foundation) and hence bound to their digitization requirements [2]. Due to the huge amount of documents and their heterogeneity, we decided to make use of bar-coded cover pages to identify groups

C. Xing, F. Crestani, and A. Rauber (Eds.): ICADL 2011, LNCS 7008, pp. 168–171, 2011.

of documents. Barcodes are also essential for automated, efficient digitization and indexing. Because of the specific type of knowledge, domain experts are needed to attach descriptive metadata to documents, based on reliable homogeneous finding aids and directories. Unifying responsibilities for barcode, cover page, and directory creation within one entity fosters efficiency and saves money.

2 The Semi-automatic Digitization Workflow

The digitization workflow altogether leads from the original archive to the digital archive and illustrates the describing process. Fig. 1 gives a general overview about the workflow and depicts how archival contents are acquired and amended in a way that generates the high degree of interconnectivity in the digital archive. This is described by the following seven steps:

(1) Some of the archive's five heterogeneous corpora already feature paginations which put individual documents in the required order within their structural units. One of the challenges was to develop unified signatures for every structural unit in the archive. This is important because corpora are paginated so diversely, that it would be quite difficult to handle every type of document in a different way. Additionally, unified signatures make sure archival contents are labeled

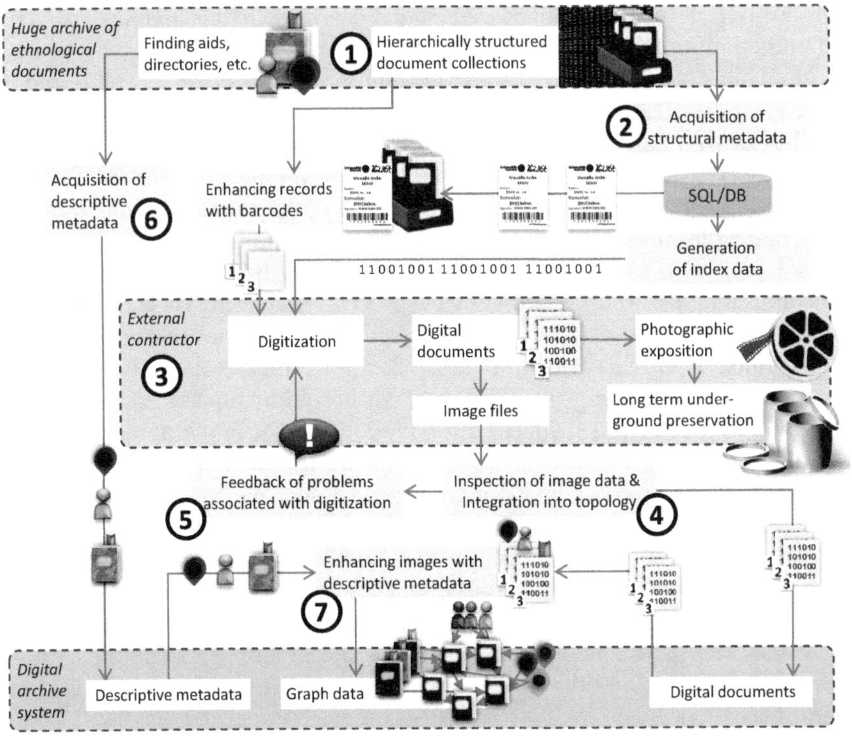

Fig. 1. Digitization workflow (according to sect. 2)

transparently. The first segment of the signature is the corpus ID. The following segments represent structural units according to characteristics of the respective corpus, starting from boxes and ending up in individual pages. Fig. 2 shows an example mapping of a particular pagination to the unified signature.

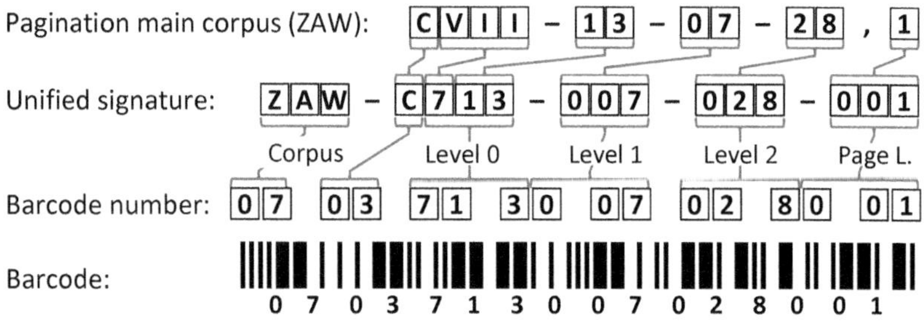

Fig. 2. Example: Pagination mapping → unified signature and conversion → barcode

(2) Preparing documents primarily includes the acquisition of structural metadata, i.e. the hierarchical archive topology itself, which is stored in a relational database. Signatures, which are the basis for barcodes and file system directory structures for digitals, are generated based on that database. To provide for a nearly automatic digitization process, barcodes have to be assigned to structural units of appropriate granularity. That does not mean to assign barcodes to individual slips of paper, but to structural units, such as boxes (level 0) and envelopes (level 1). The conversion from alphanumeric signatures to pure numeric 2-in-5-interleaved barcodes is shown in the lower part of fig. 2.
(3) Bar-coded archival units and index data are submitted to an external contractor who performs digitization according to technical requirements defined in [2]. Image files are delivered to our digital archive and 35mm-film copies are made for long term preservation at a central cultural heritage recovery site.
(4) Digital images are subject to a review process, evaluating their quality. After this, the image documents are integrated into the digital archive automatically, as determined by barcodes created in step 2. However, barcodes only cover directory units of coarse granularity. They can neither address particular documents nor individual pages. Our project staff is responsible for fine-granular structural describing of every digital image using the signature levels 2 and 3.
(5) Evaluation in step 4 sometimes requires to go back to step 3 in order to re-digitize documents, that were not digitized properly in the first place. Since quality data was recorded during evaluation, it can be queried and extracted into feedback reports accumulatively. Based on these reports the external contractor re-digitizes and replaces invalid documents in our content repository.
(6) Besides the huge number of documents, the archive also contains various types of finding aids and directories giving access to its contents. Finding aids include lists of persons, such as narrators, contributors, and scholars, as well as directories of places and locations, literature references, abbreviations, correspondences, inventory sheets, headwords, etc. Respective items are referenced

from within the individual documents. This kind of information also needs to be represented digitally, in order to be put into relation with digital documents. Hence, acquisition and preparation of descriptive metadata is a vital task that needs to be performed in advance of attaching descriptive metadata (see fig. 1). **(7)** On the basis of step 4, contentual describing is performed by manually tagging metadata descriptors to documents that have been prepared in step 6. This requires lots of labor and a great deal of know-how in order to read and understand the hand-written documents and to extract relevant descriptive metadata. Interconnectivity between documents and descriptive metadata is represented with graph data structures that can be queried and evaluated efficiently for information retrieval purposes, based on semi structured XML and graph technologies as described in publications related to our project, e.g. [6].

3 Conclusions and Future Work

In this article we have outlined the digitization workflow of our project. A special aspect was the usage of bar-coded cover pages as state-of-the-art identifiers for the digitization process. In the near future, we will provide a more direct means to access content and metadata by using barcodes and RFIDs. Since barcode pages barely include any descriptive metadata associated with documents, information from the database needs to be accessed *silently* from within the original 'paper archive'. Our favored scenario is the usage of mobile devices for scanning barcodes, to retrieve and display information about the structural units they are tagged to. This significantly helps to find specific documents amongst millions in the shelves of the archive. Investigations on how to use RFID techniques for this purpose are going on. RFID tags can store more information than barcodes and are read non-visually. A disadvantage is, that they fail to work reliably when stacked densely - this is still a scaling problem that needs to be solved.

References

1. Library Archiving and Access System. Technical report, Bavarian State Library (2006), `http://www.babs-muenchen.de`
2. DFG-Praxisregeln "Digitalisierung" (digitization requirements). Technical report, DFG (2009), `http://www.dfg.de/download/.../lis/praxisregeln_digitalisierung.pdf`
3. Goobi - An Open-Source Workflow Management Software. Technical report, intranda GmbH (2010), `http://www.digiverso.com/en/products/goobi`
4. ISO 14721:2003. Reference Model for an Open Archival Information System (OAIS). Technical report (2003)
5. Schering, A.C., Bruder, I., Schmitt, C., Meyer, H., Heuer, A.: Towards a Digital Archive for Handwritten Paper Slips with Ethnological Contents. In: Goh, D.H.-L., Cao, T.H., Sølvberg, I.T., Rasmussen, E. (eds.) ICADL 2007. LNCS, vol. 4822, pp. 61–64. Springer, Heidelberg (2007)
6. Schering, A.C., Meyer, H., Heuer, A.: Managing Highly Correlated Semi-structured Data: Architectural Aspects of a Digital Archive. In: CIKM (2007)
7. Graf, N., Inden, Y.: Best Practices Digitalization. Technical report, ETH Zurich (2009)

An Approach for Processing Large and Non-uniform Media Objects on MapReduce-Based Clusters

Rainer Schmidt and Matthias Rella

Austrian Institute of Technology, Donau-City-Straße 1, 1220 Vienna, Austria
firstname.lastname@ait.ac.at

Abstract. Cloud computing enables us to create applications that take advantage of large computer infrastructures on demand. Data intensive computing frameworks leverage these technologies in order to generate and process large data sets on clusters of virtualized computers. MapReduce provides an highly scalable programming model in this context that has proven to be widely applicable for processing structured data. In this paper, we present an approach and implementation that utilizes this model for the processing of audiovisual content. The application is capable of analyzing and modifying large audiovisual files using multiple computer nodes in parallel and thereby able to dramatically reduce processing times. The paper discusses the programming model and its application to binary data. Moreover, we summarize key concepts of the implementation and provide a brief evaluation.

1 Introduction

Over the last years, major Internet companies have made their data centers available to public users via cloud services. New resource provisioning models (like PaaS, IaaS) evolved, allowing users to create and host applications that utilize virtually unlimited back-end infrastructures. Cloud computing provides a versatile technology that may be used just for storing data or to create applications that service massive user-requests. A popular web application that uses cloud resources to render videos for user-selected digital content is provided by Animoto [1]. This application gained attention in 2008 when it scaled from about 50 to 4000 instances in less than a week in order to deal with massively increasing user requests. Although supported by the enormous power of a cloud infrastructure, it is a grand challenge to build a web application of that dimension. Most notably, one has to cope with the complexity of coordinating the involved subsystems (like application servers, computing farms, and database systems) in order to achieve scalability and robustness.

In this paper, we target a more generic approach to support the processing of large volumes of digital content in cloud-based environments. We present research on applying a widely used programming model for data-intensive computations to the problem domain of audiovisual (AV) data. The paper describes a

[1] http://animoto.com

C. Xing, F. Crestani, and A. Rauber (Eds.): ICADL 2011, LNCS 7008, pp. 172–181, 2011.

method and corresponding application for analyzing video archives based on the MapReduce programming model [1]. The application automates data compression and decomposition by utilizing native codec libraries and parallel processing based on file partitions. It can thereby take advantage of scalable computational environments in order to speed up execution times and throughput. Users are provided with full programmatic control over the processing logic (as opposed to the execution of wrapped 3rd party applications). The application utilizes Java-based abstractions for relevant concepts like containers, tracks, and frames allowing one to easily implement and incorporate custom application logic. We have implemented a set of examples that perform tasks such as applying filters (e.g. for face recognition) or performing text extraction on a collection of arbitrary input video files. A range of use-cases that support digital libraries and archives exist. Examples include error checking an repair of TV/video collections, the application of preservation actions, or the on-demand generation of access copies for different client devices.

2 Data-Intensive Computing

2.1 Execution Environments

In 2004, Dean and Ghemawat introduced MapReduce [1], a programming model and implementation that is capable of processing vast amounts of data on clusters of commodity computers. Data is shared using a distributed file system (GoogleFS) that provides a decentralized storage layer on top of local storage disks, called a shared nothing architecture. Other data-intensive computing environments include Apache Hadoop [2], Microsoft Dryad [5], and Nephele [11], which provide a range of languages, programming models, and runtime implementations for processing data on large storage and computing networks. A major benefit of utilizing such data processing environments compared to using low-level parallel libraries (like MPI) is the automated synchronization and handling of IO operations. Within their application scope, these systems can efficiently handle issues like workload distribution, coordination, and error handling. This in turn enables users to easily implement extremely robust applications that can be executed over thousands of simultaneously running nodes.

2.2 Processing Structured Data

Programming patterns like MapReduce and All-Pairs [6] are specifically designed for implementing applications that perform operations on files-based and structured content. These programming models provide a small set of abstractions allowing their users to express and execute data-intensive workloads. MapReduce provides a relatively simple programming model that is based on two interfaces; one for distributing workload among the cluster nodes, and one for aggregating the results. The programming model has proven to be applicable to a range

[2] http://hadoop.apache.org/

of problems that deal with the processing of textual data (e.g. graph processing and data mining). It has also been demonstrated that these programming models are applicable to types of scientific applications, like loosely-coupled and massively parallel problems, e.g. found in bioinformatics [4]. Data is typically accessible via an underlying distributed file system, which provide distributed and fault-tolerant storage. Database and data warehouse systems like HBase and HIVE [10] have been built on top of Hadoop's distributed file system (HDFS) providing distributed data stores. These systems relieve the users from the burden of creating applications based on the programming model primitives only and allow them to perform more familiar database operations based on a query language instead.

2.3 Application to Binary Data

A range of use-cases for processing archived binary content exist, these include tasks like scaling, transcoding, and feature extraction of images, audio, and video content. An example for handling binary data in an MapReduce application for scientific data analysis is provided by Ekanayake et al. [3]. The authors describe a MapReduce application that must combine different features from generated binary histogram files. Apache Hadoop already provides basic support for processing binary data like for example dealing with compressed input/output files on the distributed file system. When handling large compressed text files (e.g. for performing log file analysis), it is important to extract and process only parts of the data in the map task. Specific binary formats (like Hadoop's Sequence-File) can provide an efficient intermediate representation for handling large data volumes on HDFS supporting serialization, compression, and splitting. However, these binary formats are designed to provide storage representations that efficiently encode large data sets. In order to process audiovisual data, it will be important to implement suitable data abstractions as well as storage handlers to directly access and decode natively binary data on the fly. Once these abstractions are in place, users should be able to utilize the available abstraction (like MapReduce) to easily implement data-intensive operation for processing audiovisual content.

2.4 Problem Description and Strategy

2.5 Approach to the Problem

In the literature, we found existing work that describes a strategy for transcoding video data in a cloud computing scenario, called Split&Merge architecture [7]. The idea is to ensure fixed duration times for video encoding by scaling-up the number of cloud nodes used to process chunks of data depending on the size of the payload file. Here, we take a similar but more generic approach and focus on its application to a widely used programming model (and corresponding execution environment). Designing an application around a standard programming model in general, provides a number of well known benefits like encapsulation,

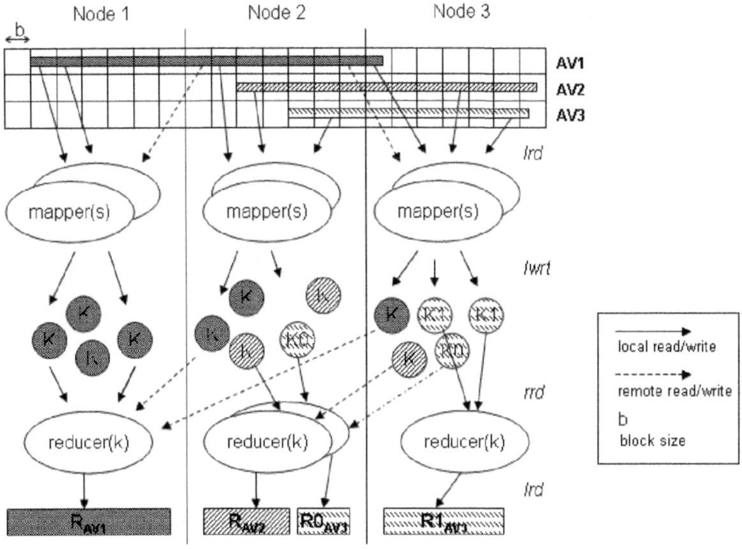

Fig. 1. A data flow for processing audiovisual data using the MapReduce model. This application takes m audiovisual (AV) input files as input and generates n AV output files. In a first phase, the MapReduce framework create a set of mappers for each input file and assigns them to file partitions (*splits*). The Mappers subdivide and process the input splits based on interpretable binary chunks (*records*). For each record, a mapper creates a set of intermediate output data streams which are grouped using keys (K), representing (parts of) the input file. Reducers remotely read the intermediate data streams and generate a result file for each key.

robustness, and portability. In large distributed systems like grids, clusters, or enterprise systems it is key to employ defined programming and deployment models in order to achieve speedup and scalability. The application presented in this paper has been implemented using the MapReduce programming model and Apache Hadoop as an execution environment.

2.6 Data Placement

Figure 1 shows a MapReduce application for processing a set of video files (AV1, AV2, AV3). The files are available on Hadoop's distributed file system (HDFS) which provides a shared storage network across the worker nodes (node1-3). Files that are stored on HDFS are automatically broken up into blocks (typically of 64MB) and stored (and replicated) on different data node based on a placement policy. During runtime, the execution environment facilitates the processing of payload data by assigning parts of the data (called splits) to worker nodes. Input splits are of the same size as storage blocks per default and Hadoop spawns one map task for each split on a worker node. It thereby tries to assign map tasks to workers that reside closely to the input splits in order to maximize local read operations. A number of input splits may be accessed remotely by map operations as the scheduling of map tasks is subject to load balancing.

2.7 Data Decomposition

In order to support the parallel processing of binary input data it is important to provide suitable mechanisms to divide the data into parts that can be interpreted by the application. Most binary formats cannot be broken into chunks at arbitrary positions or do not support splitting at all. For processing sets of relatively small files one can overcome this problem by dividing the payload on a per-file basis [9]. Audiovisual content however tends to be large and its processing can easily become too resource demanding to be performed on a single processor in a reasonable time frame. It is therefore desirable to process single video files using multiple nodes in parallel in order to speed up the overall execution time. This is in particular important for applications that perform such operations on demand based on user requests, for example when triggered via a web interface.

Digital video images (frames) provide a natural unit for decomposing video materials into independently processable parts. Video file formats (like AVI, Flash, Quicktime) are however complex, containing a range of data streams like audio, video or text tracks. The different tracks are typically compressed based on a compression format (like mp3, h264, MJPEG) and must be decoded before being interpretable. One approach to obtain video frames from input splits is to migrate the data into an uncompressed format before processing it within a parallel application [8]. However, the migration itself is a resource and storage consuming process, which hinders the application of this approach for large volumes of content.

The application presented in this paper supports the parallel processing of video content directly from the compressed original formats. The idea is to split videos at key frame positions and perform the decompression of the data chunks in parallel within the MapReduce application. Video compression combines image compression and temporal motion compensation in order to reduce the amount of data required to encode video sequences. Key frames denote independently encoded frames that do not depend on other video data to be decoded. In order to split a media byte-stream into parts it is important to identify these key frame positions within the media container. Using the Hadoop MapReduce framework, we have implemented the required concepts (like *input format*, *record reader*, *compression codec*) that support the automated splitting and parallel processing of compressed media streams (section 3). This allows us to process heterogeneous collections of audiovisual content directly from the storage location without enforcing restrictions on the container and compression formats.

2.8 User Defined Functions

The MapReduce programming model allows users to write parallel applications by implementing the functions *map* and *reduce*. Data portions are automatically generated and passed between the map/reduce functions using a generic data model based on key-value pairs. This programming model allows users to easily implement parallel applications without having to deal with cumbersome parallelization strategies. The model has been widely used for analyzing and transforming large-scale data sets like text documents, database query results, or log files that can reside on different storage systems [2]. In order to enable the development of such user-defined function for processing audiovisual content, we have identified two major requirements: (1) the framework must provide a mechanism to generate meaningful records that can be processed within the map function, and (2) the framework must provide the required abstractions and mechanisms to analyze and transform the records. This paper, however, primarily concentrates on data decomposition and record generation.

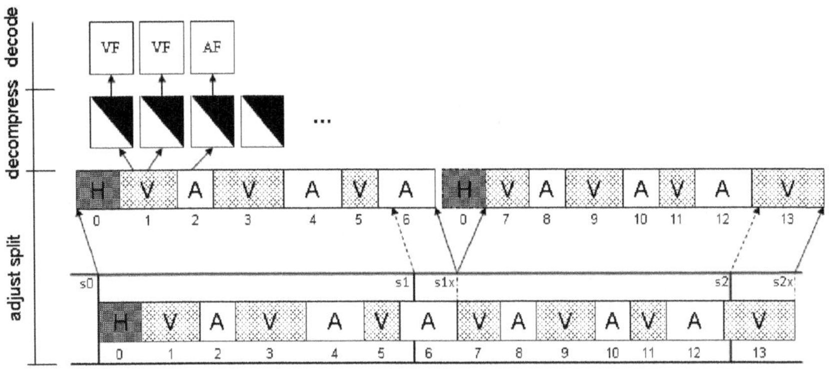

Fig. 2. A data pipeline for extracting records (audio/video frames) from splits of compressed input data. Split borders (s1, s2) are adjusted (s1x, s2x) based on key frames discovered within the multiplexed data stream. The adjusted input splits together with the container's header information are fed into a decompressor. The extracted packets are finally decoded into interpretable binary records and passed to the map function.

3 Implementation

3.1 Software Stack

The application has been implemented based on Apache Hadoop's distributed file system (HDFS) and MapReduce environment (release 0.21.0). For handling media streams from Java, we utilize the Xuggler [3] open-source library, which provides an very stable wrapper around FFmpeg's *libav* libaries [4]. In previous

[3] http://www.xuggle.com/
[4] http://www.ffmpeg.org/

versions of this application, other Java media frameworks and bindings to native applications have been evaluated (including JMF, FMJ, Theora-Java, FOBS, Jffmpeg, FFMPEG-Java). It is in general desirable to base an applications for user-defined data processing on a high-level language like Java as this can greatly simplify the readability of the code. Due to dependencies to native libraries, the video processing application demands the installation of FFmpeg on every cluster node. Other dependencies may result from the incorporation of native special purpose libraries like e.g. for optical character recognition. It is important to note that virtualization and cloud technology provide a well suited model for efficiently deploying such environments within large data centers on demand. In previous work, we have instantiated a cluster with >150 video processing nodes on Amazon's utility cloud [5] based on a single virtual machine image having the entire software stack pre-installed.

3.2 Application Design

In the following, we provide an overview of the application design, and briefly describe some of the application's basic abstractions and their implementation.

AV Splittable Compression Codec. One of the most critical issues when dealing with the parallel processing of large files is handling compression. A compression codec provides a helpful abstraction allowing one to easily read/write from/to compressed data streams. Hadoop provides codec implementations for a set of file compression codecs including gzip, bzip2, LZO, and DEFLATE as part of its API. It is however critical to consider if a file format supports *splitting* for processing it with MapReduce. Hadoop utilizes a specific interface called `SplittableCompressionCodec` to denote codecs that support the compression/decompression of streams at arbitrary positions. Codecs like bzip2 that implement this interface are highly valuable in this context as they support the partitioning and parallel processing of compressed input data. We have implemented *AV Splittable Compression Codec*, a class that supports the compression, decompression, and splitting of audiovisual files.

AV Input Stream. In order to split a binary stream, it must be possible to detect positions where the data can be decomposed into blocks. This class implements a splittable input stream for compressed audiovisual content. As shown in figure 2, split boundaries must be repositioned to key frame positions by the codec in order to support decomposition of the data stream. Hence, during execution the reader must be advanced from an arbitrary position within the data stream to the next key frame position. This is done by utilizing a key frame index that is automatically generated from the container prior to the execution. In order to produce an interpretable data stream from the adjusted file split, the stream reader appends the container's header information (kept in memory) to each data portion. It is however not required to read the entire split into memory as the payload is directly read from HDFS.

[5] http://aws.amazon.com

Frame Record Reader. Record readers are plugged into the input file format of a particular MapReduce job. They typically convert the data provided by the input stream into a set of key/value pairs (called records) that are processed within the map and reduce tasks. We utilize the concept of packets, which are logical data entities read and uncompressed from the input sequences. Packets are subsequently decoded (optionally error checked and resampled) into objects of a specific data type. For example, a *frame record reader* can utilize the above described concepts in order to obtain audio/video frames from an generic input split.

Output Generation. *Record writers* provide the inverse concept to *record readers*. They write the delivered job outputs in the form of key/value pairs to the file system. Output files are produced per reduce task and might have to be merged in a postprocessing stage. To continue the example above, a *frame record writer* writes audio/video frames to an instance of *AV Output Stream* which can be obtained from *AV Splittable Compression Codec*. The codec implementation is customizable regarding the compression formats used to encode the diverse data tracks.

4 Evaluation

In the following, we provide an evaluation that investigates the impact of input file size and compression on the application's performance and scalability.

4.1 Experiment Setup

The evaluation has been conducted on a dedicated testing infrastructure that comprises a front-end and five worker nodes (Single Core 1.86GHz Intel CPU, 1.5GB RAM) connected through Gigabit Ethernet. For benchmarking, the video processing application was configured to decode every video frame of an input file and traverse the content for a given time period. Although the application supports a very large range of formats due to its bindings to FFmpeg, we have utilized a set of homogeneous input files in order to generate comparable results. The files (shown in table 1) differ in bitrate and duration only and utilize mp3 (48KHz) and MPEG4 (25fps) as compression formats and AVI as a container. The GOP (Group of Pictures) length basically determines the amount of successive pictures between two key frames, which also influences the achievable compression ratio. The application has been executed for each file sequentially as well as on 1-5 cluster nodes.

Table 1. Payload data file sizes are depending on encoding and duration

Encoding		Input File			
GOP length	bitrate	1800 sec	3600 sec	5400 sec	7200 sec
1 frame	2380 kb/s	535 MB	1071 MB	1607 MB	2141 MB
10 frames	436 kb/s	98 MB	196 MB	294 MB	393 MB
100 frames	341 kb/s	76 MB	153 MB	230 MB	306 MB

4.2 Results and Improvements

The left part of table 2 shows performance results that have been obtained using a static input split size that corresponds to the file system's block size (i.e. the default configuration). The results show execution times that increase horizontally (with growing duration) as well as vertically (with growing compression rate). Here, a higher compression rate of the payload data has a significantly negative impact on the application throughput. This effect however is caused by an imbalanced workload distribution, as the content is split and distributed using data chunks of a fixed size. This strategy however provides only a reasonably fair workload distribution if every frame is encoded as a key frame (GOP length=1). Compression algorithms like motion compensation disrupt this even density of information within the byte stream. Hence, the size of a byte stream does not provide an adequate measure for the workload it imposes on the application. For video content, it is therefore important to balance the workload (i.e. the uncompressed frames) based on GOPs rather than fixed chunk sizes. We have implemented a simple algorithm that adjusts the split size based on the average GOP, block, and frame size, in order to achieve a better workload distribution. The results in the right part of table 2 were obtained using the dynamic input split adaption algorithm. Here, we see an overall improved throughput rate that is independent of the volumes of content. Also, higher compression rates slightly improve the throughput due to the smaller input file size. Figure 3 shows the application performance on different numbers of nodes. In the tested setting,

Table 2. Execution time [sec.] and throughput (frames per second) on 5 nodes with static split size (left) and dynamic split size adaption (right)

GOP	1800 sec	3600 sec	5400 sec	7200 sec
1	921/49	1720/52	2623/51	3480/51
10	3686/12	3853/23	3890/35	4400/41
100	4910/9	4923/18	4911/27	4944/36

GOP	1800 sec	3600 sec	5400 sec	7200 sec
1	814/55	1624/55	2441/55	3333/54
10	772/58	1499/60	2236/60	2988/60
100	754/60	1440/62	2119/64	2830/64

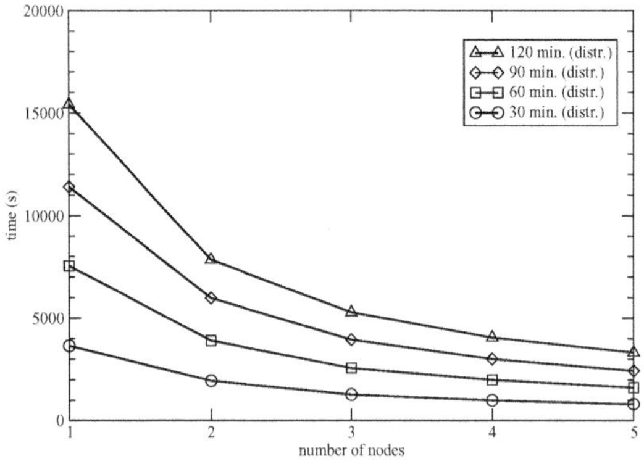

#n	Avg. Speedup	Avg. Efficiency
1	0,99	99,8%
2	1,90	95,2%
3	2,89	96,2%
4	3,74	93,6%
5	4,62	92,4%

Fig. 3. Application performance for different input files on 1-5 worker nodes

the application showed almost linear speedup, allowing one to efficiently reduce response times by increasing the number of worker nodes, e.g. important when processing content on demand and/or under SLA constraints.

5 Conclusion

We have presented an application for the parallel processing of binary media objects based on the MapReduce programming model. The implementation relies on Apache Hadoop and implements the required concepts to generate meaningful records like audio/video frames, which can be processed using common Java abstractions and user-defined logic. Furthermore, we provide insights on interpreting the compressed payload data, as this is highly important in order to assess and balance the application's workload. We motivate the employment of this approach in order to achieve minimal response times for Internet-accessible applications that maintain audiovisual content.

References

1. Dean, J., Ghemawat, S.: Mapreduce: simplified data processing on large clusters. Commun. ACM 51, 107–113 (2008)
2. Dean, J., Ghemawat, S.: Mapreduce: a flexible data processing tool. Commun. ACM 53(1), 72–77 (2010)
3. Ekanayake, J., Pallickara, S., Fox, G.: Mapreduce for data intensive scientific analyses. In: IEEE Fourth International Conference on eScience 2008, pp. 277–284 (2008)
4. Gunarathne, T., Wu, T.-L., Qiu, J., Fox, G.: Cloud computing paradigms for pleasingly parallel biomedical applications. In: Proc. of the 19th ACM Int. Symposium on High Performance Distributed Computing, HPDC 2010, pp. 460–469 (2010)
5. Isard, M., Budiu, M., Yu, Y., Birrell, A., Fetterly, D.: Dryad: distributed data-parallel programs from sequential building blocks. In: Proc. of the 2nd ACM SIGOPS/EuroSys European Conf. on Computer Systems 2007, pp. 59–72 (2007)
6. Moretti, C., Bui, H., Hollingsworth, K., Rich, B., Flynn, P., Thain, D.: All-pairs: An abstraction for data-intensive computing on campus grids. IEEE Transactions on Parallel and Distributed Systems 21, 33–46 (2010)
7. Pereira, R., Azambuja, M., Breitman, K., Endler, M.: An architecture for distributed high performance video processing in the cloud. In: Proc. of the 2010 IEEE 3rd International Conference on Cloud Computing, CLOUD 2010, pp. 482–489 (2010)
8. Schmidt, R., Rella, M.: Considering data locality for parallel video processing. ERCIM News (83) (2010)
9. Schmidt, R., Sadilek, C., King, R.: A service for data-intensive computations on virtual clusters. In: International Conference on Intensive Applications and Services, pp. 28–33 (2009)
10. Thusoo, A., Sarma, J.S., N., Jain, Z.S., P., Chakka, N.Z., S., Antony, H.L., Murthy, R.: Hive - a petabyte scale data warehouse using hadoop. In: 2010 IEEE 26th International Conference on Data Engineering (ICDE), pp. 996–1005 (2010)
11. Warneke, D., Kao, O.: Nephele: efficient parallel data processing in the cloud. In: Proc. of the 2nd Workshop on Many-Task Computing on Grids and Supercomputers, MTAGS 2009, pp. 8:1–8:10 (2009)

Risks, Benefits and Revelations: An Exploratory Study of Doctoral Students' Perceptions of Open Access Theses in Institutional Repositories

Kate V. Stanton[1] and Chern Li Liew[2]

[1] Massey University Library, Albany Campus, Auckland, New Zealand
k.v.stanton@massey.ac.nz
[2] Victoria University of Wellington, School of Information Management, Wellington, New Zealand
ChernLi.Liew@vuw.ac.nz

Abstract. This exploratory study examined doctoral students' awareness of and attitudes towards open access (OA) particularly in relation to institutional repositories (IR). Levels of students' awareness of OA and the concept of IRs, publishing behaviour, and perceived benefits and risks of OA publishing were explored. The study also examined students' willingness to comply with mandatory submission policies. The study sample was drawn from Massey University, one of the two universities in New Zealand which has had a mandatory submission policy in place since 2007. Qualitative and quantitative data was collected from doctoral students in two stages: first through a series of qualitative interviews with students from different disciplines, followed by self-completion questionnaires collected from a larger sample. In this paper, we discuss and highlight a number of potential strategies for raising awareness and for improving understanding of the benefits of OA IR to encourage acceptance and adoption.

Keywords: Electronic publishing, ETD, institutional repository, open access thesis and dissertations, scholarly communication.

1 Introduction

Through the networked electronic environment, academic staff and students have ever-increasing access to an overwhelming amount of scholarly research output, and are faced with a new range of options for disseminating their own research. Traditional modes of journal and monograph publication are being supplemented by a proliferation of open access (OA) e-journals and e-books, e-theses, document delivery services, departmental Web sites, subject-based or institutional repositories (IRs), collaborative wikis and faculty blogs. This research project focuses on one aspect of developments in electronic scholarly communication and publishing – postgraduate student thesis publication in OA IRs, and on one group of academic authors – doctoral students.

In a widely-cited SPARC (Scholarly Publication and Academic Resources Coalition) paper on IRs, Crow [1] expresses the view that IRs have the potential to

C. Xing, F. Crestani, and A. Rauber (Eds.): ICADL 2011, LNCS 7008, pp. 182–191, 2011.

provide a complement to existing modes of publication and stimulate changes to the current model of scholarly publishing (particularly mitigating the cost of accessing scholarly research); and the ability to "make visible" the research output of an institution.

In relation to IRs, the Bethesda Statement states that an OA publication is one that is "deposited immediately on publication in at least one online repository that is supported by an academic institution, scholarly society, government agency or other well established organization that seeks to enable open access, unrestricted distribution." [2]

It therefore appears that on one hand, OA to postgraduate theses is exciting – in addition to being held in print collections, student theses can be accessed and cited more easily by researchers and other students. On the other, OA publication may raise concerns over copyright, plagiarism and premature publication of findings.

In this paper, we report on an exploratory study that examines doctoral students' awareness of and attitudes towards OA publication, particularly in relation to OA IRs using a sample of doctoral students in one New Zealand university which has a mandatory IR submission policy.

2 Related Literature

Mandatory deposit policies have been adopted by many institutions to boost repository content and create a sustainable archive of research outputs [3]. These policies may apply to specific types of research output, to academic staff outputs or to postgraduate theses. While mandates take time and effort to be embedded into staff work processes [3], they have been shown to be an effective way to achieve growth and guarantee the sustainability of the repository. Although Pickton and McKnight found postgraduate students willing to comply with institutional mandates [4], other authors propose that mandates damage goodwill toward the repository, and that staff "buy-in" is more important to the sustainability of an IR than any guarantee of regular submissions of content [5].

Despite an observed reluctance to submit content to IRs, the principle and intended benefits of OA publishing and IRs do receive widespread support among the majority of academic authors [4, 6]. In Pickton and McKnight's study [4], postgraduate participants recognised the potential benefits of IR publishing, particularly greater exposure and the possibility of receiving feedback from the academic community. OA IR archiving allows for greater exposure of a work (via Google Scholar and other research discovery services), thus a greater potential for research impact [6, 7]. Increased exposure is partly achieved by allowing OA to researchers in government and private organisations or professional practitioners who would not normally have access to subscription journals in the workplace [6]. In addition, archiving allows research outputs to be preserved in digital form, making them easier to retrieve for future use, for example, in preparation for funding or promotion rounds [8].

In terms of risks, academics have expressed concern over their ability to publish if the pre-print of an article (or thesis) is already available in an IR. Confusion and concern over copyright, provenance and quality control, particularly the risk of copyright infringement or plagiarism, is also a recurring theme in the IR literature [4, 7, 9-11].

Despite their value as original research, master and doctoral theses have traditionally been considered "grey literature", due to their physical inaccessibility in closed access library collections [12]. Troman, Jacobs and Copeland [13] describe IRs as the "natural home" of electronic theses, as they offer flexibility in relation to managing submission processes, control in terms of access (e.g. managing an embargo), rates of deposit (by mandate), and with the guidance of library staff, quality control of metadata and file format. The inclusion of e-theses in IRs allows original research by emerging scholars to be visible and accessible to the research community. Statistics indicate that theses that are made available OA are more widely used and cited [13].

Two related research papers by Pickton and McKnight [4, 14] focus on the perceptions and needs of postgraduate students. The researchers found the majority of students to be enthusiastic about making their research available through IRs. The postgraduate students interviewed were most interested in disseminating their work and receiving feedback and commentary on their work. Support for the principle of OA was also a motivating factor. However, students were found to hold many of the same concerns around OA and IR publication as academic staff, particularly regarding plagiarism, quality and hindering later publication in journals [4].

3 Research Methods

This study used a mixed-methods approach consisting of two stages: qualitative, semi-structured interviews with eight doctoral students from different colleges of Massey University followed by a self-completion questionnaire that was sent to the entire doctoral student population. The analyzed interview data formed the basis of the quantitative survey to expand on and "test" the interview data against a larger sample of doctoral students. Results were examined by broad discipline (college), and mapped against a conceptual framework of criteria adopted from Diffusion of Innovations [15] and Social Exchange [16] theories, as an aid to interpretation.

4 Findings and Discussion

In this section, we discuss the key study findings related to perceived risks and benefits of OA IRs, followed by our recommendations for strategies for promoting the adoption and use of OA IRs.

4.1 Support for the Principle of OA

Consistent among interviewees and survey respondents was support for the concept of OA publishing, the importance of making research available to the wider public, and removing cost as a barrier to accessing research.

In the survey data, 86.3% of respondents either agreed or strongly agreed with the following statement on the benefit of OA: *"Open access is important because it removes cost as a barrier to accessing research and allows public access to research and information."*

The desire to provide public access to information and the perceived significance of OA should not be underestimated, and are evident in other studies in the IR literature. In Cullen and Chawner's recent survey of New Zealand academic staff, altruistic intent in making work available to the public was established as one of the "chief reasons" for making work available OA [17].

Swan and Brown's [18] author study found the principle of OA for all readers to be an important reason for publishing in OA journals. Similarly, participants in Pickton and McKnight's [4] study felt that the principle of OA was "an important motivating factor" for the deposit of theses. So, the concept of OA and equitable access is supported by the majority of authors, which is important to bear in mind when marketing IRs to the academic community.

4.2 Perceived Benefits and Risks

Many benefits of repository and OA publishing were mentioned by the interviewees and reiterated by the survey respondents. The most frequently mentioned relating to enhanced access on the web, providing public access to research, and increasing the speed of information dissemination (Table 1).

The major perceived risks described by the interview participants included the potential conflict of interest with journal publishers, concerns around copyright and plagiarism (particularly in relation to theses), and a perception that OA publication was less prestigious than traditional forms of publication.

Two interviewees expressed concern over opening themselves up to criticism by placing their thesis work in the public domain. Survey respondents had similar concerns, but overall, results were more balanced. The majority of responses fell in the "neutral" category, indicating uncertainty or ambivalence to the presented risks (Table 2). The two risks that caused the most concern included the copying or use of work without permission and plagiarism. Conflict of interest with journal publishers and issues over copyright restrictions should also be noted.

4.3 Influences on Perceptions and Publishing Behaviour

Interview participants mentioned a variety of influences on where and how they would publish their doctoral research. The influence of one's supervisor was noted most frequently. Survey respondents also identified supervisors as a major influence on deciding where to publish. 79.5% of respondents chose "agree" or "strongly agree" when presented with the following statement: "I publish where my supervisors recommend I publish".

4.4 Google Scholar – Use and Importance in the Promotion of OA and IRs

All interviewees said that they used a wide range of online catalogues and article databases to find information for the literature review stage of their research. However, one search engine stood out as the most frequently used resource: Google Scholar. Among the survey respondents, Google Scholar was also found to be a popular tool for literature searching, with 91.8% of respondents indicating that they used it during the literature review process.

Table 1. Perceived benefits associated with archiving research output in an IR

Answer Options	**Strongly Agree (benefit)**	**Agree**	**Neutral/ Not sure**	**Disagree**	**Strongly Disagree (no benefit)**	**Rating Average**	**Response Count**
I can share my work with the public more easily	35.5% (81)	**43% (98)**	16.7% (38)	3.9% (9)	0.9% (2)	1.92	228
I can share my work with other students more easily	38.9% (88)	**47.3% (107)**	11.5% (26)	0.9% (2)	1.3% (3)	1.78	226
I can share my work with my peers and the research community more easily	37% (84)	**45.8% (104)**	15.4% (35)	0.9% (2)	0.9% (2)	1.83	227
It will give my work more exposure	27.4% (62)	**51.3% (116)**	18.1% (41)	1.8% (4)	1.3% (3)	1.98	226
It will increase the number of times my work is cited	18.1% (41)	34.1% (77)	**41.2% (93)**	5.8% (13)	0.9% (2)	2.37	226
My work will be available on Google Scholar	22.8% (52)	**39% (89)**	36.8% (84)	0.9% (2)	0.4% (1)	2.17	228
My work will be available on research websites like KRIS (Kiwi Research Information Service)	15.1% (34)	28.9% (65)	**52.9% (119)**	1.3% (3)	1.8% (4)	2.46	225
I can get feedback and commentary on my research	22.2% (50)	**45.3% (102)**	25.8% (58)	4.0% (9)	2.7% (6)	2.20	225
I can publish my research findings more quickly	20.1% (45)	**40.6% (91)**	33.5% (75)	3.1% (7)	2.7% (6)	2.28	224
I can rely on the university to preserve a digital copy of my research in the long term	33.9% (77)	**40.1% (91)**	20.3% (46)	3.5% (8)	2.2% (5)	2.00	227
Open access is important because it removes cost as a barrier to accessing research (and allows public access to research and information)	**52.9% (119)**	33.8% (76)	11.1% (25)	1.3% (3)	0.9% (2)	1.64	225
Other academics in my field can find my work and contact me (professional networking)	36.2% (81)	**50.4% (113)**	11.6% (26)	1.3% (3)	0.4% (1)	1.79	224
It will enhance the research profile of my university	22.6% (51)	**45.6% (103)**	27.9% (63)	2.2% (5)	1.8% (4)	2.15	226
It will enhance my own research profile	26.2% (59)	**46.2% (104)**	24.4% (55)	2.2% (5)	0.9% (2)	2.05	225
It will help me to keep track of my research (for CV or PBRF)	23.6% (53)	**36.4% (82)**	31.6% (71)	5.8% (13)	2.7% (6)	2.28	225

Table 2. Perceived risks associated with archiving research output in an IR

Answer Options	Strongly Agree (risk)	Agree	Neutral/ Don't Know	Disagree	Strongly Disagree (not risk)	Rating Average	Response Count
There may be a conflict of interest with journal publishers	17.8% (40)	37.8% (85)	**38.7% (87)**	4.4% (10)	1.3% (3)	2.34	225
If I publish my work in an IR I may not be able to publish elsewhere due to copyright or other restrictions	22.9% (52)	36.6% (83)	**37% (84)**	2.6% (6)	0.9% (2)	2.22	227
People may copy or use my work without permission	24.1% (55)	**38.6% (88)**	25.5% (59)	10.1% (23)	1.3% (3)	2.26	228
My work may be plagiarized	24.2% (55)	**34.4% (78)**	28.2% (64)	11.5% (26)	1.8% (4)	2.32	227
I am concerned about confidentiality concerning the participants' information	11.1% (25)	21.7% (49)	**39.4% (89)**	21.2% (48)	6.6% (15)	2.91	226
I am concerned about confidentiality concerning the researcher's details	8.4% (19)	24.0% (54)	**38.2% (86)**	24.4% (55)	4.9% (11)	2.93	225
I am concerned that the research in an IR is not perceived as prestigious when compared to print journals or books etc	19.6% (44)	32% (72)	**34.7% (78)**	12% (27)	1.8% (4)	2.44	225
I am concerned that work that is published OA is not perceived as prestigious when compared to print journals or books etc	17.9% (40)	29.6% (66)	**32.3% (72)**	17.0% (38)	3.1% (7)	2.58	223
People will not be able to find my work	6.8% (15)	15.8% (35)	**43.7% (97)**	28.4% (63)	5.4% (12)	3.10	222
My work will not be cited	7.6% (17)	18.2% (41)	**42.2% (95)**	28.0% (63)	4% (9)	3.03	225
I am concerned about the long-term preservation of my work	9.4% (21)	14.3% (32)	**37.2% (83)**	32.7% (73)	6.3% (14)	3.12	223
My thesis research might not be good enough to put in the public domain	6.6% (15)	14.2% (32)	31.9% (72)	**34.5% (78)**	12.8% (29)	3.33	226
Publishing my thesis open access will allow people to criticize my work	6.3% (14)	26.7% (59)	**33.0% (73)**	26.7% (59)	7.2% (16)	3.02	221

Respondents indicated that IR-specific search tools were used far less frequently. Kiwi Research Information Service (KRIS) was the least frequently used research resource, with just 3.7% of respondents indicating that they had used it during their literature search.

Other thesis databases such as the Electronic Theses Online Service (EThOS) and the Australian Digital Theses Program (ADT) were used more frequently, at 25.8% and 6.1% respectively. However, only one of the eight interviewees was aware that material in IRs appears in (is indexed by) Google Scholar. This finding was surprising, given the wide use of Google Scholar, and the potential "pay-off" for the student – placing work in an IR allows the work to be available to the international research community via Google Scholar, which in turn is very widely used by other students. Making students, academic staff and library staff aware of this critical connection between repository content and Google Scholar is a key strategy for raising the profile of the IR and OA publishing.

4.5 Library Outreach: Research Management Courses and Research Consultations

Questions regarding students' attendance of library courses or research consultations were not investigated in the questionnaire due to the length of the survey instrument. However, those interviewees who had attended a library course or research consultation were found to have a deeper awareness and understanding of the potential benefits of IRs and the concept of OA than those who had not. This difference in awareness further reinforces the importance of the role of postgraduate and staff-focused library outreach programmes in raising awareness of the repository and its role in the research process.

4.6 Willingness to Comply with Mandatory Submission Policies

Five (of eight) interviewees and over 90% of those survey respondents who were covered by the mandatory submission policy were found to be willing to comply and submit their theses to the IR. Of the other three interviewees not covered by the policy, all stated that they would voluntarily submit a copy of their theses. In addition, 83% of survey respondents not covered by the policy were also willing to deposit their theses in an IR.

Among the survey respondents, 77.3% either agreed or strongly agreed with the following statement: *"Except in the case of an embargo, all doctoral theses should be made openly available online through an institutional repository."*

5 Conclusion and Implications for IR Managers

5.1 Awareness of OA and IR, and Use of OA Materials

The study shows that awareness of OA and IRs is still fairly low, even in a research-focused community of doctoral students. While the majority of interview and survey respondents indicated that they support the concept of OA, the majority of interview transcripts indicated a lack of deep understanding of OA. The mandatory thesis submission policy has been in effect at Massey University since 2007. All students are given information on repository submission via the Doctoral Handbook, policy information is available online via the Graduate Research School and Library websites, and the university's IR (Massey Research Online) is linked from the library homepage. Yet awareness of the university IR remains disappointingly low. Just over half of survey respondents indicated that they were aware of the university IR.

Results also indicated that only a very small minority of the respondents used research services like KRIS, ADT and EthOS. However, the overwhelming majority of respondents used Google Scholar, so it is likely that they access OA material without realising it. In relation to perceived benefits and risks, respondents had similar perceptions and concerns to those of academic staff and UK postgraduate students already documented in the IR literature. The respondents perceived that repository publication of theses was beneficial both for themselves and the research community, but also had concerns regarding copyright, potential for plagiarism, and the effect IR deposit may have on their ability to publish their thesis work.

5.2 Key Strategies for Promoting IRs and OA Publishing

Support for the principle of OA – OA advocates will be pleased by the respondents' overwhelming support for the principle of OA and enabling public access to research, and motivation to share their work with the public and research community. Knowledge sharing for the public good was clearly a key point of principle for the doctoral students.

The role of Google Scholar in promoting IR publication - The ubiquity of Google Scholar as a research tool cannot be ignored. Over 90% of students surveyed and all interviewees used Google Scholar during the literature gathering stage of their research, making it an essential research tool. Making students aware that work deposited in an IR will be made available to an international audience via Google Scholar is another useful strategy. One interviewee described finding out about this aspect of OA IR deposit "a *revelation*". The difficulty, of course, is how to make students aware of this critical connection between depositing a work in an IR and having it indexed by Google Scholar, thus making that work available to an international audience.

Research mangement courses for postgraduates and staff – "Knowledge Management in Research" (KMIR) is a course developed and delivered by Massey University Library staff [19]. KMIR aims to equip doctoral students and academic staff with the skills to undertake a major research project. Making Your Research Count, a module of KMIR, identifies practical ways for researchers to enhance the impact of their publications and advocates use of the IR as a means of increasing citation. Greater levels of awareness of the wider information landscape and support for the IR among those interviewees who had attended KMIR indicate that this is an effective forum for communicating information about the IR, OA and the mandatory deposit policy to postgraduate students (and staff) within a wider research and information context.

Don't forget supervisors and colleagues! - Communicating directly with postgraduate students is only part of the picture when attempting to raise levels of awareness and use of OA and IRs. Supervisors and academic colleagues were found to have the greatest influence on students' perceptions, and their decision on where and how to disseminate their work.

Mandatory thesis deposit policies - This study indicates that a mandatory thesis deposit policy is a useful strategy to build the core collection of an IR, and one that is supported by the majority of those doctoral students surveyed. Over 90% of survey respondents who were covered by the mandatory deposit policy were found to be willing to comply. It must also be noted that 83% of those survey respondents who were not covered by the policy were willing to voluntarily deposit a copy of their thesis in the repository.

A key recommendation for the library and information profession is ensuring that the mandatory submission policies and the benefits of repositories are communicated to students through a variety of channels including supervisors, official research school communications, handbooks, library workshops and other liaison or outreach services, in order to ensure that each student understands the purpose and benefit of OA IR publication.

References

1. Crow, R.: The case for institutional repositories: A SPARC position paper (2002), `http://scholarship.utm.edu/20/1/SPARC_102.pdf`
2. Bethesda Statement on open access publishing (2003), `http://www.earlham.edu/~peters/fos/bethesda.htm`
3. Sale, A.: The acquisition of open access research articles. First Monday 11(9) (2006), `http://firstmonday.org/issues/issue11_10/sale/index.html`
4. Pickton, M., McKnight, C.: Research students and the Loughborough institutional repository. Journal of Librarianship and Information Science 38(4), 203–219 (2006)
5. Palmer, C.L., Teffeau, L.C., Newton, M.P.: Strategies for institutional repository development: A case study of three evolving initiatives. Library Trends 57(2), 142–167 (2008)
6. Kingsley, D.: Those who don't look don't find: Disciplinary considerations in repository advocacy. OCLC Systems & Services: International Digital Library Perspectives 24(4), 204–218 (2008)
7. Cullen, R., Chawner, B.: Institutional repositories and the role of academic libraries in scholarly communication. Paper Presented at the Asia-Pacific Conference on Library Education and Practice, Japan, March 6-8 (2009a), `http://hdl.handle.net/10063/866`
8. Performance-Based Research Fund Sector Reference Group (n.d.). Performance-Based Research Fund Sector Reference Group review: Managing nominated research outputs. TEC, Wellington, NZ
9. Davis, P.M., Connolly, M.J.L.: Evaluating the reasons for non-use of Cornell University's installation of DSpace. D-Lib Magazine 21 (2007), `http://www.dlib.org/dlib/march07/davis/03davis.html`

10. Gadd, E., Loddington, S., Oppenheim, C.: A comparison of academics' attitudes towards the rights protection of their research and teaching materials. Journal of Information Science 33(6), 686–701 (2007)
11. White, B.: Minding our ps and qs: Issues of property, provenance, quantity and quality in institutional repositories. Paper Presented at the International Association of Technological University Libraries (IAUTL) Conference 2008 (2008), http://hdl.handle.net/10179/645
12. Jones, R., Andrew, T.: Open access, open source and e-theses: The development of the Edinburgh Research Archive. Program: Electronic Library and Information Systems 39(3), 198–212 (2005)
13. Troman, A., Jacobs, N., Copeland, S.: A new electronic service for UK theses: Access transformed by EThOS. Interlending & Document Supply 35(3), 157–163 (2007)
14. Pickton, M., McKnight, C.: Is there a role for research students in an institutional repository? Some repository managers' views. Journal of Librarianship and Information Science 39, 153–161 (2007)
15. Rogers, E.M.: Diffusion of innovations, 4th edn. The Free Press, New York (1995)
16. Homans, G.C.: Social behavior as exchange. The American Journal of Sociology 63(6), 597–606 (1958)
17. Cullen, R., Chawner, B.: National Survey: Perceptions of New Zealand academic staff towards institutional repositories (2009b), http://www.oarinz.ac.nz/oarinzwiki/index.php?n=OARiNZ.NationalSurveyPerceptionsOfNewZealandAcademicStaffTowardsInstitutionalRepositoriesSeptember2009
18. Swan, A., Brown, S.: Authors and open access publishing. Learned Publishing 17, 219–224 (2004)
19. White, B., Gendall, R., Naidoo, K.: No end in sight: Information skills for academics and researchers. Paper Published in the Proceedings of the 3rd International Lifelong Learning Conference, Yeppoon, Queensland, Australia (2004), http://hdl.handle.net/10179/965

A Social Tagging Based Collaborative Filtering Recommendation Algorithm for Digital Library

Zhenming Yuan*, Tianhao Yu, and Jia Zhang

College of Information Science and Engineering,
Hangzhou Normal University,
310036 Hangzhou, China
{Zmyuan,zhangjia}@hznu.edu.cn,
yutianhao@stu.hznu.edu.cn

Abstract. Recommendation is one of the new personalized services in the digital library. This paper proposes a new collaborative filtering recommendation algorithm based on the social tagging, which try to settle the semantic gap and the cold start problems of traditional collaborative filtering. Firstly, the communities with the similar habits are detected in the social network of the digital library. Then the candidate tags are derived from the user-book-tag correlation model. Finally, the books with highest posterior of the tags are recommended by the naïve Bayes classifier. Experiments results show that the proposed algorithm improves the performance of the collaborative filtering algorithms. And it has been a core recommendation algorithm in China Academic Digital Associative Library (CADAL).

Keywords: recommender system, digital library, social network, collaborative filtering.

1 Introduction

Personalized service is more and more popular in the applications of the digital library, which provides the recommendation ability according to the habit and preference of the readers. The recommendation systems provide the users interested books from enormous digital library in response to the challenge which come from the enormous amount of data [1].

Though the recommendation systems have made great progress, they still face great difficulties, among which the recommendation algorithm is one of the core problems [2]. The content-based filtering and the collaborative filtering are the most successful personalized recommendation algorithms which have been used in digital library. The content-based filtering [3] is focus on the content of books interested by the users. The book profile and the user profile are represented in the same feature space. However, it has the semantic gap of modeling the book profile using the term

* This research is supported by the Natural Science Foundation of China (No. 60773051), the 211 Key Project of Ministry of Education of China (Project of China Academic Digital Associative Library, No. 201003010).

C. Xing, F. Crestani, and A. Rauber (Eds.): ICADL 2011, LNCS 7008, pp. 192–201, 2011.

frequency which does not reflect the content of book effectively. Collaborative filtering method [4] predicts the rating of items for a particular user based on the ratings from other users with similar preference. Nevertheless, it has two serious limitations. One is the sparsely problem; the other is the cold start problem. The collaborative filtering commendation system is unable to make high quality recommendation to a new user with few activities and comments.

This paper proposes a new collaborative filtering algorithm based on collaborative tagging and community detection in the social network of the digital library. Firstly, the communities of the social network are detected to find the readers who have the similar preference. Then the user-book-tag triple model is built to calculate the best candidate tags. Finally, we use the naïve Bayes classifier to recommend books. The proposed algorithm derived from collaborative tagging, which can solve the problem of the semantic gap [6,8]. It is also able to settle the cold start problem attribute to using the community detection.

2 Social Tagging Based Collaborative Filtering

Suppose there are a set of m tags, $T=\{t_1,t_2,\ldots,t_m\}$, a list of k users, $U=\{u_1,u_2,\ldots,u_k\}$, a list of n books, $B=\{b_1,b_2,\ldots,b_n\}$ in the digital library. They are used to build the user-book-tag triple model, which composed of the user-book matrix $\boldsymbol{R}$, the user-tag matrix $\boldsymbol{A}$, and the tag-book matrix $\boldsymbol{Q}$ [9].

Definition 1: User-Book binary matrix, R. A list of users $\boldsymbol{U}=\{u_1,u_2,\ldots u_k\}$and a list of books $\boldsymbol{B}=\{b_1,b_2,\ldots b_n\}$ are used to build the User-Book binary matrix. If the user u has read or tagged a book b, $R_{u,b}=1$, otherwise $R_{u,b}=0$, as in Fig.1.

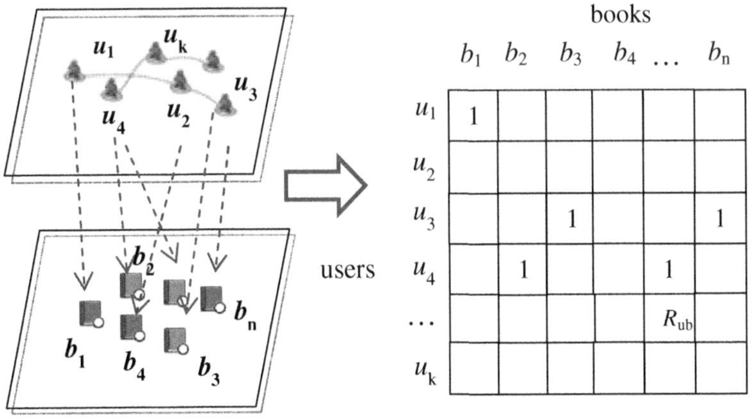

Fig. 1. Build the User-Book binary matrix

Definition 2: User-Tag matrix, A. A list of users $\boldsymbol{U}=\{u_1,u_2,\ldots,u_k\}$ and a set of tags $\boldsymbol{T}=\{t_1,t_2,\ldots,t_m\}$ to are used to build the User-Tag matrix. Each $A_{u,t}$ in the matrix represents the frequency of tags which a user has tagged with a tag t, as in Fig.2.

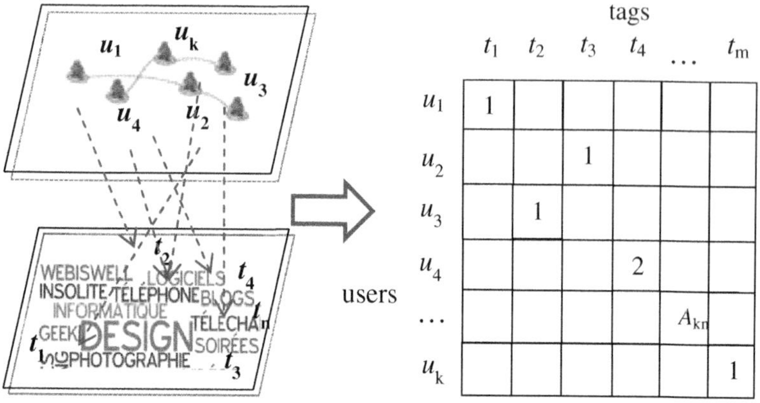

Fig. 2. Build the User-Tag matrix

Definition 3: Tag-Book matrix, *Q*. A set of tags ***T***=$\{t_1,t_2,\ldots,t_m\}$ and a list of books ***B***=$\{b_1,b_2,\ldots,b_n\}$ are used to build the Tag-Book matrix. Each $Q_{,b,t}$ in the matrix state the number of tag *t* tagged for book *b* by user, as in fig.3.

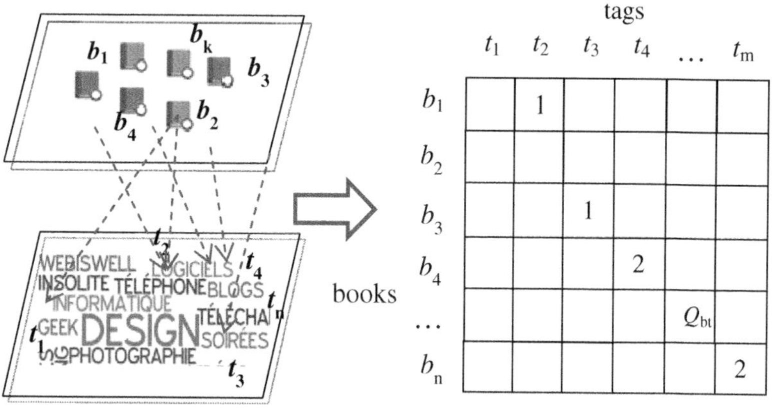

Fig. 3. Build the Book-Tag matrix

2.1 Community Detection

Considering there is a social network of digital library, *G*, the users with similar interests or preferences trends to gather in the same group. So we can conclude that the users in the same community would tag the same book with the similar tags [10]. To discover such user groups, the community detection is used to find the community users who have the same habits in the social network of digital library. In the proposed algorithm, Q-based GN algorithm [7] is employed to detecting the community structures in the social network of digital library. The algorithmic procedure for the community detection is as follows:

Step 1: The edge betweenness is calculated in the social network of digital library. Edge betweenness of an edge means the number of shortest paths between pairs of vertices that run along it.

Step 2: Finding the largest the edge betweenness in the social network of digital library, then remove it from the digital library social network. So we can get k communities.

Step 3: Recalculating betweenness for all edges affected by the removal in the social network of digital library. Repeat Step 2 and Step3 until get the highest value of modularity, Q. Then we can get p final groups.

Defined a $k\times k$ matrix $\boldsymbol{E}=\{e_{ij}\}$, whose element e_{ij} is the fraction of all edges in the digital library social network that links the users in the community i to the users in the community j. The trace, T, of the matrix is the ratio of the edge numbers in one community to the whole social network.

$$T=\sum_i e_{ii} \tag{1}$$

The modularity Q is defined by equation (2).

$$Q=\sum_i (e_{ii}-(\sum_j e_{ij})^2)=T-\| e^2 \| \tag{2}$$

where $\|e\|$ is the sum of the elements of the matrix. If the number of within-community edges is no better than random, Q equals to 0. The maximum of the modularity Q is 1, which indicates strong community structure. Finally, the social network of digit library G is divided into p groups $\{g_1,g_2,\ldots,g_p\}$.

2.2 Candidate Tags Extraction

In the proposed algorithm, the community detection plays the important role on the candidate tags selection and the books recommendation. The users in the same community have similar preferences, which is the basic to derive the candidate tags of the target user in the same community based on the user-book-tag correlation model. Suppose a recommend target user, u, belongs to the community g_p in the social network of digital library. If the target user has no books or tags, we make the measurement of how much the user, u, prefers a tag, t, is given by:

$$S_u=\sum_{o\in g_p}(A_{o,t}) \tag{3}$$

Otherwise, we make the measurement of how much the user, u, prefers a tag, t, is given by:

$$S_{u,t}=\sum_{o\in g_p}(A_{o,t})*sim(u,o) \tag{4}$$

where $sim(u,o)$ is given by the cosine similarity between the user u and the user o.

$$sim(u,o) = cos(u,o) = \frac{\sum_{t \in T} A_{u,t} * A_{o,t}}{\sqrt{\sum_{t \in T} (A_{u,t})^2} \sqrt{\sum_{t \in T} (A_{o,t})^2}} \quad (5)$$

The higher score a user has the more similar he is to the target user. S_u *and* $S_{u,t}$ represent the prediction value of target user's interest for tag t in the social network of digital library. We make the sum of tag frequency $A_{o,t}$ weighted by $sim(u,o)$ which leads to a higher prediction value for more similar user to the target user. Finally, w tags are selected for candidate tags according to the highest-order of prediction value S_u or $S_{u,t}$. Because of using the community detection to find the the community users who have the same habits in the social network of digital library, it could settle the new user cold start problem to some extent.

2.3 Active Recommendation

After extracting w candidate tags from the community which the target user belongs to, the books with highest posterior of the tags are recommended by the naïve Bayes classifier. Given a tag instance t_j in w candidate tags, the naïve Bayes classifier is used to calculate the posterior $p(B=b_y|t_j)$ and the priori probability $p(B=b_y)$ which tag t_j was tagged for book b_y for each possible book in the set of all book $B=\{b_1,b_2,\ldots,b_n\}$.

$$p(B = b_y) = \frac{\sum_{u=1}^{k} R_{u,y}}{\sum_{y=1}^{n} \sum_{u=1}^{k} R_{u,y}} \quad (6)$$

$$p(t_j | B = b_y) = \frac{1 + Q_{j,y}}{m + \sum_{t=1}^{m} Q_{t,y}} \quad (7)$$

where $R_{u,y}$ and $Q_{j,y}$ denote the binary value of *u-th* user for *y-th* book in User-Book matrix and the frequency of *j-th* tag for *y-th* book in the Tag-Book matrix of the user-book-tag model respectively. Finally, the ordered set of books $TopN_u$ with the highest $p_{u,y}$ is recommended for the target user[11]. The framework of the proposed algorithm is shown in the fig.4.

$$p_{u,y} = p(B = b_j) \prod_{j=1}^{w} p(t_j | B = b_y) \quad (8)$$

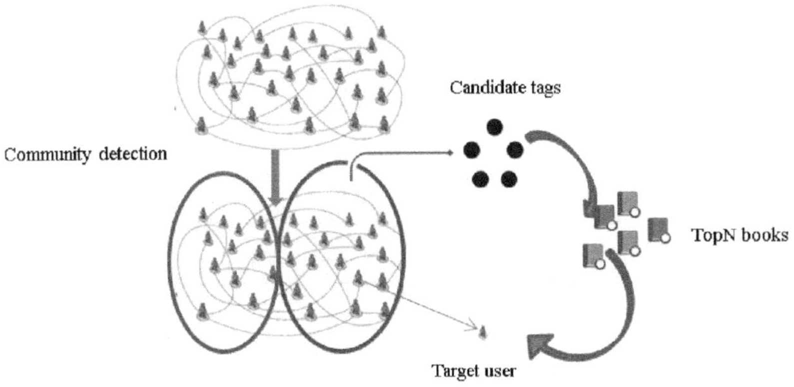

Fig. 4. The framework of the social-tag collaborative filtering based recommendation algorithm

2.4 Cold-Start Problem

There are two types of the cold-start problems. The first one is to recommend for a new reader who has not browse or borrow any book, and has not written any tag for books. In this case, we assume the reader joined the digital library social network invited from his/her friends already in the social network. In addition, many potential friends also are recommended and linked according to the new reader's initial personal information. Therefore, the Q-based GN community detection algorithm is used to find out which community the new reader belongs to. Consequently, we utilize the relation properties among the users in the social network of the digital library to recommend books by above active recommendation algorithm.

The second type of the cold-start problem is to recommend the new books without tags. Actually, when each new book collected in the digital library, the meta data, such as the cataloging in publication data (CIP), can be regarded as the potential initial clues to refine the initial tags by the modified nearest neighbor voting algorithm [13].

3 Experiments and Analysis

In order to investigate the performance of the proposed algorithm, the Douban.com [5] data is used as the test dataset. We compare our algorithm with the content-based recommendation, item-based collaborative recommendation algorithm and user-based collaborative recommendation algorithm. The system prototype is implemented by C# language. The experiments were performed on Intel(R) Pentium® Dual 1.79GHzCPU, 2GB memory.

3.1 Dataset and Evaluation Metrics

Douban.com [5] is an online book-sharing service which contains hundreds of millions of books. It is uploaded, tagged and commented by more than 8.5 million social network users. We collected our dataset from Douban.com and constructed a 268×36836 User-Book matrix, a 268×8526 User-Tag matrix, 8526×36836 Tag-Book matrix. The dataset was divided into two parts: a training set containing 29469 books (80%) and a test set containing 7367 books (20%).

Table 1. The dataset from Douban

readers	books	tags	tagging
268	36836	8526	106460

The hit-ratio and recall are used to measure the performance of our proposed algorithm [12]. The performance is measured by looking at the number of books in the test set which are included in $TopN_u$ recommended for the target user. The recall and hit-ratio scheme is given by Equation (9) and (10).

$$hit\text{-}ratio(u) = \frac{|Test_u \cap TopN_u|}{Test_u} \tag{9}$$

$$recall = \frac{\sum_{u=1}^{k} hit\text{-}ratio(u)}{k} \times 100 \tag{10}$$

where $Test_u$ represents a set of books tagged by the target user u in the test set. $|Test_u \cap TopN_u|$ denotes the value of the hit-ratio.

3.2 Experiments with Different Sizes of Candidate Tags

In the proposed algorithm, we expect that the number of the candidate tags having significant impact on the quality of result from collaborative filtering. So we evaluate our algorithm by measuring recall in the light of tags with a character length of 3, 5, 10, 50. The $TopN_u$ is set to 10. The fig.5 shows the various recalls based on different sizes of the candidate tags.

By analyzing the experimental data in the fig.5, we can find that the size of the candidate tags has significant impact on the performance of result from collaborative filtering. The fig.5 shows that the performance tends to better as the size of the candidate tags is increased. We get the best quality of our proposed algorithm when the size of the candidate tags equal to 10. It can be indicated that there is noise in the candidate tags which do not represent user's preference, because the quality of our proposed algorithm is worse when the size is equal to 50.

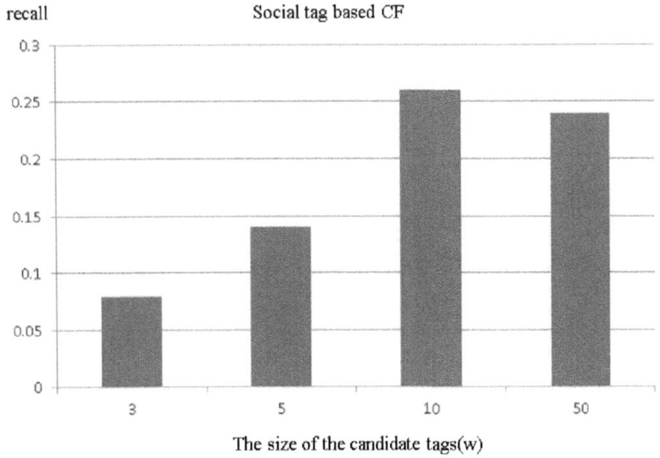

Fig. 5. Recall ratio comparison based on different size of the candidate tags

3.3 Comparison with Different Algorithms

In order to compare the performance of our proposed algorithm with those of content-based recommendation, item-based CF and user-based CF, the fixed number and the varied number of the returned books decided two types of the comparison. As in the table.2., we compare the recall and computing time among the four different algorithms when the size of the candidate tags and the number of returned books top n are set to 10 and 20 respectively.

In fig.6., we compare the recall ratio under different sizes of the returned books among our social tag-based algorithm, the item-based collaborative filtering (CF) and user-based CF algorithms. The size of the candidate tags is set to 10.

By analyzing the experiment data in the fig.6 and table.2, we can conclude that the proposed algorithm performed better than item-based recommendation, user-based CF and item-based CF. But the collaborative filtering recommendation cost more time than content-based recommendation.

Table 2. Comparisons of recall and cost time with different algorithms

Recommendation algorithm	Recall ratio	time
Content-based	0.158	2(s)
User-based CF	0.195	5(s)
Item-based CF	0.225	7(s)
Social Tag-based CF	0.270	9(s)

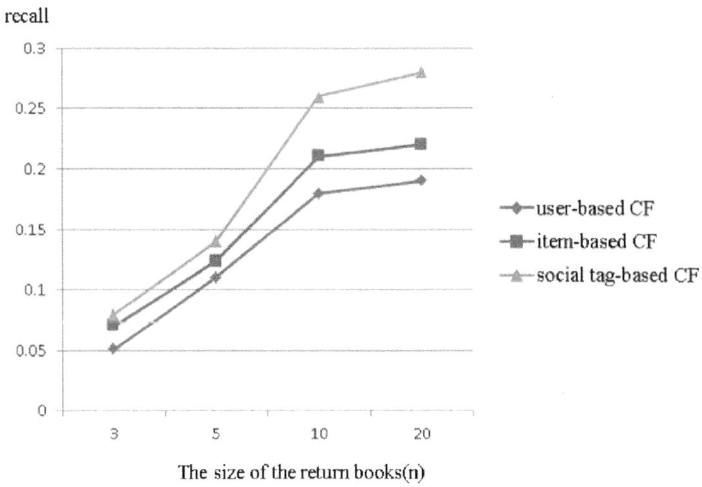

Fig. 6. Various comparison of recall with different size of returned books

4 Conclusions

While Web 2.0 technology and the social network are applied increasingly to the digital library, the recommender systems have emerged in response to the challenge of searching and sharing the user oriented interesting digital resources from the massive and heterogeneous digital library. In this paper, we propose a new collaborative filtering algorithm based on the social tagging and the community detection. Firstly, the communities of the social network are detected to find the readers who have the similar preference. Then the candidate tags are derived from the user-book-tag correlation model. Finally, the books with highest posterior of the tags are recommended by the naïve Bayes classifier. The experiments results show that the proposed algorithm improves the performance of the collaborative filtering algorithm. And it has been a core recommendation algorithm in China Academic Digital Associative Library (CADAL).

References

1. Chen, S.-Y., Teng, Y.-Y., Ke, H.-R.: Social Tagging in Digital Archives. In: Buchanan, G., Masoodian, M., Cunningham, S.J. (eds.) ICADL 2008. LNCS, vol. 5362, pp. 414–415. Springer, Heidelberg (2008)
2. Cantador, I., Bellogín, A., Vallet, D.: Content-based Recommendation in Social Tagging Systems. In: Proceedings of the 4th ACM Conference on Recommender Systems (2010)
3. Amato, G., Straccia, U.: User profile modeling and applications to digital libraries. In: Abiteboul, S., Vercoustre, A.-M. (eds.) ECDL 1999. LNCS, vol. 1696, pp. 184–197. Springer, Heidelberg (1999)

4. Rohini, U., Ambati, V.: A collaborative filtering based re-ranking strategy for search in digital libraries. In: Fox, E.A., Neuhold, E.J., Premsmit, P., Wuwongse, V. (eds.) ICADL 2005. LNCS, vol. 3815, pp. 194–203. Springer, Heidelberg (2005)
5. Douban.com, `http://www.douban.com`
6. Gupta, M., Li, R., Yin, Z., Han, J.: Survey on social tagging techniques. SIGKDD Explor. Newsl. 12, 58–72 (2010)
7. Newman, M.E.J.: Fast algorithm for detecting community structure in networks. Phys. Rev. E 69, 066133 (2004)
8. Milicevic, A.K., Nanopoulos, A., Ivanovic, M.: Social tagging in recommender systems: a survey of the state-of-the-art and possible extensions. Artif. Intell. Rev. 33, 187–209 (2010)
9. Jing, P., Daniel, Z., Huimin, Z., Feiyue, W.: Collaborative Filtering in Social Tagging Systems Based on Joint Item-Tag Recommendations. In: Proceedings of the 4th ACM Conference on Recommender Systems (2010)
10. Changyan, C., Qinying, L., Yingxin, P., Shiwan, Z.: Smarter Social Collaboration at IBM Research. In: Proceedings of the 4th ACM Conference on Recommender Systems (2011)
11. Deshpande, M., Karypis, G.: Item-based Top-N recommendation algorithms. ACM Transactions on Information Systems (TOIS) 22(1), 143–177 (2004)
12. Sarwar, B., Karypis, G., Konstan, J., Riedl, J.: Analysis of recommendation algorithms for E-commerce. In: Proceedings of the Second ACM Conference on Electronic Commerce, pp. 158–167 (2000)
13. Li, X., Snoek, C.G.M.: Learning Social Tag Relevance by Neighbor Voting. IEEE Transactions on Multimedia 11(7), 1–14 (2009)

Co-Ranking Multiple Entities in a Heterogeneous Network: Integrating Temporal Factor and Users' Bookmarks*

Ming Zhang[1], Sheng Feng[1], Jian Tang[1], Bolanle Ojokoh[1,2], and Guojun Liu[1]

[1] Institute of Networking and Information Systems,
School of Electronics Engineering and Computer Science,
Peking University, 100871, Beijing, China
[2] Department of Computer Science, Federal University of Technology,
Akure, Nigeria
{tangjianpku,fengsheng.pku}@gmail.com, bolanleojokoh@yahoo.com,
{lgj,mzhang}@net.pku.edu.cn

Abstract. In this paper, we present a novel approach that models the mutual reinforcing relationship among papers, authors and publication venues with due cognizance of publication time. We further integrate bookmark information which models the relationship between users' expertise and papers' quality into the composite citation network using random walk with restart framework. The experimental results with ACM dataset show that 1) the proposed method outperforms the traditional methods; 2) by incorporating the temporal factor, the ranking result of latest publications can be greatly improved; 3) the integration of user generated content further enhances the ranking result.

1 Introduction

With the increasing growth in scientific publications, automatic and effective ranking of scientific publications is becoming more and more essential to the academic groups. It can be very useful because it provides a means for making decisions concerning matters of appointment, promotion and award assignments, in addition to qualitative search. Earlier approaches focused on counting the number of publications and/or citations [4]. Traditional link analysis approaches[5] like PageRank [2] and HITS [7] algorithm were also applied with success. Lempel and Moran [8] combined the ideas of HITS and *PageRank* and proposed the SALSA algorithm. More recently, Du et al. [3] proposed *PaperRank*, an extension of the *HITS* and *PageRank*. Several authors have expanded on the heterogeneity concept to include publication venues information [10], [13] and very recently, affiliation [16]. Some other authors have worked on the inclusion of temporal information [9], [12]. Li and Tang [9] proposed a temporal random walk model

* This study is partially supported by the HGJ Grant (No. 2011ZX01042-001-001) as well as the Specialized Research Fund for the Doctoral Program of Higher Education of China under Grant ("FSSP" Grant No.20100001110203).

C. Xing, F. Crestani, and A. Rauber (Eds.): ICADL 2011, LNCS 7008, pp. 202–211, 2011.

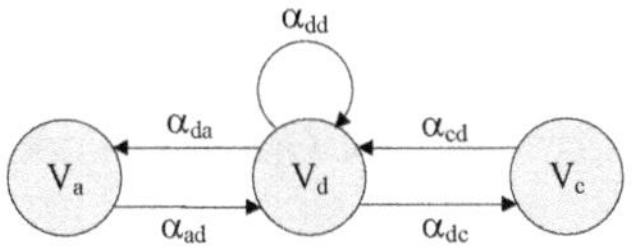

Fig. 1. Transition probability

for expertise search, dealting with a heterogeneous social network but emphasized on categorizing authors' publications based on publication time. Sayyadi et al [12] also combined information from the authorship network and publication time to predict future ranking of publications. User generated content (UGC) such as tags have been used to evaluate users' expertise in a folksonomy [11], [1]. To the best of our knowledge, UGC has not been utilized in ranking publications.

This paper has designed a hybrid approach for ranking scientific publications, with the following contributions:

- Consideration of the mutual reinforcing relationship, and the publication time to construct the academic network for random walk.
- Modelling of user generated content to help evaluate the quality of papers.
- Utilization of the random walk with restart framework, combining the analyses of academic network and user generated content.

In Section 2, we present our proposed model. The experiment are presented in Section 3. In section 4, we conclude the paper.

2 Our Proposed Method

2.1 Random Walk on the Academic Network

The random walk method is used to jointly rank the papers, authors and venues on the academic network, shown in Fig.1. We adopt a similar strategy as in the *PopRank* algorithm [10] by adding a *population propagation factor (PPF)* to each link pointing to an object. The propagation factors vary for different kinds of objects, because of the variation in the influence of one type of object on another.

The random surfer will have the α_{dd} probability to stay in the paper citation network, and will have α_{da} and α_{dc} probabilities to respectively find authors and the publication venues related to the paper.

Firstly, we define the weighted adjacency matrix W(G) of the academic network in equation(1).

$$W(G) = (w(i,j))_{i,j \in V} = \begin{bmatrix} W_{dd} & W_{da} & W_{dc} \\ W_{ad} & 0 & 0 \\ W_{cd} & 0 & 0 \end{bmatrix} \quad (1)$$

Now we give the detailed initial definition of each sub matrix in accordance with their respective sub graphs in the equations that follow:

- Paper citation network G_{dd}

The weighted adjacency matrix W_{dd} of G_{dd} is defined as

$$W_{dd}(i,j) = \begin{cases} \left|\{d \mid t(d) < t(d_j) \wedge d \in \Gamma_{dd}(d_i)\}\right| + 1, & d_j \in \Gamma_{dd}(d_i) \\ 0 & \text{, otherwise} \end{cases} \tag{2}$$

$t(d)$ is the publication time of paper d, and $\Gamma_{dd}(d_i)$ is the references of paper d_i.

In equation 2, the cell $W_{dd}(i,j)$ is equal to 1 plus the number of references of d_i which are published earlier than d_j. Hence, if d_j is the most recent reference, $W_{dd}(i,j)$ will be equal to the total number of references of paper d_i. If d_j is the oldest reference of paper d_i, $W_{dd}(i,j)$ will be equal to 1.

- Author-paper network $G_{ad} = (V_a \cup V_d, E_{ad})$

The weight matrix W_{ad} of the edges from authors to papers is defined as

$$W_{ad}(i,j) = \begin{cases} \left(\left|\{d \mid t(d) < t(d_j) \wedge d \in \Gamma_{ad}(a_i)\}\right| + 1\right) \times f(a_i, d_j), & d_j \in \Gamma_{ad}(a_i) \\ 0 & \text{, otherwise} \end{cases} \tag{3}$$

$$f(a,d) = \frac{1}{k + \lambda} \tag{4}$$

$$W_{da}(j,i) = \left|\Gamma_{da}(d_j)\right| - k + 1 \tag{5}$$

In equation 3, $\Gamma_{ad}(a_i)$ is the publication set of author a_i. In equation 4, a is the k-th author of paper d. In equation 5, $\Gamma_{da}(d_j)$ is the author set of paper d_j, and a_i is the k-th author of paper d_j.

We apply a weight function $f(a,d)$ attached to the former weight to model the different contribution each author has made.

- Venue-paper network $G_{cd} = (V_c \cup V_d, E_{cd})$

$$W_{dc}(i,j) = \begin{cases} 1, & d_i \textit{ published at } c_j \\ 0, & \text{otherwise} \end{cases} \tag{6}$$

$$W_{cd}(i,j) = \begin{cases} \left|c_{ik} \mid t(c_{ik}) < t(c_{im})\right| + 1, & d_j \in \Gamma_{cd}(c_{im}) \\ 0 & \text{, otherwise} \end{cases} \tag{7}$$

A paper can only be published in one venue, so in equation 6 we give a weight of 1 for $W_{dc}(i,j)$ if the paper d_i is published in venue c_j and 0 otherwise.

In equation 7, we bring in the time factor. c_{ik} is one proceeding of the conference c_i, or one volume of the journal c_i. $\Gamma_{cd}(c_{im})$ is the paper set which are published on c_{ik}, and $t(c_{im})$ is the corresponding time of c_{im}. We have $\Gamma_{cd}(c_{ik}) = \{d \mid t(d) = t(c_{ik}) \wedge d \in \Gamma_{cd}(c_i)\}$, where $\bigcup_k \Gamma_{cd}(c_{ik}) = \Gamma_{cd}(c_i)$. Obviously, $\Gamma_{cd}(c_{ik}) \cup \Gamma_{cd}(c_{il}) = \emptyset, \forall k, l, t(c_{ik}) \neq t(c_{il})$.

Lasting, we apply a scoring function Φ to weight matrix W:

$$W(i,j) = \Phi\big(W(i,j)\big) \tag{8}$$

Note that Φ should be an increasing function with a decreasing first derivative: $\Phi''(x) > 0$ and $\Phi'(x) \leq 0$. In other words, the function should retain the ordering of the scores in W so that new references still score higher than old references but it should reduce the differences between scores which are too high. We conduct our experiments with $\Phi(x) = \sqrt{x}$.

Now we give the sub transition probability matrix of each sub graph, after which we give the final transition probability matrix of the academic network.

- Paper citation network G_{dd}

The transition probability matrix from paper to paper is M_{dd},

$$M_{dd}(i,j) = P(d_j \mid d_i) = \frac{W_{dd}(i,j)}{\sum_k W_{dd}(i,k)}(i,j \in V_d) \tag{9}$$

- Author-paper network G_{ad}

The transition probability matrix from author to paper is M_{ad},

$$M_{ad}(i,j) = P(d_j \mid a_i) = \frac{W_{ad}(i,j)}{\sum_k W_{ad}(i,k)}(i \in V_a, j \in V_d) \tag{10}$$

The transition probability matrix from paper to author is M_{da},

$$M_{da}(j,i) = P(a_i \mid d_j) = \frac{W_{da}(j,i)}{\sum_k W_{da}(j,k)}(i \in V_a, j \in V_d) \tag{11}$$

- Venue-paper network G_{cd}

The transition probability matrix from paper to venue is M_{dc},

$$M_{dc}(i,j) = P(c_j \mid d_i) = \frac{W_{dc}(i,j)}{\sum_k W_{dc}(i,k)}(i \in V_d, j \in V_c) \tag{12}$$

The transition probability matrix from venue to paper is M_{cd},

$$M_{cd}(j,i) = P(d_i \mid c_j) = \frac{W_{cd}(j,i)}{\sum_k W_{cd}(j,k)}(i \in V_d, j \in V_c) \tag{13}$$

So the transition matrix of academic network is:

$$M(G) = (P(i,j))_{i,j \in V} = \begin{bmatrix} \alpha_{dd} M_{dd} & \alpha_{da} M_{da} & \alpha_{dc} M_{dc} \\ M_{ad} & 0 & 0 \\ M_{cd} & 0 & 0 \end{bmatrix} \tag{14}$$

Here, $\alpha_{da} + \alpha_{dc} + \alpha_{dd} = 1$.

Let $\mathbf{d}$ be the ranking vector for papers, $\mathbf{a}$ be the ranking vector for authors, $\mathbf{c}$ be the ranking vector for venues. We set vector π as $\pi = [\mathbf{d}^{\mathbf{T}}, \mathbf{a}^{\mathbf{T}}, \mathbf{c}^{\mathbf{T}}]^{\mathbf{T}}$, then the random walk on the academic network could be defined as:

$$\pi_{t+1} = \mathbf{M}^{\mathbf{T}} \pi_t \tag{15}$$

2.2 Analysis of User Generated Content

The user-paper bookmark system on the social sharing and tagging web sites is defined as B $= (U, D, T, R)$, where U is the user set, D is the paper set, T is the timestamp set, and $R \subseteq U \times D \times T$ stands for the bookmark relationship set. $(u, d, t) \in R$ means that the user u bookmarked paper d at time t.

User-paper bookmark network is a bipartite graph. The vertices in this graph are the users and the papers. The edges indicate the bookmark relationship between users and papers. Inspired by [11], we propose an algorithm based on the HITS. Because only users can bookmark papers, in this case, users will only receive hub score which shows the expertise, and the papers will only receive the authority score which shows the quality. We define $\mathbf{e}$ as a vector of expertise scores of users, and $\mathbf{q}$ as a vector of quality scores of papers.

The algorithm is based on the two assumptions:

- an expert should be one who tends to identify high quality papers before other users discover them,
- there should be a relationship of mutual reinforcement between the expertise of a user and the quality of a paper.

Based on these assumptions, we define the adjacency matrix:

$$A(i,j) = \left| u \mid (u, d_j, t), (u_i, d_j, t_{ij}) \in R \wedge t < t_{ij} \right| + 1 \tag{16}$$

We also apply a scoring function $\Psi(x) = \sqrt{x}$ to adjust the value of A:

$$A(i,j) = \Psi\big(A(i,j)\big) \tag{17}$$

Based on this matrix, the calculation of expertise and quality scores is an iterative process similar to that of the HITS algorithm:

$$\mathbf{q} = \mathbf{e} \times A \tag{18}$$

$$\mathbf{e} = \mathbf{q} \times A^T \tag{19}$$

2.3 Random Walk with Restart

The basic idea of random walk with restart[14] on our academic network can be expressed as (combined with equation 15):

$$\pi_{t+1} = c\mathbf{M}^{\mathrm{T}}\pi_t + (1-c)\mathbf{Q} \tag{20}$$

The restart vector $\mathbf{Q}$ is built from the analysis result of user generated contents. Since the user bookmark behaviour cannot cover all the papers, we use the following formula to form the restart vector:

$$\mathbf{Q} = \begin{cases} e^{\mathbf{q}(i)} & , i \in V_d \wedge i \text{ is bookmarked by some user} \\ 1 & , \text{otherwise} \end{cases} \tag{21}$$

the vector $\mathbf{Q}$ needs to be normalized in order to satisfy $\sum_{i \in V} \mathbf{Q}(i) = |V|$.

Then the process iteratively runs equation 20 until convergence. After that, we can get the ranking result of papers, authors and venues.

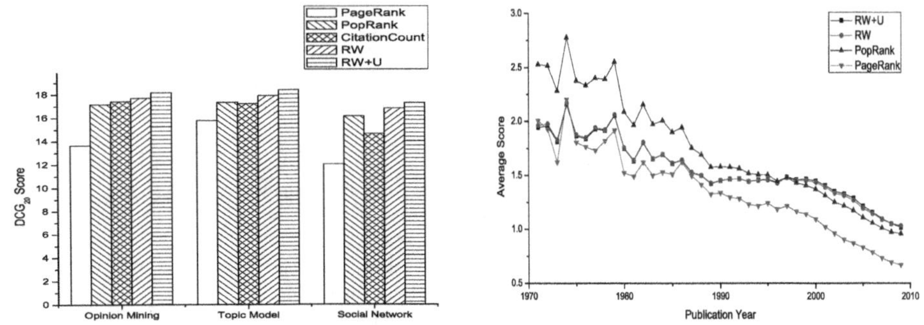

(a) (b)

Fig. 2. (a)Paper ranking result;(b)Average score

3 Evaluation

3.1 Datasets

We collected 603,530 ACM papers in computer science, published until 2009. There are 476,562 authors and 5,851 venues related to the paper dataset.

We also collected about 1,801,254 user bookmarks from citeulike.org from November 4, 2004 to May 18, 2009, which are related to 1,537,998 scientific papers and 47,431 users. After matching with our paper set, we finally obtained 65,949 papers in our paper set, which are related to 128,874 user bookmarks.

3.2 Evaluation Preparation

We chose three domains: "Opinion Mining", "Topic Model" and "Social Network" to evaluate the ranking result of papers and authors.

3.3 Paper Rankings

We used a well-known metric, the Discounted Cumulated Gain(DCG)[6], in order to compare the five different rankings of papers, as following:

$$DCG_p = score_1 + \sum_{i=2}^{p} \frac{score_i}{\log_2 i} \tag{22}$$

$score_i$ is the user given score of the i-th item in the ranking result.

Top 20 papers according to each ranking are merged in a single list, shuffled and submitted for human judgment. 7 CS graduate student volunteers participated in evaluation and there are 3 feedbacks for one query in average. Scores of 0, 1, 2, and 3(the higher, the more agreeable) are collected to reflect the judges' opinion with regard to whether a paper is ranked top 20 in a certain field.

Table 1. Top documents in the topic *opinion mining*

	Title	Authors	Venue	Year
1	Thumbs up Sentiment Classification using Machine Learning Techniques	Bo Pang, Lillian Lee and Shivakumar Vaithyanathan	EMNLP	2002
2	Mining the Peanut Gallery: Opinion Extraction and Learning Techniques	Kushal Dave, Steve Lawrence, David M.Pennock	WWW	2003
3	Mining and summarizing customer reviews	Minqing Hu and Bing Liu	KDD	2004
4	A Sentimental Education: Sentiment Analysis Using Subjectivity Summarization Based on Minimum Cuts	Bo Pang and Lillian Lee	ACL	2004
5	Opinion Observer: Analyzing and Comparing Opinions on the Web	Bing Liu, Minqing Hu and Junsheng Cheng	WWW	2005
6	Recognizing Contextual Polarity in Phrase-Level Sentiment Analysis	Theresa Wilson, Janyce Wiebe and Paul Hoffmann	EMNLP	2005
7	Determining the sentiment of opinions	Soo-Min Kim and Eduard Hovy	COLING	2004
8	Seeing stars: Exploiting class relationships for sentiment categorization with respect to rating scales	Bo Pang and Lillian Lee	ACL	2005
9	Opinion Mining and Sentiment Analysis	Bo Pang and Lillian Lee	FTINR	2008
10	Sentiment Analyzer: Extracting Sentiments About A Given Topic Using Natural Language	Jeonghee Yi, Tetsuya Nasukawa,et al.	ICDM	2003

Fig.2(a) shows that our proposed method, RW+U, outperforms all other baseline methods. This is because user generated content reflects the mutual reinforcing relationship between users' expertise and papers' quality, which is modeled by an algorithm similar to HITS. RW+U makes use of random walk with restart framework to incorporate the previous paper quality result by setting it as restart vector. This result also gives us an understanding of the impact of incorporating the user generated content (found by the difference in the performance of RW and RW+U) and that of incorporating the temporal factor (emphasized by the difference in the performance of PopRank[10] and RW).

Fig.2(b) shows the average scores of papers over a year. Our method gives the highest average score to the papers from 2000 to 2010. This is because our method takes temporal factor into consideration while all other methods rely largely on the citation analysis, which is unfair for latest publications.

In table 1, we list the top 10 papers ranked by our proposed method on the topic opinion mining. Our proposed method can enhance the ranking of latest publications by taking temporal factor into consideration and thus the result is reasonable.

To verify that our ranking results are appropriate, we used the well known "ACM SIGMOD Test of Time Award". The evaluation steps are as follows:

1. Get the papers which are in proceedings of SIGMOD conferences,
2. Sort the rank sub lists grouped by year,
3. Check the position of the awarded papers in the above rank sub list.

To get an overall view of the performance of the ranking methods, we sum the positions of the awarded publications in the last row. The smaller this sum, the better the performance of the ranking method, shown in table 2.

Table 2. Rank positions of papers awarded with the SIGMOD Test of Time Award

Year	Title	Pa	Po	CC	R	RU
2010	NiagaraCQ: A Scalable Continuous Query System for Internet Databases. Jianjun Chen, David J. DeWitt, Feng Tian, and Yuan Wang	1	1	1	1	1
2009	Approximate Computation of Multidimensional Aggregates of Sparse Data Using Wavelets. Jeffrey Scott Vitter and Min Wang	4	4	2	4	4
2008	Integration of Heterogeneous Databases without Common Domains Using Queries Based on Textual Similarity. William W. Cohen	8	6	9	6	6
2007	Online Aggregation. Joseph M. Hellerstein, Peter J. Haas, and Helen J. Wang	2	1	1	2	2
2006	BIRCH: An Efficient Data Clustering Method for Very Large Databases. Tian Zhang, Raghu Ramakrishnan, and Miron Livny	1	1	1	1	1
	Implementing Data Cubes Efficiently. Venky Harinarayan, Anand Rajaraman, and Jeffrey D. Ullman	2	2	2	2	2
2005	View Maintenance in a Warehousing Environment. Yue Zhuge, Hector Garcia-Molina, Jennifer Widom, Joachim Hammer	5	6	6	5	5
2004	From Structured Documents to Novel Query Facilities. Vassilis Christophides, Serge Abiteboul, Sophie Cluet, and Michel Scholl	5	5	4	4	3
	Shoring Up Persistent Applications. Michael J. Carey, David J. DeWitt, Michael J. Franklin et al.	2	2	1	2	2
2003	Mining Association Rules Between Sets of Items in Large Databases. Rakesh Agrawal, Tomasz Imielinski, and Arun Swami	1	1	1	1	1
		31	29	28	28	**27**

3.4 Author Rankings

Fig.3(a) shows the evaluation result. This reveals the fact that the proposed method outperforms the other ranking methods. Among all the algorithms, publication count performs the worst. This coincides with the actual case.

We list the top 5 authors ranked by our proposed method on the topic *opinion mining*: Bing Liu, Lillian Lee, Janyce M. Wiebe, Bo Pang and W. Niblack. Based on our domain knowledge, the result is reasonable.

3.5 Venue Rankings

The modified impact factor(mIF) [15] was adopted. It is defined as follows:

$$mIF_X = \frac{C}{D} \tag{23}$$

where C is the number of times that articles of X were cited, and D is the total number of articles in X.

We use $P@N$ as the evaluation method:

$$P@N = \frac{|E \cap F|}{N} \tag{24}$$

where E are the top N journals of the proposed method, and F are the top N journals of mIF.

From Fig.3(b) we can see that PageRank performs the worst. The performance of RW is almost the same with PopRank, this is because temporal factor does

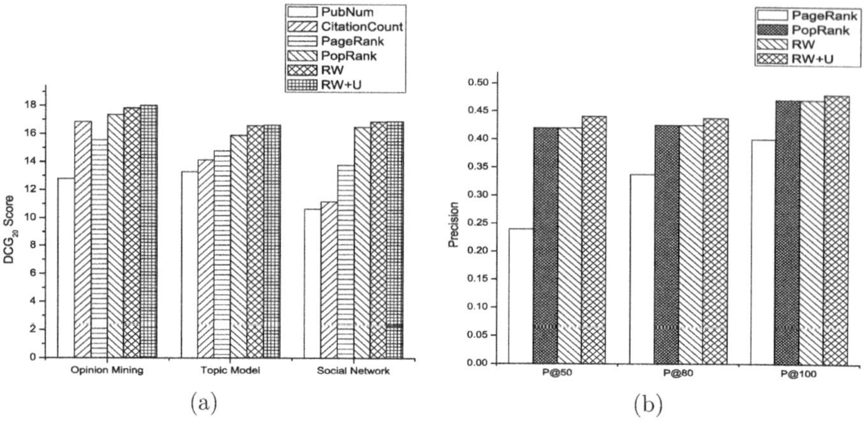

Fig. 3. (a)Author ranking result; (b)Journal ranking result

not influence the result of venue ranking as all the venues we evaluate has been in existence for a long time. RW+U algorithm performs better than RW, because user generated content provides information about publication quality, which directly influences the quality of venues in our method. The precisions of the four algorithms are all quite low, this is because all these methods are link-based and this is quite different from the citation based mIF method.

4 Conclusions and Future Work

In this paper, we have presented an approach that jointly ranks publications, authors and venues. We first constructed a heterogeneous academic network which is composed of publications, authors and venues. Then we performed random walk over the network hence yielding global ranking result of the objects on the network. When considering the transitional probabilities among these objects, we took the temporal factor into account. Besides, we modeled the mutual reinforcing relationship between user expertise and publication quality based on users' bookmarks. We adopted the random walk with restart framework to integrate the publication quality. Experimental results with ACM dataset show that the performance of our proposed method outperforms all other baseline algorithms, such as Citation Count, PageRank, and PopRank. In the future, we intend to incorporate more user generated content information such as users' click behavior and users' opinions (which can be analyzed from online reviews towards publications) into our framework.

References

1. Bateman, S., Brooks, C., Mccalla, G.: Applying collaborative tagging to e-learning. In: Proc. of ACM WWW, vol. 3(4) (2007)
2. Brin, S., Page, L.: The anatomy of a large-scale hypertextual Web search engine. Computer Networks and ISDN Systems 30(1-7), 107–117 (1998)

3. Du, M., Bai, F., Liu, Y.: PaperRank: A Ranking Model for Scientific Publications. In: Proceedings of the 2009 WRI World Congress on Computer Science and Information Engineering, pp. 277–281 (2009)
4. Garfield, E.: Citation analysis as a tool in journal evaluation. Science 178, 471–479 (1972)
5. Gollapalli, S.D., Mitra, P., Giles, C.L.: Ranking authors in digital libraries. In: JCDL, pp. 251–254. ACM, New York (2011)
6. Järvelin, K., Kekäläinen, J.: IR evaluation methods for retrieving highly relevant documents. In: Proceedings of the 23rd ACM SIGIR, pp. 41–48 (2000)
7. Kleinberg, J.M.: Authoritative sources in a hyperlinked environment. Journal of the ACM 46(5), 604–632 (1999)
8. Lempel, R., Moran, S.: The stochastic approach for link-structure analysis (salsa) and the tkc effect. ACM Transactions on Information Systems, 387–401 (2000)
9. Li, Y., Tang, J.: Expertise Search in a Time-Varying Social Network. In: Proceedings of the Ninth International Conference on Web-Age Information Management, pp. 293–300 (2008)
10. Nie, Z., Zhang, Y., Wen, J.R., Ma, W.Y.: Object-level ranking: bringing order to Web objects. In: Proceedings of the 14th WWW, pp. 567–574 (2005)
11. Noll, M.G., et al.: Telling experts from spammers: expertise ranking in folksonomies. In: Proceedings of the 32nd ACM SIGIR, pp. 612–619 (2009)
12. Sayyadi, H., Getoor, L.: FutureRank: Ranking Scientific Articles by Predicting their Future PageRank. In: Proceedings of the 9th SIAM International Conference on Data Mining, pp. 533–544 (2009)
13. Tang, J., Jin, R., Zhang, J.: A Topic Modeling Approach and Its Integration into the Random Walk Framework for Academic Search. In: Proceedings of ICDM, pp. 1055–1060 (2008)
14. Tong, H., Faloutsos, C., Pan, J.Y.: Fast Random Walk with Restart and Its Applications. In: Proceedings of the Sixth ICDM, pp. 613–622 (2006)
15. Yan, S., Lee, D.: Toward alternative measures for ranking venues: a case of database research community. In: Proceedings of the 7th ACM/IEEE-CS Joint Conference on Digital Libraries, pp. 235–244 (2007)
16. Yang, Z., Hong, L., Davison, B.D.: Topic-driven multi-type citation network analysis. In: The 9th International Conference on Adaptivity, Personalization and Fusion of Heterogeneous Information, RIAO 2010 (2010)

On Modeling Virality of Twitter Content*

Tuan-Anh Hoang, Ee-Peng Lim, Palakorn Achananuparp, Jing Jiang, and Feida Zhu

School of Information Systems, Singapore Management University

Abstract. Twitter is a popular microblogging site where users can easily use mobile phones or desktop machines to generate short messages to be shared with others in realtime. Twitter has seen heavy usage in many recent international events including Japan earthquake, Iran election, etc. In such events, many tweets may become viral for different reasons. In this paper, we study the virality of socio-political tweet content in the Singapore's 2011 general election (GE2011). We collected tweet data generated by about 20K Singapore users from 1 April 2011 till 12 May 2011, and the follow relationships among them. We introduce several quantitative indices for measuring the virality of tweets that are retweeted. Using these indices, we identify the most viral messages in GE2011 as well as the users behind them.

1 Introduction

1.1 Objectives

Events that attract social media attention often see their information becoming highly popular through word-of-mouth. These highly popular information are *viral*. Examples of viral information include viral videos, viral blog posts, viral games, etc.. Viral information may have accounted for most of the internet traffic consumption as they get pass on from people to people. Businesses are also trying to tap on viral information to promote brand and product awareness and this has resulted in several viral marketing techniques.

In this paper, we focus on viral content in Twitter. Twitter has seen very heavy usage in many countries due to its very low barrier to messaging posting and realtime dissemination. In Twitter, users *follow* other users to receive short text messages also known as *tweets* from the latter users. Tweets can be posted via mobile phone or desktop machine. To study viral content in Twitter during the Singapore's GE2011 event, we developed the *Palanteer* search engine[1] that offers tracking and search of GE2011 tweets. Palanteer performs daily crawl of tweets from 1 April 2011 till 12 May 2011 generated by users located in Singapore. The election date of GE2011 was May 7, 2011. Since GE2011 is a socially interesting

* We would like to acknowledge that this research was carried out at the Living Analytics Research Centre (LARC), sponsored by Singapore National Research Foundation and Interactive & Digital Media Programme Office, Media Development Authority.

[1] http://palanteer.sis.smu.edu.sg

C. Xing, F. Crestani, and A. Rauber (Eds.): ICADL 2011, LNCS 7008, pp. 212–221, 2011.

event, we expect viral information to exists in the Twitter data generated by the event.

In this paper, we thus propose the following research objectives related to viral content and highlight the corresponding contributions:

- **Measurement of viral content:** One may measure viral information by their popularity, i.e., the number of users adopting the piece of information. Popularity however does not consider the word-of-mouth, or user-to-user propagation. It is therefore important to consider other quantitative measures of viral content. One major contribution of this paper is to introduce several new measures that use retweeting structure to quantify the virality of tweet information.
- **Analysis of viral content:** Using our proposed virality measures, we analyze the GE2011 Twitter data. We looks out for viral tweets propagated among Singapore based users, as well as viral politicians and topics found in these tweets. This allows us to understand the foci of attention of users in the event.
- **Measurement and analysis of viral behavior:** Viral information can be associated with users who generate them and users who propagate them. Users who generate viral information are likely influential in the Twitter network. In this analysis, we focus on measuring and analyzing these users.

There are several useful applications of virality measures and analysis. Viral marketing techniques can apply virality to measuring the success of viral advertisements or new product diffusion within the Twitter network. By tuning the advertisements and changing product designs, one may improve virality which in turns improves the marketing outcome. Search engines can utilize virality to rank information so as to return more socially interesting search results. Finally, we may use virality to identify influential users in the Twitter network related to some social events. These may also be the users one should track closely for future events.

2 Related Works

The term virality originated from medical science to describe the ability of viruses to spread among organisms. Emerged from economics and social studies, viral behavior of an object is one that indicates how good it is at "network-enhanced word of mouth" [5]. Marketing community later created viral marketing strategies that focus on designing messages that grow awareness of products or advertisements within a targeted customer community[4,11]. These previous works focus on largely qualitative study instead of quantitative modeling. The effects of network is also often neglected.

Leskovec, Adamic and Huberman analysed the viral effect of customers recommending products in a large user-user recommendation network over time[6]. They observed that most customers do not propagate product recommendation after purchases. Only very few products enjoy active recommendation within small communities. The recommendation network is quite different from Twitter network which involves little cost for users to generate and propagate tweets. The work also did not introducing any quantitative measure. Ienco, Bonchi and Castillo proposed to rank messages for users to consume so as maximize the overall user activeness in the network[4]. They however focused on the optimization of the selection of a small subset of messages instead of measuring the message virality.

3 Twitter Dataset

Data crawling. Using the *Palanteer* search engine, we collected around three weeks of Twitter data before the Singapore's general election on May 7, 2011. The data crawling selected a seed user set U_0 which contains 58 Twitter users who have dominant interests in GE2011. They include the political party accounts (e.g., `PAPSingapore` and `wpsg`), politician accounts (e.g., `georgeyeo`) and political commentators (e.g, `temasekreview`). We then derived the followers and followees of users from U_0 and created a larger set of users denoted by U. All tweets generated by U during the three weeks are crawled.

Dataset selection. As there are tweets in the crawled data that are not related to GE2011, we hence extracted a set of tweets generated by U containing the 5 most frequently mentioned politician names during the election, namely, George Yeo, Tin Pei Ling, Nicole Seah, Chiam See Tong and Low Kia Thiang. The first two are members of the ruling party PAP while the latter three belong to opposition parties NSP, SPP and WP respectively. The statistics of this dataset (known as the 5-Politician dataset) is shown in Table 1:

Table 1. Statistics of 5-Politician Dataset

# tweets	# users	# retweets	# retweeting users
32,696	4718	11,436	2994

Inferred Retweets. In the 5-Politician dataset, we realised that many retweets are in response to tweets not covered in the dataset. For example, our dataset contains a tweet with content, ``RT @djkoflow: NSP at lavender foodcourt!!! Nice will i see nicole seah?'' which is a retweet of another tweet by `@djkoflow` not found our database. In this case, we inferred that the tweet ``NSP at lavender foodcourt!!! Nice will i see nicole seah?'' actually exists based on the former tweet. There is also an inferred retweeting relationship between these two tweets. By introducing inferred retweets, we increased the number of retweets by 48% from 11,436 to 16,932. We define an *original* tweet to be one that is not a retweet. We obtained 4096 original tweets that have been retweeted. The increase in retweets allows us to have a more complete set of data for modeling the virality of tweet messages.

4 Virality Models

In this section, we present some models for measuring the virality of tweets, topics, and users. Each model is derived based on a different set of heuristic principles. Each model assigns a numerical virality value to an entity of some type. We use $mscore(m)$, $tscore(t)$, and $uscore(u)$ to denote the virality value of the tweet m, topic t, and user u respectively. A standard set of notations used in the definition is given in Table 2.

Table 2. Notations

$M/T/U$	set of tweets/topics/users
$u(m)$	original author of the tweet m
$U_R(m)$	set of users retweeting the tweet m
$W(m)$	set of non-stop words in tweet m
$M(t)$	set of (original) tweets in topic t
$M(u)$	set of (original) tweets tweeted by user u
$M_R(u)$	set of (original) tweets retweeted by user u
$Follow(u)$	set of followers of user u
$mscore(m)$	viral score of tweet m
$tsore(t)$	viral score of topic t
$usore(u)$	viral score of user u

4.1 Viral Tweets

Virality by retweet count. Like viruses, a viral tweet should be one having ability of being replicated on as many people as possible. In the Twitter context, retweet is analogous to replication. Therefore, it is natural to use the frequency of a tweet being retweeted to measure its virality. Since Twitter allows a user to retweet a tweet at most once, the number of times a tweet is retweet is simply the number of users retweeting it. That is,

$$mscore_{rtc}(m) = |U_R(m)| \tag{1}$$

Virality by retweet likelihood. The retweet count model tells us the popularity of a tweet, but does not tell us how good the tweet is propagated through the social network structure. Consider the two original tweets, m_1 by user u_1 and m_2 by user u_2. Suppose that only 4 out of 20 followers of u_1 retweeted m_1, and all 3 followers of u_2 retweeted m_2. By retweet count model, m_1 is more viral than m_2. However, one may consider m_2 to be more viral since all of its receivers have retweeted while only a small fraction of users receiving m_1 actually retweet m_1.

We therefore propose an alternative model that measures the virality of a tweet m by its retweet likelihood which is computed by the fraction of users retweeting m over all users receiving m.

$$mscore_{rtl}(m) = \frac{|U_R(m)|}{|\bigcup_{u \in U_R(m) \cup \{u(m)\}} Follow(u)|} \tag{2}$$

Note that our dataset was crawled from a subset of Twitter users at most one hop away from the set of 58 core users in the follow graph. The dataset hence does not have complete follower information for all the non-core users. Given this incomplete follow information and that follow links are rather weak relationships, we decide to use the set of users interested in u instead of the set of u's followers. Furthermore, a user can may receive a tweet by u through public timeline and generate a retweet even when he/she does not follow u.

We say that user u_1 is interested in user u_2 if u_1 mentions "@u_2" in at least k of u_1's tweets. In our experiments, we set $k = 1$. This means u_1 mentions u_2 at least once or u_1 retweets at least one tweet from u_2.

The Equation 2 is now modified as follows:

$$mscore_{rtl}(m) = \frac{|U_R(m)|}{|\bigcup_{u \in U_R(m) \cup \{u(m)\}} \{u' : u' \text{ interested in } u\}|} \tag{3}$$

Virality by a combined model. The virality by retweet likelihood model has a shortcoming since it favors tweets published by users having fewer followers while ignoring the tweet popularity among retweeting users. Therefore, we propose a combined model incorporating both retweet count and retweet likelihood.

$$mscore_c(m) = mscore_{rtc}(m) \cdot mscore_{rtl}(m) \tag{4}$$

4.2 Viral Topics

Topic construction. A topic is a group of tweets of similar content. We assume that each tweet contains only one topic and adopt a clustering approach to construct topics. We take a modularity based clustering algorithm [3] on a *tweet graph* so as to derive subgraphs representing topics. The tweet graph consists of original tweets as nodes, and pairs of tweets with overlapping terms as edges. Each edge is weighted by the similarity of the two corresponding tweets. To compute similarity between two tweets, we use the "bag-of-words" representation of each tweet after removing tweet-encoding terms, e.g. @, *RT*, and *via*, common stop words and internet slang words[7]. The remaining words form a word vocabulary W. The similarity between tweet m_1 and tweet m_2 is computed by Jaccard coefficient as follows. At the end of clustering, we obtained 69 topics.

$$similarity(m_1, m_2) = \frac{|W(m_1) \cap W(m_2)|}{|W(m_1) \cup W(m_2)|} \tag{5}$$

Note that $similarity(m_1, m_2)$ returns a value between 0 and 1. We remove from the tweet graph all edges with weight less than 0.25 to reduce noise in topic construction.

Modularity $q(C)$, proposed by Newman[9], is an objective function that measures the quality of cluster results C of a network. $q(C)$ falls within the range $[-1, 1]$, and the larger $q(C)$ the better is the resultant clusters C. Maximization of $q(C)$ is proven to be a $NP-hard$[1] problem. Some top-down and bottom-up heuristic and approximate algorithms for this problem were proposed by

Newman et. al [9], [3], [8]. In our work, we choose the agglomerative method to determine clusters as topics since it outperforms the top-down method as reported in [3].

Topic virality. We employed a simple principle to measure virality of each topic. Topic with a lot of viral tweets is viral. Then, the virality of topic t is simply the sum of virality of all tweets in t. That is

$$tscore(t) = \sum_{m \in M(t)} mscore(m) \tag{6}$$

Similar to modeling the virality of tweet, we derive three models to measure topic virality based on three different tweet virality models. Due to space constraint, we shall just show the results using the combined model, i.e., $mscore = mscore_c$.

4.3 Viral Users

There are two basic modeling approaches for viral users, namely the *author-take-all* and the *shared contribution* models.

"Author take all" model. This approach measures the virality of a user by how viral her tweets are. In this model, we consider the virality of a tweet can be fully attributed to its original author but none to the users retweeting the tweet. In other words, retweeting users does not get any share of the virality of the tweet. With this assumption, the virality of user u is the sum of virality of all original tweets of u. That is,

$$uscore(u) = \sum_{m \in M(u)} mscore(m) \tag{7}$$

Depending on how we measure tweet virality, we can have three corresponding ways to define the virality of user. In this paper, we shall only present the one using combined virality model.

$$uscore_c(u) = \sum_{m \in M(u)} mscore_c(m) \tag{8}$$

"Shared contribution" model. Sometimes, a tweet may not be viral until it is retweeted by an influential user who attracts a large number of subsequent retweets by her followers. To account for the contribution of retweeting users to tweet virality, we propose the *Share contribution* model that measures the virality of a user u by including u's contribution to all tweets' virality (including retweets). Given an original tweet m and a user u, the contribution of u to virality of m is proportional to number of retweets on the tweet (if u is author of m) or on the u's retweet of m (if u retweets m). That is,

$$uscore(u) = \sum_{m \in M(u) \cup M_R(u)} \left(\frac{mscore(m)}{1 + |U_R(m)|} \cdot (1 + |U_R(m) \cap Follow(u)|)\right) \tag{9}$$

Similarly, we have three ways to compute the virality of users corresponding to three models mearsuring tweet virality. Furthermore, as mentioned in Section 4.1, we slightly modify Equation 9 by using the users interested in u instead of u's followers. In the following, we show the one using the combined model.

$$uscore_c(u) = \sum_{m \in M(u) \cup M_R(u)} \frac{mscore_c(m)}{1 + |U_R(m)|} \cdot (1 + |U_R(m) \cap \{u' : u' \text{ interested in } u\}|) \quad (10)$$

5 Data Analysis

We now apply our virality measures on the 5-Politican dataset. Our aims is two-fold. Firstly, we want to analyze the differences between measures for tweet messages, topics and users. This comparison helps us to understand the characteristics of these measures. Secondly, we want to determine the highly viral messages, topics and users in GE2011 and provide some empirical analysis on them. The analysis will lead us to understand the socio-political issues that interest the Singapore electorates. In some cases, these issues may not appear within the radar of traditional mainstream media.

5.1 Tweet Analysis

According to the combined virality measure $mscore_c$, the following are the most viral tweets that have been both retweeted by many users and have experienced high retweet likelihood. The combined virality measure is adopted due to its balanced consideration of both retweet count and likelihood. The $mscore_c$, number of retweets each of them had generated and the retweet likelihood are shown in normal typeface, boldface and italized numbers respectively before each tweet.

1. (7.50, **127**, *0.67*): `Retweet if you want George Yeo to remain in Parliment. #sgelection`
2. (7.41, **206**, *0.41*): `Join us in our salute to George Yeo and Chiam See Tong for their service to Singapore. RT this. #sgelections #sgelection`
3. (6.46, **172**, *0.43*): `Retweet if you are sad that Tin Pei Ling is elected more than you are sad that Nicole Seah is not elected. #sgelections`
4. (5.98, **161**, *0.42*): `Tin Pei Ling is in Parliament..God Bless Singapore`
5. (4.58, **138**, *0.38*): `Retweet this if you love Mr Chiam See Tong, the people's humble leader who sat in his aluminum cubicle all these years for his people.`
6. (3.61, **48**, *0.86*):`If only we can exchange Chiam See Tong and George Yeo for TPL and WKS:( #sosingaporean`
7. (2.43, **69**, *0.40*): `I'm happy that Tin Pei Ling in our govt. Now everyones truly represented. Even the intellectually disabled :) #sgel ...`
8. (2.21, **85**, *0.30*): `To all those regretting the loss of George Yeo, Chiam ST, the election of TPL, remember: it's the GRC system that caused ...`
9. (1.86, **108**, *0.20*): `With extra $16000 per month in her pocket now Tin Pei Ling can upgrade from Kate Spade to Hermas. #GeorgeYeoInTinPeiLingOut #SGElections`
10. (1.68, **28**, *0.68*): `@Xiaxue my 4yr old daughter is gonna be so sad when she knows that George Yeo, the uncle that gave her a lollipop, lost.. ;(`

All the above viral tweets were generated right after election results were known and they expressed the sentiments from the users, including giving encouragement to George Yeo and Chiam See Tong, the two popular politicians who lost their parliamentary seats after election, and satirical remarks to Tin Pei Ling who suffered from online flaming throughout her election campaign. In particular, the topmost viral message shows support to George Yeo despite his loss.

Among the top 10 tweets, one can find some with not many retweets (28 retweets) but high retweet likelihood. This suggests that such tweets attracted attention of followers of every user who tweets or retweets them. For example, the 10th most viral tweet, ``@Xiaxue my *(omitted)*'', has observed a total of 41 users receiving it, out of which 28 have retweeted. This is a significantly high retweet likelihood making the tweet very viral.

On the other hand, we can also find another tweet ``WP's 5-man Aljunied team have been garlanded at Hougang Stadium. Low Thia Khiang thanking supporters #sgelections'' having similar number of retweets (33 retweets) but much lower retweet likelihood. This resulted in the tweet being ranked 1217th, much lower than the earlier example. The low retweet likelihood is caused by a total of 806 users receiving the tweet but only 33 retweeted. In other words, users are generally not interested to pass on this retweet to their followers.

5.2 Topic Analysis

We next analyze the viral topics in 5-Politician dataset using the model proposed in Section 4. The combined message virality model is used in deriving the topic virality measure. We rank all the 69 constructed topics by decreasing order of virality values and show the top 10 viral topics in Table 3. The top viral topics are dominated by topics with large number of original tweets and retweets.

Even among the top viral topics, the top five appear to have significantly larger virality scores compared with the lower five. This observation is also supported by the number of original tweets and number of retweets. The topmost topic is largely about encouraging George Yeo to stand as a Presidential candidate upon the loss of his parliamentary seat. A total of 1115 original tweets followed by 4600 retweets were generated on this topic. The second most viral topic covers the suggestion to replacement of Tin Pei Ling who won the election by George Yeo. Again, this topic generates many original tweets and retweets. Similar observations can be made for the remaining viral topics.

5.3 User Behavior Analysis

Now, we analyse the virality behaviors of users in our dataset. Table 4 shows the top 10 viral users by virality scores using author-takes-all (ATA) and shared contribution (SC) models. The two models return different but overlapping sets of top users. The ATA model favors users who contribute many highly viral messages while the SC model favors users who also contribute to retweets of viral messages. The highly viral users include information aggregator users (e.g., yahoosg, sgelection), mainstream media (e.g., stcom, todayonline, etc.),

Table 3. Top 10 Viral Topics

	Virality Score	# Orig Tweets	# Re tweets	Topic Words
1	23.65	1115	4600	lost,yeo,wp,parliament,chiam,foreign,minister,george,tpl,president, pap,aljunied,grc,tong,sad
2	14.94	828	4077	ling,yeo,pap,lings,mp,goh,spade,george,say,pei,dont,kate,tong,tin,parliament
3	10.92	812	2905	ling,rally,nicole,party,pap,nsps,facebook,tpl,pei,nsp,tin,marine,grc,seah,parade
4	8.62	442	2009	tong,lost,chiam,mr,pasir,you,singaporeans,potong,residents,payoh,respect, crowd,spps,bishantoa,spp
5	2.01	253	1127	wps,thia,low,wp,lim,mao,rally,khiang,yeo,hougang,george,chen,sylvia,grc, aljunied
6	0.95	36	98	did,yeo,salute,mr,ukiss,you,proud,mp,minister,respect,george,election, grc,elections,aljunied
7	0.75	16	132	ling,truly,government,disabled,represented,happy,bwahaha,tt,pei,tin,govt,every, im,everyones,intellectually
8	0.41	29	101	selfblinged,statement,hes,speech,thing,poor,minister,press,yeos,conference,mr, george,commentary,yeo's,why
9	0.29	7	90	really,yeo,wp,politics,hua,lim,quit,announced,george,me,consider,hwee,needs, important,decision
10	0.28	2	89	bts,yeo,wong,getting,kss,still,mistakes,george,tpl,sad,mah,weight,parliament

Table 4. Top 10 Viral Users

User	Rank/Score (ATA)	Rank/Score (SC)
yahoosg	1/90.1	6/246.4
firefliesinajar	2/85.7	
thenooselite	3/80.1	
temasekreview	4/73.7	
fakemoe	5/70.0	2/535.9
merylzhanyee	6/54.5	
stcom	7/49.1	4/284.8
todayonline	8/44.6	1/768.8
sosingaporean	9/41.1	
fake_pmlee	10/39.1	
sgelection		3/419.1
tocsg		5/251.6
ongweizhong		7/237.2
eisen		8/184.3
mrbrown		9/179.5
calvin_lee		10/136.5

satirical users (e.g., `fakemoe` and `fake_pmlee`), and activist users (e.g., `mrbrown`, `thenooselite`, `ongweizhong`, `calvin lee`, etc.). The user `firefliesinajar` ceased to exist at the point this paper was written. Interestingly, the mainstream media users have demonstrated the ability to stay viral compared with other users by generating viral tweets.

6 Conclusion

In a social network, messages passed among users may become viral especially in an event that attracts wide interests. To offer a systematic approach to study virality of content, this paper proposes a few models to measure virality at the tweet, topic and user level. Other than considering the number of retweets of a tweet, we also introduce retweet likelihood as another component in these models. Using the Palanteer search engine, we collected a sizeable collection of Twitter data and applied our proposed models on it. The empirical analysis results demonstrated that the virality models can effectively find the viral content and users. As part of future work, one can extend the virality models for searching Twitter content. This will allow viral content to be returned as search results to meet the users' demand for trending content in social media. Viral messages may be linked to influential users in the network although this relationship has yet to be well studied. Several researchers have studied different ways of calibrating influential users [10,2]. It is thus interesting to study the relationship between influential users and users contributing to viral content.

References

1. Brandes, U., Delling, D., Gaertler, M., Goerke, R., Hoefer, M., Nikoloski, Z., Wagner, D.: On modularity clustering. IEEE Trans. Knowledge and Data Engineering 20(2), 8577–8582 (2008)
2. Cha, M., Haddadi, H., Benevenuto, F., Gummadi, K.P.: Measuring user influence in twitter: The million follower fallacy. In: ICWSM (2010)
3. Clauset, A., Newman, M.E.J., Moore, C.: Finding community structure in very large networks. Phys. Rev. E 70 (2004)
4. Ienco, D., Bonchi, F., Castillo, C.: The meme ranking problem: Maximizing microblogging virality. In: ICDM Workshops (2010)
5. Jurvetson, S.: What exactly is viral marketing? Red Herring 78, 110–112 (2005)
6. Leskovec, J., Adamic, L.A., Huberman, B.A.: The dynamics of viral marketing. ACM TWeb 1(1) (2007)
7. Manning, C., Raghavan, P., Schatze, H.: Introduction to information retrieval (2008)
8. Newman, M.E.J.: Finding community structure in networks using the eigenvectors of matrices. Phys. Rev. E 74 (2006)
9. Newman, M.E.J.: Modularity and community structure in networks. Proc. Natl. Acad. Sci. USA, 8577–8582 (2006)
10. Weng, J., Lim, E.-P., Jiang, J., He, Q.: Twitterrank: Finding topic-sensitive influential twitterers. In: ACM WSDM (2010)
11. Wilson, R.F.: The six simple principles of viral marketing. Web Marketing Today (2005)

Creating a Handwriting Recognition Corpus for Bushman Languages

Kyle Williams and Hussein Suleman

Department of Computer Science,
University of Cape Town,
Private Bag X3, Rondebosch, 7701
{kwilliams,hussein}@cs.uct.ac.za

Abstract. Handwriting recognition systems rely on the existence of a corpus for training recognition models and evaluating accuracy. Creating a handwriting recognition corpus for the Bushman languages of southern Africa is difficult due to the complexities of the script used to represent them and the fact that this script cannot be represented using Unicode. To solve this problem, a semi-automatic Web-based tool was developed to segment, capture and encode the Bushman text. A case study demonstrated how the tool could be used to create a Bushman handwriting corpus with few errors.

Keywords: Corpus creation, transcription, digital libraries.

1 Introduction

The Bleek and Lloyd Collection [1] contains notebooks that document Bushman language, culture and belief. These notebooks have been digitised and digital library systems have been built to preserve them and make them available online [1]. Transcriptions of the text that appears in the notebooks would allow for enhanced ways of interacting with the collection by allowing for the text to be indexed, searched and compared and could be used in speech-to-text applications and allow for the Bushman text to be reprinted in books. However, manual transcription is a time consuming and costly process and is generally not a viable option, thereby motivating the need for an automatic solution. In order to perform automatic transcription, a corpus is needed for training and evaluating automatic recognition systems. A problem arises in that the Bushman languages are complex due to the diacritics that they contain and cannot be represented using Unicode. To solve this problem, a custom Web-based tool was built that relied on automatic algorithms as well as user interaction to create a corpus.

The rest of this paper is structured as follows. Section 2 discusses some related work, followed by a description of the Bleek and Lloyd Collection in Section 3. The Web-based tool that was created to segment, capture and encode the Bushman text is described in Section 4 and a case study in Section 5 demonstrates the use of this tool to create a corpus. The quality of the data captured during the case study is analysed and evaluated in Section 6 and, finally, conclusions are drawn in Section 7.

C. Xing, F. Crestani, and A. Rauber (Eds.): ICADL 2011, LNCS 7008, pp. 222–231, 2011.

2 Related Work

The existence of a high quality corpus is necessary for any system that performs automatic recognition of text, whether handwritten or machine printed. Corpora for English or other well-understood and well-studied languages are relatively easy to access and make use of. Examples of widely-used corpora are the Lancaster-Oslo/Bergen corpus [2] and the George Washington manuscripts [3]. However, for less well-studied languages and scripts, it is often necessary to create a new corpus that can be used for experimentation since it is unlikely that a suitable one already exists.

Creating corpora for modern languages is relatively easy compared to creating corpora for historical languages. Usually, creating corpora for modern languages involves getting a group of users to write words on forms that were specifically designed for creating the corpora. For instance, Agrawal et al [4] used HP TabletPCs to collect data and create a corpus for complex Indic scripts and Al-Ma'adeed et al [5] used paper forms on which specific Arabic words were written and then scanned to create an Arabic database.

The creation of corpora for historical texts is, however, more difficult than that of modern texts. For instance, it is not possible to make use of forms to easily and accurately capture specific data. Historical texts also introduce a number of difficult problems related to the segmentation of lines and words due to poor handwriting and the effects of age, such as ink-bleed and paper degradation. Fischer et al [6] note that, in many cases, transcriptions of historical texts can only be performed by language specialists, whereas for modern texts it can be performed by lay-persons. This is of course open to debate since it is possible to train lay-persons in transcribing the text or, alternatively, use a crowd-sourcing approach, which has, in many cases, been shown to be more accurate than specialists [7].

Fischer et al [6] describe a tool for the creation of a corpus for the IAM Historical Handwriting Database (IAM-HistDB). The specific manuscripts that they make use of all appear to have been scribe-written and are presented in a neat fashion. The tool described by Fischer et al [6] contains algorithms that automatically segment text on a page and that allows for users to manually correct any errors. The transcription of the text is not performed as transcriptions for the text already exist. Instead, transcription alignment is automatically performed and users are able to perform alignment correction.

Setlur et al [8] describe the complexities of the Devanagari script, which is the basis of many Indian languages and the lack of consistency among researchers in terms of the representation of the script, specifically in terms of what the granularity of a character should be. They describe a Java-based tool for creating a corpus for Devanagari script recognition, which automatically segments words and lines and also allows for user interaction. Once lines and words have been segmented, the Devanagari script is then captured using a simulated keyboard or by typing transliterated text on the keyboard. The text is stored in Unicode and the Devanagari output can be previewed by rendering the Unicode using a Devanagari font.

3 The Bleek and Lloyd Collection

The Bushman people are South Africa's oldest human inhabitants [9] and it is likely that they would have had a unique view of the world. However, over time much of their culture and language has been lost. Fortunately though, some of it was recorded and written down by linguists Wilhelm Bleek, Lucy Lloyd and others [1] in the 19th century and this historical collection of notebooks, dictionaries and pieces of art have collectively come to be known as the Bleek and Lloyd Collection [1]. Systems have and are being built for preserving the notebooks [1] and dictionaries [10] as well as for enhancing interaction with the collection [11]. An example of part of a notebook page is shown in Figure 1.

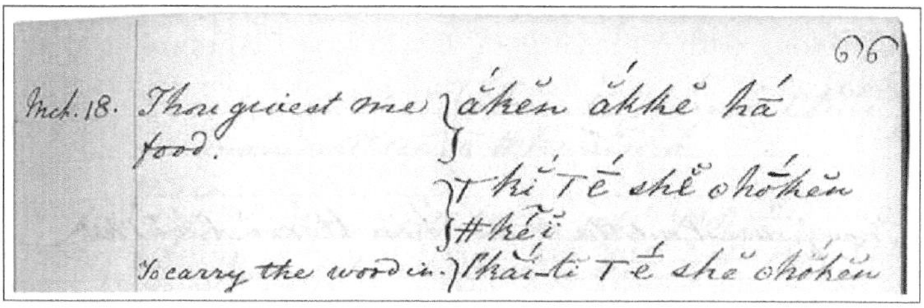

Fig. 1. An example of part of a page in the Bleek and Lloyd Collection

The script used to represent the Bushman text is complex due to the diacritics and the fact that the text cannot be represented using Unicode. Thus far, about 137 different combinations of single and stacked diacritics have been discovered that appear above, below and above and below characters. These diacritics need to be supported by a tool for capturing Bushman text. Table 1 shows some of the types of diacritics that appear in the text, starting with simple diacritics that appear above characters, then diacritics that appear above and below characters and then diacritics that are stacked and that span multiple characters.

3.1 Bushman Text Representation

Since the Bushman text cannot be represented using Unicode, a custom solution for representing the characters was developed. The custom solution involves the use of LaTeX and the TIPA package [12] for processing International Phonetic Alphabet symbols. The TIPA package does not support many of the diacritics that appear in the Bushman text by default. However, it does allow for the creation of custom macros that can be used to create nested and stacked diacritics.

Each Bushman diacritic is represented using a backslash (\) followed by a command that specifies the diacritic and braces ({}) that contain the characters that the diacritics should be added to. Using the custom macros, the Bushman text can be encoded and represented. While the encoding is generally correct, the visual representation is just an approximation. Future work includes the possibility of creating a custom font for Bushman script. Table 1 shows how

Table 1. Encoding and visual representation of Bushman text using custom TIPA macros

Text Line	Encoding	Visual Representation
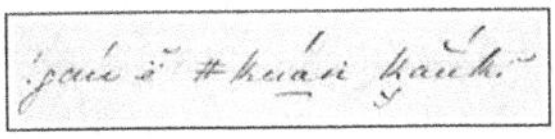	!ga\dialine{u} \twodotsu{e} \texthash{}ku\barblinet{a}n \ybelow{k}a\uline{u}k\u{i}	!gaủ ĕ #kua̱n ḵaůkĭ
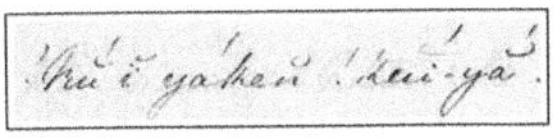	!k\uline{u} \twodotsu{i} y\dialine{a}ke\u{n} !ku\uline{i}-y\uline{a} .	!kŭ ĭ yảkeň !kui̱-yă .
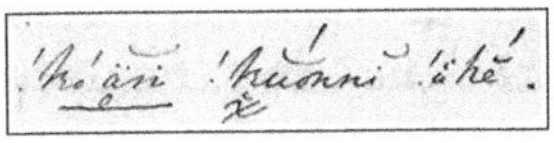	!k\barbelow{\dialine{o}} \circbtwodotst{a}\onedot{n}} \xcbelow{k}\uline{uo}nn\u{i} !\uu{u}h\uline{e} .	!kọ̈ảṅ ḵŭonnĭ !ŭhĕ .

Bushman text can be encoded using LaTeX and the custom TIPA macros and also shows the approximate visual representation.

4 A Specialised Tool for Capturing Bushman Text

A custom tool called xở'xở, which means *to write* in the |xam Bushman language, was created to assist in creating the corpus for Bushman handwriting recognition. xở'xở is an AJAX Web-based tool that allows multiple users to assist in the creation of the corpus simultaneously. xở'xở contains a segmentation component that allows for a page of Bushman text to be segmented into individual lines and words using automatic segmentation algorithms and user interaction to correct errors and a text capture component that allows users to capture the Bushman text using a special input form.

xở'xở also acts as a job management and monitoring system. Each user has an account on the system, which assigns segmentation and transcription jobs and keeps track of the jobs that they have completed. xở'xở will be discussed in detail in this section, starting with a description of the preprocessing and segmentation steps and then a discussion of how the Bushman text is captured.

4.1 Preprocessing

Preprocessing involves preparing the notebook pages for segmentation. The first step is the separation of Bushman text from English text since they appear alongside in the notebooks (see Figure 1). The user is able to use a rectangle selection tool to draw a box around the Bushman text, which is then automatically cropped. In cases where the page contains no Bushman text, the user is able to click a button and the page will be ignored. The text is then thresholded using a method based on a local adaptive approach [13] using a sliding window. Figure 2 shows an example of Bushman text that has been separated from the English text and thresholded.

4.2 Line Segmentation

To segment a page into the individual lines, an approach based on the horizontal projection profile is used [14] where each horizontal pixel-row is analysed to determine the number of foreground-background transitions and a horizontal projection profile is created. In the projection profile, minima represent the spaces between individual lines, which is smoothed using a Gaussian filter to remove false minima.

Before the page is segmented into lines, the candidate segmentation locations are presented to the user who is given the option to move them and add or delete segmentation locations. Furthermore, the user is able to change the size of the Gaussian kernel. Figure 2 shows an example of the interface when the line segmentation candidates are automatically identified by the line segmentation algorithm. The lines are segmented once the user has made any changes that they feel are necessary and the system keeps track of each page, the lines that it was segmented into and which user performed the segmentation.

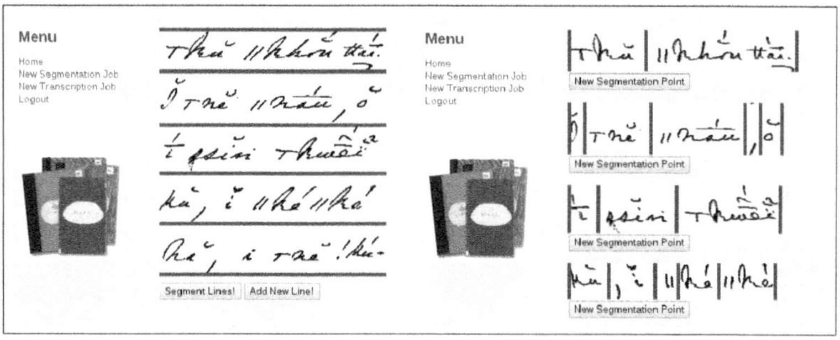

Fig. 2. Interface allowing user to view line and word segmentation candidates and change them

After the text lines have been segmented the slant is automatically corrected using a method based on the vertical projection profile [15]. The slant-corrected text lines are then presented to the user who has the option to crop them, re-apply the slant correction or even delete the line entirely.

4.3 Word Segmentation

Word segmentation is performed on each of the individual text lines segmented in the previous step using a technique based on the distance between connected components (CCs) [4]. In this method, the CCs in each line are first identified and labelled [16] and then they are then sorted by their x-coordinate. The horizontal distances between every pair of adjacent CCs is calculated and when CCs overlap vertically the distance is set to 0. If the distance D between two adjacent

CCs CC_i and CC_{i+1} is greater than the threshold T then CC_i and CC_{i+1} are considered as belonging to two separate words; otherwise, CC_i and CC_{i+1} are considered as belonging to the same word. The threshold T is calculated as:

$$T = \frac{\sum_{i=1}^{n-1} Distance(CC_i, CC_{i+1})}{n} / 2 \quad (1)$$

where n is the number of CCs in the image. As was the case with line segmentation, word segmentation candidates are presented to the user and the user has the option to move segmentation points as well as add and delete segmentation points. The words are segmented once the user is satisfied with the segmentation points and the system automatically keeps track of each word that a line was segmented into and the user who performed the segmentation. Figure 2 shows an example of the interface for word segmentation.

4.4 Text Capture

In order to capture the text, a custom input form was created and is shown in Figure 3. The text input interface is made up of several components. The first of these contains the image that is to be captured and is displayed at the top of the interface. Below this is the text input box where the user can input the text that they see, which can then be copied to the LaTeX box below. From the LaTeX box users are able to highlight text and add diacritics by clicking on the diacritics that they would like to add from the right side of the interface. The interface also allows the user to preview the LaTeX representation and keeps track of each transcription and the user who captured it.

The interface caters for diacritics that appear above, below and above and below diacritics. The diacritics that it supports were pre-determined by scanning through a subset of the full Bleek and Lloyd Collection. However, it is possible that some diacritics might be encountered that the interface does not support. If this occurs, the user is able to mark a line as not being supported by the interface and at a later stage these lines can be reviewed and previously unseen diacritics can be added to the interface. Text lines can also be marked when users feel that they are not suitable for transcription.

5 Corpus Creation Case Study

A workshop was held in which twenty-nine data capturers used xȍä'xȍä to assist in creating a corpus of Bushman handwriting. The workshop began with an introduction to the problem, a discussion of what the goals of the workshop were and what the data created would be used for and the data capturers were encouraged to talk to one another during the workshop. For the workshop a total of 900 pages in the collection, written by two authors, were randomly sampled and inserted into the job management system as segmentation jobs.

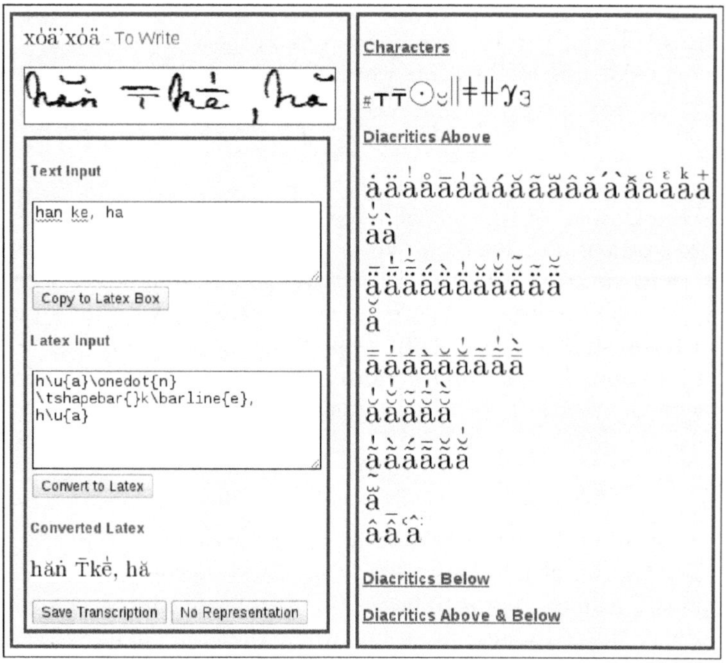

Fig. 3. Interface allowing user to capture the Bushman text

5.1 Segmentation

The segmentation part of the workshop started off with a short demonstration of how the segmentation was to be performed. Segmentation jobs were then randomly allocated to users by the job management system, which kept track of who the jobs were allocated to and what changes each individual user made. Text lines were the standard input method for capturing the Bushman text and therefore the segmented lines were added to a list of transcription jobs.

In total there were 900 segmentation jobs, of which 729 were completed and segmented into 7950 text lines. There are a number of reasons why the other 171 jobs might not have been completed, such as them not containing Bushman text. These incomplete jobs will be reviewed in a follow-up workshop.

5.2 Text Capture

Transcriptions were captured for each of the individual text lines. The data capturers were given a demonstration of how to perform transcription and the various special cases that arise in the Bushman text were highlighted. The data capturers were encouraged to collaboratively determine what the characters were and the importance of consistency was stressed.

Of the 7950 transcription jobs, 1547 were completed and 452 were marked as having no representation. Future work will investigate each of the text lines that was marked as having no representation and appropriate action will be taken, such as adding the diacritics to the interface.

There are two reasons for 5951 transcription jobs not being completed. The first is that there was insufficient time to complete all of the jobs in one sitting and the other is that some users were more efficient at completing transcriptions.

6 Evaluation

The previous section described a case study where xȍä'xȍä was used to created a handwriting recognition corpus. In this section, the output from the case study is evaluated in terms of the quantity and quality of contributions of each data capturer and the relationship between them. This information is valuable in planning for a follow-on workshop to expand the corpus as well as to give an idea of the quality of the data produced during the workshop.

6.1 Contributions

The first issue considered is the contribution to the corpus made by each data capturer as it is important in planning for follow-on workshops. Table 2 shows the segmentation and transcription jobs completed by the data capturers and, as can be seen, there is wide variation between the different data capturers.

Table 2. Number of segmentation and transcription jobs completed by users

Segmentation Jobs		Transcription Jobs	
Jobs	**Number of Users**	**Jobs**	**Number of Users**
0-9	0	0-9	0
10-19	5	10-19	1
20-29	9	20-29	3
30-39	7	30-39	4
40-49	5	40-49	9
50-59	0	50-59	4
60-69	0	60-69	6
70-80	0	70-80	2

6.2 Data Quality

In the case of a corpus for handwriting recognition, quality is represented by the accuracy and correctness of the transcriptions of the data. To evaluate quality, three transcriptions by each data capturer were randomly sampled and reviewed by a post-graduate research assistant who noted the errors encountered in each text line. It was found that, on average, each text line transcription had 0.48 errors, where an error represents a character or diacritic that was transcribed incorrectly. Given that most text lines contain more than 10 characters, it is felt that this number is acceptable for the purpose of handwriting recognition and

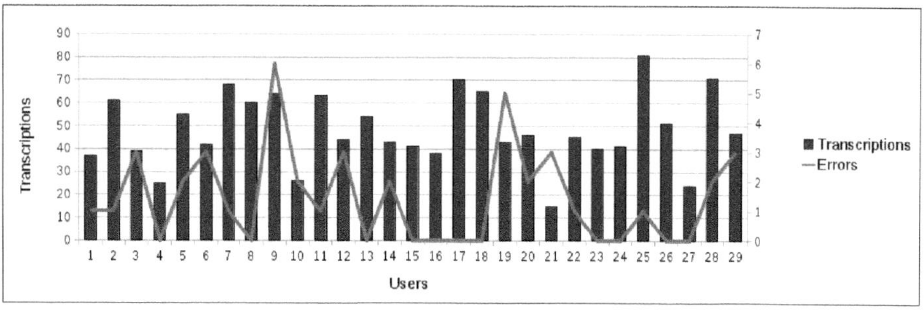

Fig. 4. Comparison of number of transcriptions and errors

shows that lay-persons are able to produce high quality data with relatively low levels of perceived errors.

An investigation was also conducted to see if there was a relationship between the number of transcriptions made by each user and the error rate. Figure 4 provides a comparison between the number of transcriptions and number of errors made by each data capturer and shows that there appears to be no relationship between the two.

The evaluation has shown how the different data capturers contributed to the creation of the corpus in terms of efficiency. It has also shown that accuracy and efficiency appear to be unrelated. Both of these findings are useful in planning for future corpus creation workshops.

7 Conclusions

The creation of corpora for historical texts is often difficult due to the effects of degradation and the complexities of historical scripts. This paper has discussed the creation of a corpus for the Bushman languages using a Web-based tool that was specifically built to cater for the complexities of the Bushman script. The tool was semi-automatic and involved automatic algorithms as well as operations that required user interaction. A case study demonstrated the use of the tool in a workshop to create a corpus and an evaluation of the data showed that data capturers were able to create a corpus with few perceived errors. The creation of transcriptions is a first step towards making the Bushman texts more accessible and it is hoped that it will allow for the development of new techniques for exploring and interacting with the collection.

Acknowledgements. This research is supported by the University of Cape Town, the Telkom/NSN/Telesciences/THRIP Centre of Excellence and the National Research Foundation.

References

1. Suleman, H.: Digital libraries without databases: The Bleek and Lloyd collection. In: Kovács, L., Fuhr, N., Meghini, C. (eds.) ECDL 2007. LNCS, vol. 4675, pp. 392–403. Springer, Heidelberg (2007)
2. Marti, U., Bunke, H.: A full English sentence database for off-line handwriting recognition. In: Proceedings of the Fifth International Conference on Document Analysis and Recognition, pp. 705–708. IEEE, Washington, DC (1999)
3. Rath, T.M., Manmatha, R.: Word spotting for historical documents. Int. J. Doc. Anal. Recognit. 9, 139–152 (2007)
4. Makridis, M., Nikolaou, N., Gatos, B.: An efficient word segmentation technique for historical and degraded machine-printed documents. In: Proceedings of the Ninth International Conference on Document Analysis and Recognition, pp. 178–182. IEEE, Washington, DC (2007)
5. Al-Ma'adeed, S., Elliman, D., Higgins, C.A.: A data base for arabic handwritten text recognition research. In: Proceedings of the Eighth International Workshop on Frontiers in Handwriting Recognition, pp. 485–489. IEEE, Washington, DC (2002)
6. Fischer, A., Indermühle, E., Bunke, H., Viehhauser, G., Stolz, M.: Ground truth creation for handwriting recognition in historical documents. In: Proceedings of the 9th IAPR International Workshop on Document Analysis Systems, pp. 3–10. ACM, New York (2010)
7. Surowiecki, J.: The wisdom of crowds: why the many are smarter than the few. Abacus (2005)
8. Setlur, S., Kompalli, S., Ramanaprasad, V., Govindaraju, V.: Creation of data resources and design of an evaluation test bed for Devanagari script recognition. In: 13th International Workshop on Research Issues in Data Engineering: Multilingual Information Management, pp. 55–61. IEEE, Washington, DC (2003)
9. Lee, R.A., Balick, M.J.: Indigenous use of hoodia gordonii and appetite suppression. EXPLORE: The Journal of Science and Healing 3(4), 404–406 (2007)
10. Williams, K., Manilal, S., Molwantoa, L., Suleman, H.: A visual dictionary for an extinct language. In: Chowdhury, G., Koo, C., Hunter, J. (eds.) ICADL 2010. LNCS, vol. 6102, pp. 1–4. Springer, Heidelberg (2010)
11. Williams, K., Suleman, H.: Translating handwritten bushman texts. In: Proceedings of the 10th Annual Joint Conference on Digital Libraries, pp. 109–118. ACM, New York (2010)
12. Rei, F.: Tipa: A system for processing phonetic symbols in LaTeX. TUGboat 17(2), 102–114 (1996)
13. Sezgin, M., Sankur, B.: Survey over image thresholding techniques and quantitative performance evaluation. J. Electron. Imaging 13(1), 146–168 (2007)
14. Marti, U., Bunke, H.: On the influence of vocabulary size and language models in unconstrained handwritten text recognition. In: Proceedings of the Sixth International Conference on Document Analysis and Recognition, pp. 260–265. IEEE, Washington, DC (2001)
15. Pastor, M., Toselli, A.H., Vidal, E.: Projection profile based algorithm for slant removal. In: Campilho, A.C., Kamel, M.S. (eds.) ICIAR 2004. LNCS, vol. 3212, pp. 183–190. Springer, Heidelberg (2004)
16. Shapiro, L., Stockman, G.: Computer vision. Prentice Hall, Englewood Cliffs (2001)

User Value Oriented Functional Architecture and Implementation of Regional Digital Library: The Case of ZADL Project

Jiaping Qian, Hong Li, Huazhang Tong, and Jindi Ma

Zhejiang University Libraries,
Hangzhou, China
qianjp@zju.edu.cn

Abstract. To construct the functional framework of digital library based on utilization value of target user is the core idea of Zhejiang Academic Digital Library (ZADL) project started in 2008. This paper highlights objectives of the project, functional design of the system architecture driven by the user-oriented services and the implementation strategy and discusses the experience learnt from the project.

Keywords: Functional Framework, Regional Digital Library, Service Functions.

1 Introduction

There are 84 tertiary institutions, including 28 universities and 56 colleges in the Zhejiang Province, an eastern coastal province of China. Those institutions are located in the major cities all over the province. The libraries of those universities and colleges vary from one to another in many aspects including their available resources, service models, especially the information service capabilities under digital environment. In 2008, the province launched a project called Zhejiang Academic Digital Library (ZADL). Through the ZADL Portals, all academic libraries within the province are able to effectively share the academic resources and services and to improve their services for learning, teaching, researching and other social services. This paper highlights objectives of the project, functional design of the system architecture driven by the user-oriented services[1,2], and the implementation strategy and discusses the experience learnt from the project .

2 Objectives

The ZADL project that is based on China Education and Research Network (CERNET) is to build a regional academic digital library in Zhejiang province. The aim of the ZADL is to establish a framework of a reliable and extensive document retrieval and delivery system for tertiary education and scientific research in the Zhejiang province. The ZADL will be gradually implemented over 3 years. Specifically, the objectives of the project include:

C. Xing, F. Crestani, and A. Rauber (Eds.): ICADL 2011, LNCS 7008, pp. 232–236, 2011.

1) to develop a first-class digital information resource library shared by all Zhejiang colleges and universities,
2) to establish a digital library service system accessed by all the Zhejiang colleges and universities, and
3) to train a team of qualified librarians to provide the digital library management and service and to further develop new services.

3 Functional and Implementation Framework

3.1 Operational Frame

As a provincial digital library, the ZADL system consists of one provincial management center and five sub-centers. Fig. 1 shows the relationship among the six centers. Each center, including the provincial center, has its own member libraries and users. Such a structure unifies the library stuff training program and promotion of the use of the ZADL services. Moreover, it enables simple and quick services to the users at low cost compared to multiple separated systems. Also users can access the resources that the present system is not subscribed to from CALIS via ZADL service.

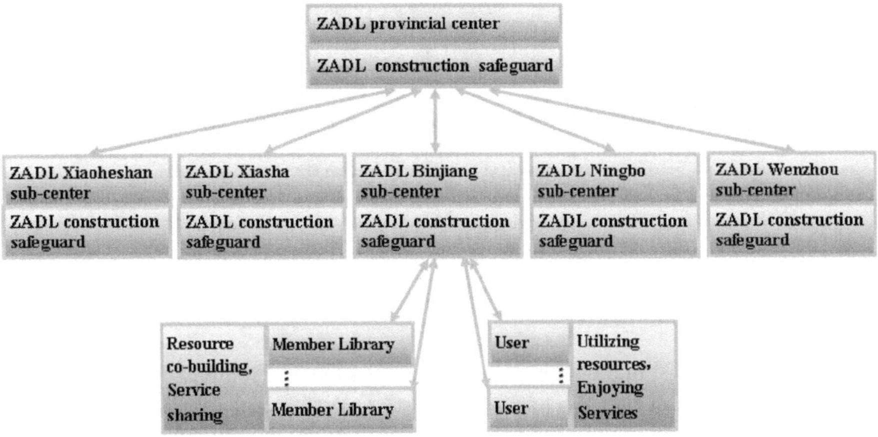

Fig. 1. Service system architecture scheme of ZADL

3.2 Service Functions

The primary principle of designing the ZADL functions is its use-value for library users[3]. User value includes the following three aspects: 1) the needs of users, 2) functions for searching and retrieving target documents, and 3) efficient access. Based on a study and analysis of user requirements, the ZADL defines its core function. That is to provide all in one-step service in finding resource and retrieving target documents. By navigating through the Zhejiang Academic Library Union Catalogue (ZALUC) or using the Unified Search Entry (USE), all collections that are separately stored in member libraries including e-resources can be found. To facilitate users, both ZALUC and USE should integrate services to get resources seamlessly, which are to provide download links to the users who have full text resource access authorities, and a document delivery portal to those who don't have full text resource access authorities. Furthermore, bibliographic data should be submitted automatically. If users encounter problems during searching resources or utilizing services of digital library, they can acquire help from librarians by virtual reference service system in a timely manner. Fig. 2 shows the functional and implementation framework.

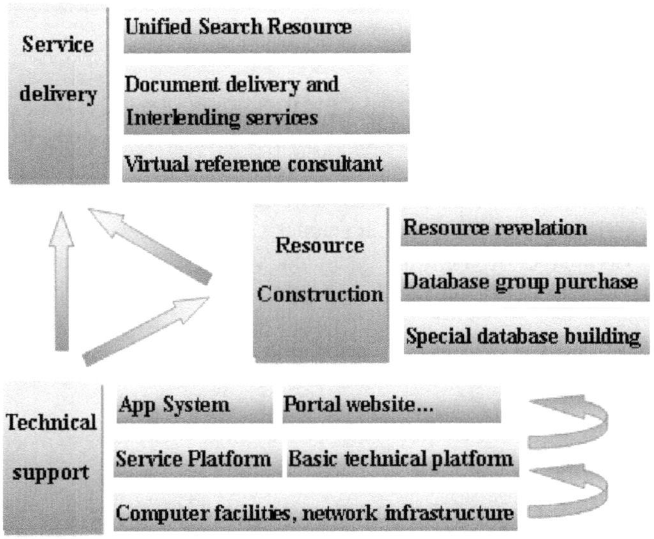

Fig. 2. ZADL Functionality and Implementation Framework

3.3 Implementation Strategy

The ZADL project is implemented by three divisions including Service Delivery, Resource Construction and Technical Support. Tasks of each division are defined specifically and clearly, one of the tasks for service division is to collect the needs of those end users in utilizing recourses and the related services, including the specific demands to portals. The resource division is mainly responsible for the database joint purchasing, special database self building and the organization and revelation of all the resources, while the technical division engages in designing the application framework of services, setting the network infrastructure, and making the procurement plans.

As a long-term development strategy, the project emphasizes on the flexibility of the functional framework during the framework design and implementation. As a result, the ZADL system allows its interface to integrate with other digital platform in the future when needed. Therefore, the project is implemented in two steps. First step is a trial run of the system after the basic system platform is completed. Then, an improvement is to be identified and implemented after consulting with users about the basic system platform.

4 Progress and Results

The ZADL project has been being on trial running since the end of 2010. The following works have been completed and related services have been made available to users:

* Establishment of web-based digital library portals. The web portals have achieved integration of key functional applications at portal levels, including unified search, union catalogue and document delivery service system.

* Establishment of a union catalog across the college libraries of the whole province called ZALUC, which includes the paper bibliographic records, electronic resources from the libraries of colleges and universities in the province, as well as important open access resources from the internet.

* A single search platform that combines meta-data based database search and union resource search system, and can provide user two choices with respective advantages.

* Implementation of a unified identity authentication by combining the central authentication and union authentication.

* Seamlessly integrated resource access services, including full text URL and Document Delivery service system, into resource discovery service system. For example, meta-data based search platform provides over 270 databases, 380 million records resources search and access that includes 3.1 million e-books search and 2.4 million e-books delivery. From ZALUC and two unified search interface, library users can effectively discover and access resources distributed in the colleges and universities in the province.

As an ongoing project, based on the trial run, demo and advertisement, the ZADL has received positive response from the libraries of colleges and universities of the province. For instance, ZALUC, started in the end of 2010, has 49 libraries members and integrated 5, 5000 magazines, 428 databases and over 1, 000, 000 books from member libraries, and has opened to the users of member libraries. Besides, the service statistics shows that the resources and services are actively used by the members. Taking document delivery as an example, as of now there are 49 active document delivery members, with over 5000 registered users, and over 25, 000 documents delivered.

5 Conclusion and Further Work

The ZADL functionality framework is built on target user needs and values, mainly considering how to simplify user operations, reducing user time instead of only focusing on the advance of technology. Positive responses from the users have shown that the design achieved its initial objectives. As an ongoing project, the ZADL adopts multiple steps strategy: trial run → improvement based on feedback → demonstration. The website and service system needs a continuous improvement and enrichment. Our work is to continue focusing on user needs meanwhile taking advantage of new technology actively, and making the ZADL a reliable digital library to further better service teachers and students in colleges and universities.

References

1. Saracevic, T., Kantor, P.B.: Studying the Value of Library and Information Services. Part I. Establishing. J. Am. Soc. Info. Sci. 48(6), 527–542 (1997)
2. Flint, D.J., Larsson, E., Gammetgaard, B., Mentzer, J.T.: Logistics innovation: A customer value-oriented social process. Journal of Business Logistics 26(1), 113–148 (2005)
3. Becker, A., Buxmann, P., Widjaja, T.: Value potential and challenges of service-oriented architectures - A user and vendor perspective. In: ECIS 2009 Proceedings (2009), http://aisel.aisnet.org/ecis2009/88

Comparative Evaluation of Interfaces for Presenting Location-Based Information on Mobile Devices

Dion Hoe-Lian Goh, Chei Sian Lee, and Khasfariyati Razikin

Wee Kim Wee School of Communication and Information,
Nanyang Technological University, Singapore
{ashlgoh,leecs,khasfariyati}@ntu.edu.sg

Abstract. Location-based information can now be easily accessed on mobile devices and is commonly presented in the form of lists of items, maps and more recently, augmented reality (AR). Each approach (list, map, AR) has its strengths and weaknesses. In this paper, we investigate the three approaches for searching and browsing location-based information, taking into account performance and users' perceptions of usability. Participants were issued with Android smartphones pre-loaded with a specific interface and performed a set of browsing and searching tasks. Results suggest that all three interfaces performed similarly for search tasks, but for browsing, the map performed worst. In terms of usability, participants appeared to rate the list better than the other interfaces for presentation location-based information.

Keywords: Augmented reality, maps, lists, user interfaces, evaluation, mobile information access.

1 Introduction

Location-based information can now be easily accessed on mobile devices with the integration of features such as Global Positioning System (GPS) and wireless Internet capabilities. The information presented on mobile devices is commonly in the form of lists of items or on maps [5], mirroring that of desktop applications. However, the limited screen size in mobile devices and their mobile nature of use brings about a new set of challenges and opportunities for designers because visualizations that have been successfully employed in desktop applications may not be suitable for mobile devices [15]. In addition, the information needs of the mobile user are different from that of a desktop user as the mobile user is on the move [2, 6, 14]. For example, mobile users may require information based on the context that they are in (e.g. the train or bus schedule), gather information with a higher level goal (e.g. decide on a restaurant in the area that fulfills certain criteria) or obtaining information serendipitously (e.g. looking at the upcoming events in the area).

Recently, with the increasing capabilities of mobile devices, augmented reality [1] has emerged as a means to present information in location-based applications. Essentially, augmented reality (AR) interfaces superimpose location-based information over a live camera view of one's surroundings on the mobile device. By doing so, the premise is that users are better able to understand one's environment by

C. Xing, F. Crestani, and A. Rauber (Eds.): ICADL 2011, LNCS 7008, pp. 237–246, 2011.

accessing this information. The growing interest surrounding AR has been spurred by easy-to-use applications such as LAYAR (http://layar.com), allowing users to experience AR from popular smartphone platforms.

With the live camera view showing the information available in the users' environment, AR interfaces may reduce the cognitive effort needed by the user to relate the information to the surroundings [11]. However, the camera view only focuses on a limited area, making it potentially difficult for users to orient themselves within the larger vicinity they are in. A list-based presentation, in contrast, is easy to understand and information is usually ranked according to some relevance criteria. It may however require more effort by the user to relate the information presented to the surroundings [5]. Finally, a map graphically displays information items according to their respective locations, allowing users to quickly grasp what is around them. A disadvantage is that map reading requires spatial awareness skills to orientate the map to one's surroundings [16].

Despite the diversity of presentation methods available for location-based information and their different strengths and weaknesses, a limited number of studies have been conducted to compare between them, to the best of our knowledge. For example, [20] compared a digital map interface with a physical map that includes AR features that superimposes content on the camera view. Findings suggested that the latter had a better performance. Further, [5] compared a map with a text-based interface for searching location-based information. Results were mixed as maps were not always the interface of choice among users. However, as yet, there have been no studies conducted to compare the three interfaces for presenting location-based information, namely, lists, maps and AR.

In this study, we fill this research gap by comparing the three approaches for searching and browsing location-based information. We take into account both task performance as well as users' perceptions of these interfaces. Our work is driven by the following research questions: (1) Which interface allows users to complete search and browsing tasks faster; and (2) Which interface for presenting location-based information is preferred by the user?

This paper is organized as follows. The next section gives an overview of related work. This will be followed by a description of the evaluation as well as the interfaces that were used. Next, results are presented. The paper ends by discussing the implications of the results, limitations as well as areas of future work.

2 Related Work

List-based interfaces have been the starting point for displaying information on many mobile applications. Magitti [3] is a context-aware mobile recommender system designed to recommend locations of interest in a tourism setting. The primary interface is a list of recommended items that match a user's current profile and context. This list is updated whenever the context changes. A field evaluation suggests that the system was useful as a guide for recommending leisure activities. Next, CityFlocks [4] is a mobile application that provides location-based information within a vicinity. A search interface is provided for information about places of interest. As in Magitti, results are presented as a list and ranked according to the likelihood of matching the searcher's interests. Findings from an evaluation showed that users found the content to be useful for navigating new environments.

Besides lists, maps provide an alternative presentation of location-based information on mobile devices. MobiTOP [13] offers a map-based interface for creating, sharing and seeking location-based content called annotations. Annotations are represented on a map by markers, and by selecting a marker, its associated content is displayed. In an evaluation of the application's usability in a geography field study setting, users reported to be satisfied with the map for information access as they were able to relate the information with the location. However, some found it difficult to view the map on small mobile screen. Next, SocialSearchBrowser [5] is a mobile application that uses a map to visualize location-based queries posted by its users. An evaluation was conducted to compare the application's map interface against a text-based version that presented queries in a list. Results were not conclusive, as of the 34 participants who took part, 16 participants preferred the map interface, 14 preferred the list, while four had no preference. Further, a more detailed analysis suggests that different interfaces are preferred in different situations, and that both options should be made available to users.

While AR interfaces have been developed and used over a number of years, it is only with the recent advent of well-equipped smartphones that they have now seen greater utilization. For example, the LOCUS project [18] presents the results of location-based searches using AR. A qualitative expert evaluation of the system suggests that LOCUS had an advantage over maps for displaying information because of the additional information that can be superimposed on objects being viewing, a feature which maps cannot offer. In a separate study, [17] compared a mobile AR map versus a standard mobile digital map to play a location-based game. Results were mixed in terms of usability as each interface type scored well on some constructs while doing poorly on others. Qualitative observations however suggest that the AR interface seem to encourage group discussion and pubic problem-solving.

In summary, evaluations for presenting location-based information tend to focus on one or at most two interface types (e.g. maps alone, or maps and AR). Further, such prior work usually involves small numbers of users (30 or less) and studies either performance measures or user perceptions. Thus, key contributions of our present research are that we investigate performance and perceptions across three different interfaces for presenting location-based information: lists, maps and AR.

3 Methodology

3.1 Interfaces Developed for the Study

Three interfaces – list, map and AR – were developed as separate mobile applications using the Android platform. These interfaces had a similar look and feel in order to maintain consistency across the applications. Furthermore, all three applications made use of the same location-based user-generated content from the MobiTOP [13] system. Each interface was designed to display information items around the vicinity of the user. The application first obtains the current location of the user through GPS and queries the database for information that is nearby. Users can specify a search radius so that only information within it is retrieved. Further, each application allows users to search for information by keywords. The location-based information, also known as annotations, describes places of interest, and each contains a title, tags, description, location coordinates, one or more images of the location, as well as other metadata.

The list interface (Figure 1a) presents annotations with the nearest item at top of the list. Each list item shows the title, tags, description snippet, one image, date of contribution and the distance between the location of the annotation and the users' present location. Full details of the annotation can be viewed by tapping on it in the list. The map-based interface (Figure 1b) presents annotations as markers on a map. The user's current location is identified by a circle at the centre of the map, which can be navigated by panning and zooming. Tapping on a marker shows the corresponding annotation's full details. Finally, the third interface (Figure 1c) employs AR. Utilizing the browser interface provided by Wikitude API (http://www.wikitude.org), the interface displays nearby annotations by superimposing them over a live camera view. As users move around, the set of annotations displayed changes. Similar to other interfaces, users view an annotation's details by tapping on its marker.

3.2 Participants and Procedure

A total of 90 participants were recruited for the study. They comprised 47 females and 43 males between the ages of 17 and 39 and a mean age of 25. There was an equal mix of students and working adults. Participants primarily had Computer Science/Information Technology (47) and Engineering (27) backgrounds. The rest were from other disciplines such as banking and finance, construction, education, and marketing. Participants were frequent mobile phone users and a majority used their phones to share multimedia with their friends via social media services. Many were also frequent users of mapping and navigation services.

Participants were randomly divided into three groups of 30, with each using a specific interface. Each group performed the same set of tasks for browsing and searching. These tasks were modeled after typical information seeking activities that mobile users undertake [7] and involved real-world locations. At the same time, these tasks varied in their complexity of the information seeking process. The browsing task required participants to examine annotations via a particular interface to locate two places of interest (e.g. restaurants, buildings and shops). The searching task required participants to issue keyword searches then use a particular interface to locate two places of interest.

Each participant was provided with an Android smartphone pre-loaded with the assigned interface. All phones were of the same model to ensure a consistent experience. The researchers first introduced the study and the application. Participants also performed various familiarization activities. Following this, the study began and participants performed the searching and browsing tasks. Next, participants completed a questionnaire that captured demographic data (reported earlier) and perceptions of their assigned application. Participants were also requested to provide qualitative comments about whether they liked their assigned interface.

(a) List-based interface

(b) Map-based interface

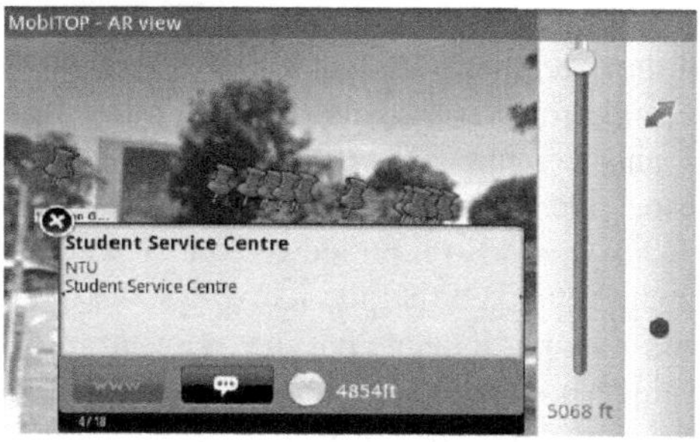

(c) Augmented reality-based interface

Fig. 1. Interfaces used in the evaluation

3.3 Measures

Measures covered two aspects. The first was the time taken to correctly complete the searching and browsing tasks. Each participant was accompanied by a researcher who recorded the completion time. As each searching and browsing task required the retrieval of two places of interest, the time taken was averaged, and these averages were used in subsequent analyses.

Secondly, participants' perceptions of the usability of the interfaces were also collected. Questions were rated on a scale of 1 (strongly disagree) to 5 (strongly agree) and consisted of the following constructs:

- **Ease of use:** the interface is perceived by the user as to not require a lot of effort to accomplish tasks [12].
- **Learnability:** the interface is easy for the user to learn and subsequently complete tasks [12].
- **Efficiency:** the interface enables the user to complete tasks quickly [19].
- **Effectiveness:** the interface enables the user to complete tasks accurately [19].
- **Satisfaction:** measures the degree to which the interface appeals to a user [9].

4 Results

4.1 Searching

Participants that used the map seemed to take the longest time (M = 62.12s, SD = 33.13) compared to those who used the list (M = 48.26s, SD = 34.10) and AR (M = 53.38s, SD = 22.27) interfaces. However, an ANOVA test suggested that these differences were non-significant.

In terms of perceptions, Table 1 presents the results of the various usability constructs. Separate ANOVA tests on the different constructs were conducted, and only ease of use yielded statistically significant differences across the interface types [$F(2, 87) = 4.02$, $p < 0.05$]. Post-hoc tests showed that the perceived ease of use for list and AR were significantly different, with the list interface performing better. There were no significant differences between list and map, or map and AR. Qualitative comments from participants appear to support our findings. Participants felt that using the list enabled them to "*locate the point of interest within a short span of time*". Another who used AR highlighted that they found it "*hard to find the location*" especially if they were not familiar with the area.

The three interfaces performed equally for the other usability constructs. For example, most participants separately felt that they only needed a moderate amount of time to learn their respective interfaces as they were "*easy to learn even to a new user*". All three interfaces were also reported to be efficient for locating information. For instance, a participant remarked that the list was a "*fast way to find information*". In terms of effectiveness, the participants felt that relevant information was somewhat easy to obtain for all three interfaces. An AR participant commented that "*it did not require a lot of effort to complete the task accurately*". Overall, participants were moderately satisfied with three interfaces for searching. For example, an AR participant remarked that the interface was a "*brilliant idea*", while the list was "*simple and easy to use*".

Table 1. Usability constructs results based on searching tasks for the different interfaces

Constructs	List (N = 30)	Map (N = 30)	Augmented Reality (N = 30)
Ease of Use*	4.04 (0.57)	3.94 (0.55)	3.64 (0.55)
Learnability	4.44 (0.51)	4.33 (0.58)	4.10 (0.64)
Efficiency	4.08 (0.53)	4.01 (0.66)	3.80 (0.65)
Effectiveness	3.84 (0.61)	3.89 (0.71)	3.54 (0.77)
Satisfaction	3.69 (0.61)	3.73 (0.70)	3.57 (0.70)

* Statistically significant differences between the three interfaces

4.2 Browsing

For browsing, participants who used the map again took the longest time to complete the tasks (M = 76.89s, SD = 46.61) than the list (M = 32.87s, SD = 20.59) and AR (M = 49.52s, SD = 33.45) interfaces. This time, there were significant differences in time taken [$F(2, 87) = 11.97$, $p < 0.001$]. Post-hoc tests indicated that there were significant differences between map and list, and map and AR.

Table 2 lists results for the perceptions of the usability constructs for the browsing tasks. Participants who used the list found that it was easier to use than those who had used the map or AR. These differences were statistically significant [$F(2, 87) = 4.83$, $p < 0.05$]. Post-hoc results indicated differences between list and augmented reality interfaces. AR Participants commented that it was "*inconvenient to hold the phone and move around*" while almost all of those using the list interface had remarks with "*easy to use*" or its variants. There were no significant differences between list and map, or map and AR.

Table 2. Usability constructs results based on browsing tasks for the different interfaces

Constructs	List (N = 30)	Map (N = 30)	Augmented Reality (N = 30)
Ease of Use*	4.06 (0.59)	3.86 (0.58)	3.61 (0.53)
Learnability	4.50 (0.52)	4.29 (0.62)	4.12 (0.65)
Efficiency*	4.14 (0.69)	3.81 (0.78)	3.72 (0.61)
Effectiveness	3.90 (0.61)	3.86 (0.74)	3.58 (0.74)
Satisfaction	3.77 (0.62)	3.63 (0.71)	3.57 (0.69)

* Statistically significant differences between the three interfaces

Unlike for searching, differences in perceptions of efficiency were also statistically significant [$F(2, 87) = 3.52$, $p < 0.05$]. Specifically, post-hoc tests suggest that the list was perceived to be faster for task completion than AR. There were no significant differences between list and map, or map and AR. This could be due to the fact that the participants who had used AR found it difficult to browse through the annotations especially when there were "*too many results and the pins (on the screen) overlapped one another*". In contrast, the list interface provided "*fast response*" times possibly because it required the least graphically intensive among the interfaces.

Beyond these, the three interfaces performed similarly for the other usability constructs with respect to browsing. Put differently, participants found the interfaces easy to learn, and moderately effective in task completion. Participants were also relatively satisfied with their assigned interfaces.

5 Discussion and Conclusion

To reiterate, the objective of our work was to compare users' performance and perceptions of three interfaces for searching and browsing location-based information on mobile devices. Our study yielded the following findings.

First, for searching, performance was similar across all three interfaces but for browsing, the map performed significantly worse than the list and AR interfaces. A key difference between the two tasks could have contributed to this finding. Specifically, the search tasks had an implicit assumption that participants had at least a vague idea of what they were looking for through the issuance of keyword queries. The queries effectively served as a filter to remove non-relevant annotations, thus leading to a less cluttered interface. This gave participants a better chance of ascertaining which annotations could fulfill the tasks' requirements. In contrast, the browsing tasks required more exploration of annotations for successful completion. Without the ability to execute keyword queries, participants had to manually sift through the annotations on their respective interfaces to determine which were relevant. Here, the list interface provided comparatively more informational cues than the map or AR interfaces, and hence likely contributed to the better performance. The AR interface presented markers without any other information, but because only markers associated with the current view were displayed, this indirectly served as a filtering mechanism, and participants could deduce which annotations were likely to be relevant by panning to a particular direction. The map however gave the fewest number of cues since all markers surrounding the participants' current location were presented, hence requiring the most time to complete the tasks.

Second, given the performance results, it was unsurprising that the list was rated best in terms of ease of use for search and browsing by participants. What was interesting however was that the AR interface was always ranked last despite its better performance when compared to the map. This divergence between perception and reality could be attributed to the relative newness of the AR concept in comparison to lists and maps, and the effort needed to use the interface. For the latter, participants had to physically move and hold up the mobile device in order to view the annotations [17]. This could be uncomfortable as fatigue could eventually set in. This pattern of results was also similar to the perception of efficiency for completing the browsing tasks. Again, the list ranked higher in terms of efficiency because participant had more informational cues than the map or AR interfaces. However, the map ranked higher in efficiency when compared to AR likely because the former was a more familiar concept. As a result, participants felt that the map was "*not cluttered*", and was therefore easier to locate annotations of interest. This perception arose because participants could easily zoom and pan within the map to reduce the number of annotations presented. While the AR interface had similar features, participants did not seem to use them as adeptly.

Third, although the list interface appears to have a slight edge over the other interfaces for presenting location-based information, the results when taken as a whole, are inconclusive. For example, while the list appears appropriate for browsing tasks, this advantage disappears for searching. In many information seeking situations, a combination of browsing and searching would be used [10]. Thus when these tasks are interleaved, the advantages of the list may become diffused. Further, differences between learnability, effectiveness and satisfaction were non-significant. This suggests that in the short term, users may find the list interface easier to use and more efficient, but as users learn to use the various interfaces, this usability perception gap may narrow.

Put differently, it appears that each of the interfaces have complementary affordances to meet the needs of users [5, 8]. Consequently, designers may consider incorporating combinations of the three interfaces in location-based applications. For instance, a list could provide an overview of the information within a vicinity, while an AR interface could help orientate a user by superimposing the information on the live camera view. Further, several design issues should be considered when integrating the interfaces:

- Enable users to select their default interface of choice for browsing and searching to cater to individual differences. Additionally, allow users to switch between interfaces easily to suit the task at hand.

- Map and AR interfaces should provide more informational cues to help users better discern relevant information around them. This is challenging given limited screen sizes but one approach is to embed information on the markers themselves. For instance, stronger hues on a marker could indicate more relevant content. The use of shapes and sizes could also be explored.
- To help manage information overload, filtering mechanisms need to be put in place, especially for map and AR interfaces. These should go beyond a coarse approach such as setting a limit for up to N nearest items to be displayed. Instead, techniques that consider a user's context should be included as well.

Despite the insights uncovered, there are some shortcomings that could be addressed in future work. First, our study was cross-sectional in nature and may favor familiar interfaces such as the list. A study that involves longer-term repeated use would be helpful in validating our findings. Second, participants' familiarity of the study's vicinities could be a factor in task completion time. Future work could involve a greater number of locations, both familiar and unfamiliar, for better comparison of results. Finally, individual differences such as spatial ability and experience with different interfaces should be investigated to attain a better understanding of the strengths and weaknesses of each user interface.

Acknowledgments. This work was supported by the Singapore National Research Foundation Interactive Digital Media R&D Program grant NRF2008IDM-IDM0004-012.

References

1. Alessandro, M., Dünser, A., Schmalstieg, D.: Zooming interfaces for augmented reality browsers. In: 12th International Conference on Human Computer Interaction with Mobile Devices and Services, pp. 161–169. ACM, New York (2010)
2. Amin, A., Townsend, S., van Ossenbruggen, J., Hardman, L.: Fancy a Drink in Canary Wharf?: A User Study on Location-Based Mobile Search. In: Gross, T., Gulliksen, J., Kotzé, P., Oestreicher, L., Palanque, P., Prates, R.O., Winckler, M. (eds.) INTERACT 2009. LNCS, vol. 5726, pp. 736–749. Springer, Heidelberg (2009)
3. Bellotti, V., Begole, B., Chi, E.H., Ducheneaut, N., Fang, J., Isaacs, E., King, T., Newman, M.W., Partridge, K., Price, B., Rasmussen, P., Roberts, M., Schiano, D.J., Walendowski, A.: Activity-based serendipitous recommendations with the Magitti mobile leisure guide. In: Proceedings of the 2008 Annual SIGCHI Conference on Human Factors in Computing Systems, pp. 1157–1166. ACM, New York (2008)
4. Bilandzic, M., Foth, M., De Luca, A.: CityFlocks: Designing social navigation for urban mobile information systems. In: 7th ACM Conference on Designing Interactive Systems, pp. 174–183. ACM, New York (2008)
5. Church, K., Neumann, J., Cherubini, M., Oliver, N.: The Map Trap? An evaluation of map versus text-based interfaces for location-based mobile search services. In: 19th International Conference on World Wide Web 2010, pp. 261–270. ACM, New York (2010)
6. Church, K., Smyth, B.: Understanding the intent behind mobile information needs. In: 14th International Conference on Intelligent User Interfaces, pp. 247–256. ACM, New York (2009)

7. Cui, Y., Roto, V.: How people use the Web on mobile devices. In: Proceedings of the 17th International Conference on World Wide Web, pp. 905–914. ACM, New York (2008)
8. Fröhlich, P., Simon, R., Baillie, L., Anegg, H.: Comparing conceptual designs for mobile access to geo-spatial information. In: 8th Conference on Human-Computer Interaction with Mobile Devices and Services, pp. 109–112. ACM, New York (2006)
9. Frøkjær, E., Hertzum, M., Hornbæk, K.: Measuring usability: Are effectiveness, efficiency, and satisfaction really correlated? In: SIGCHI Conference on Human Factors in Computing Systems, pp. 346–352. ACM, New York (2000)
10. Heimonen, T.: Information needs and practices of active mobile Internet users. In: Proceedings of the 6th International Conference on Mobile Technology, Application & Systems, Article 50. ACM, New York (2009)
11. Hile, H., Liu, A., Borriello, G., Crzeszczuk, R., Vedantham, R., Koêcka, J.: Visual navigation for mobile devices. IEEE Multimedia 17, 16–25 (2010)
12. Hornbaek, K.: Current practice in measuring usability: Challenges to usability studies and research. International Journal of Human Computer Studies 64, 79–102 (2006)
13. Kim, T.N.Q., Razikin, K., Goh, D.H., Nguyen, Q.M., Theng, Y.L., Lim, E.P., Sun, A., Chang, C.H., Chatterjea, K.: Exploring hierarchically organized georeferenced multimedia annotations in the MobiTOP system. In: Proceedings of the Sixth International Conference on Information Technology: New Generations ITNG 2009, pp. 1355–1360. IEEE Press, Los Alamitos (2009)
14. Low, G.H., Goh, D.H., Lee, C.S.: Looking for a good time: Information seeking in a mobile content sharing and gaming environment. In: Proceedings of the 2010 International Conference on Cyberworlds, pp. 231–238. IEEE Press, Los Alamitos (2010)
15. Maurer, M.-E., Hausen, D., De Luca, A., Hussmann, H.: Mobile or desktop websites?: Website usage on multitouch devices. In: 6th Nordic Conference on Human-Computer Interaction, pp. 739–742. ACM, New York (2010)
16. Montello, D.R., Lovelace, K.L., Golledge, R.G., Self, C.M.: Sex-related differences and similarities in geographic and environmental spatial abilities. Annals of the Association of American Geographers 89, 515–534 (1999)
17. Morrison, A., Oulasvirta, A., Peltonen, P., Lemmela, S., Jacucci, G., Reitmayr, G., Nasanen, J., Juustila, A.: Like bees around the hive: A comparative study of a mobile augmented reality map. In: Proceedings of the 2009 Annual SIGCHI Conference on Human Factors in Computing Systems, pp. 1889–1898. ACM, New York (2009)
18. Mountain, D., Liarokapis, F.: Mixed reality (MR) interfaces for mobile information systems. Aslib Proceedings: New Information Perspectives 59, 422–436 (2007)
19. Peevers, G., Douglas, G., Jack, M.A.: A usability comparison of three alternative message formats for an SMS banking service. International Journal of Human Computer Studies 66, 113–123 (2008)
20. Rohs, M., Schoning, J., Raubal, M., Essl, G., Kruger, A.: Map navigation with mobile devices: Virtual versus physical movement with and without visual context. In: Proceedings of the 9th International Conference on Multimodal Interfaces, pp. 146–153. ACM, New York (2007)

Who, What, Why: Examining Annotations in Mobile Content Sharing Games*

Dion Hoe-Lian Goh, Chei Sian Lee, and Guanghao Low

Wee Kim Wee School of Communications and Information,
Nayang Technological University, Singapore
{ashlgoh,leecs,lowg0008}@ntu.edu.sg

Abstract. Recently, mobile content sharing applications incorporating gaming features have attracted much interest. This paper investigates motivations for using these games. We used the TAG framework to analyze motivations based on the recipients of the content, the type of content created, and the goals behind content creation. Participants maintained a one week long diary, and selected participants were called for an in-depth interview. Results suggest that motivations for creating content include knowledge creation, self-expression, creation/maintenance of social relationships, self-presentation, competition and achievement. Additionally, we found that games and mobile content sharing are mutually reinforcing. Implications of our work are also discussed.

Keywords: Mobile content sharing, games, motivations, user study.

1 Introduction

Mobile content sharing applications allow users to view and share multimedia content on mobile devices and are becoming popular in part due to people's increasing reliance on mobile phones and their myriad uses beyond voice calling [1]. Research has also shown that the appeal of these applications can be attributed to the ability to help users achieve certain motivational needs. These include sharing information, personal satisfaction, meeting of emotional and hedonic needs, and making social connections [2, 3, 4]. While achieving these motivations yield benefits, executing these tasks can be tedious and boring [5]. For instance, the repetitive actions of creating and viewing annotations on mobile phones could dull enthusiasm overtime.

One promising approach to address the issue of tedium is to introduce games into mobile content sharing applications. Such applications provide entertainment and content is created as a result of gameplay [6]. This approach appears feasible due to the burgeoning of the mobile gaming industry. We term applications that blend mobile content sharing with gaming as mobile content sharing games. Like those for pure entertainment, these games have attracted commercial interest, and one example is Foursquare (http://foursquare.com/). The game enables users to earn points for discovering new locations and sharing location-based information. Users indicate

* This work was supported by the Singapore National Research Foundation Interactive Digital Media R&D Program, under research Grant NRF NRF2008IDM-IDM004-012.

C. Xing, F. Crestani, and A. Rauber (Eds.): ICADL 2011, LNCS 7008, pp. 247–256, 2011.

their presence at a location by "checking in". The user who most frequently checks into a location will be made its "mayor". This title encourages sustained use. Further, the content created can be harnessed by others to learn more about various locations.

Mobile content sharing games have also garnered research interest. An example is the Gopher Game [7]. Gophers represent missions to be completed and are carriers of information. As players navigate their physical surroundings, they pick up gophers and help them complete their missions by supplying images and textual content. By doing so, content describing specific locations are created, and this can be shared since other users may pick up these gophers and view their associated content.

Current research on mobile content sharing games mostly focuses on design and usability [8, 9]. While these issues are pertinent, research into usage motivations is critical, as a deeper understanding will help designers create compelling applications. However, to our knowledge, there is a dearth of such work, and prior research deals with mobile content sharing applications [2] and games for pure entertainment [10]. Relying on these is insufficient because mobile content sharing games are different as their objectives are to generate content from entertainment. The interaction between these two objectives may result in motivations that differ from other applications [11]. Further, [6] studied gratifications from using a mobile content sharing game, but from the perspective of perceived use in a lab setting, and not actual usage.

We therefore adopt an investigation of motivations for creating content in SPLASH, our implementation of a mobile content sharing game. This was accomplished using a diary study and interview. Data was analyzed using the TAG framework (Target, Annotation, Goal), employed previously for studying motivations for mobile content sharing applications [12].

2 Related Work

2.1 Mobile Content Sharing Games

Recent research has suggested that games could be used to motivate users to accomplish tasks such as content sharing [11]. Here, we review some of this work on mobile platforms. The Gopher Game [7], mentioned earlier, is one example. Next, CityExplorer [8] fosters the labeling of location-based images. The game subdivides a geographic area into segments. Within each segment, players need to label as many points of interest with category names as possible using their mobile devices. A player who creates the most number of labels in a segment wins credits. In MobiMissions [13], content sharing is achieved through the completion of missions, which are sequences of digital photographs and text annotations associated with specific locations. To complete a mission, a player has to capture up to five photographs and add up to five text annotations. This content can be accessed by other players to meet their information needs. Finally, PhotoCity [9] invites players to contribute photos to places of interest, with the goal of facilitating 3D model construction. A flag represents a place of interest, and to capture it, players have to earn points by taking as many as high-quality photos as possible. The player that scores the highest captures the flag.

2.2 Motivations for Content Sharing

A number of studies have been conducted on motivations for sharing photos via mobile devices. For example, [13] revealed five motivations for capturing photos: (1) a record of personal and collective experiences; (2) influence others' views of oneself; (3) voice one's views; (4) creating and maintaining relationships; (5) task support. A similar study [2] found motivations falling into creation/maintenance of relationships, reminding of experiences, self-presentation, self-expression and task performance.

Beyond photo sharing, [15] conducted a study on GeoNotes which allows users to create location-based notes. Two main motivations were to socialize with others, and to orientate to one's surroundings. Further, [12] conducted a study on motivations involved MobiTOP, a location-based mobile annotation system. Motivations uncovered included creation/maintenance of social relationships; reminding of individual and collective experiences; self-presentation; and task performance. Finally, [6] uncovered motivations for using Indagator, a mobile content sharing game, and these include socializing, entertainment, and information discovery.

In sum, understanding motivations are important for creating user-centric applications. Despite the insights that our review of related literature offers, there are two limitations. One, the applications do not combine content sharing and gaming, which we have argued, may influence research outcomes [11]. Two, the study on the gratifications of mobile content sharing games [6] was conducted in a lab and no actual use of the application was performed.

2.3 The TAG Framework

Mobile content has been analyzed from different perspectives in the past and have provided many useful insights for designers and developers [2, 14, 16]. In our previous research [12], we consolidated these perspectives and proposed a framework to facilitate a more comprehensive analysis of mobile content. This is known as the TAG (Target, Annotation, Goal) framework and is summarized in this section.

Target (T) looks at the recipient of the annotations (mobile content). Past work has suggested that an annotation may be created for the general public, family and friends or for individual use [14, 16, 17]. Next, the content of the annotation (A) identifies the different types of annotations that users create. Studies in photo sharing (e.g. [2, 17]) suggest a diverse range of categories, and may apply to mobile annotations as well. Finally, the goal for creating the annotation (G) refers to the motivations for doing so [4]. This dimension investigates the thoughts behind users' actions, has been extensively studied, albeit primarily in photo sharing [14, 16]. Put differently, the TAG framework provides a more holistic analysis of mobile annotation behaviors. This framework will be applied to understand the motivations for using a mobile content sharing game.

3 A Brief Introduction to SPLASH

SPLASH (Seek, PLAy, and SHare) is a mobile application that seamlessly blends content sharing and gaming. In this section, we briefly highlight SPLASH's features. More details about the application, including its design rationale can be found in [18].

SPLASH supports the sharing and seeking of location-based content (or annotations) comprising textual comments, media resources and tags (Figure 1b). Each real-world location is visualized on a map (Figure 1a) and represented by a virtual space (Figure 1c) where players interact, share content and play games. These spaces are designed to establish a sense of community among players, which has been suggested to foster content sharing [6]. Specifically, virtual spaces offer:

- **Content access.** Annotations associated with a particular location are accessed via a comment board found in each virtual space. Besides creating and viewing, users may also rate annotations to provide feedback on quality.
- **Gaming.** This is manifested in three ways. Information mini-games utilize nearby content to help users learn about a particular location. An example is a game that randomly selects a captured image around a user's current location to create a jigsaw puzzle. Mini-HCGs (human computation games [11]) elicit information from users about the current location for the purposes of sharing. One example is a game about pets that reside in the virtual rooms. Pets need to be fed with information in order to thrive. Such information is then made available to users. Finally, casual mini-games offer pure entertainment, such as a shooting game.
- **Accumulating rewards.** To motivate use, virtual credits are earned by playing games and contributing annotations. Further, badges are awarded to users when significant tasks are achieved (e.g. contributing a given number of annotations or accumulating an amount of wealth). These credits and badges are displayed on scoreboards which rank users against others.
- **Socializing.** A "friend" function allows users to add other users as friends. This allows them to quickly view updates of annotations posted by their friends, and to also send private messages to them. Additionally, users' have personalized avatars which can be customizable by purchasing items from a virtual store. These avatars are shown on users' profiles and in the virtual spaces (Figure 1d).

4 Methodology

We recruited 25 male and 15 female participants with ages between 20 and 45 years (average of 26.3). They were a mix of undergraduates, postgraduates and working adults. More than half had computer science or IT-related backgrounds. Participants were proficient with mobile phones, mobile applications, and mobile games.

Each participant was provided with a HTC Desire mobile phone for a week and tasked to maintain a diary. Participants were required to indicate the location where they created the annotations, the situation that they were in, the intended audience, the reason for creating annotations, and additional comments. Due to the mobile nature of the study, employing direct observation methods to capture user intentions may not be feasible [19]. Obtaining responses via post-study interviews only presents issues where interviewees may have filtered out or forgotten certain information. In contrast, a diary study non-intrusive, *in situ*, and is able to obtain a more personal and natural account which may be difficult to observe by another party [19].

Participants were first trained in using SPLASH. Thereafter, they were given instructions and scenarios for the study. Further, 20 participants who contributed the most comments attended a post-study interview where they elaborated on their motivations and diary entries. Finally, we manually analyzed annotations, diary entries, and interview comments to ascertain motivations using the TAG framework.

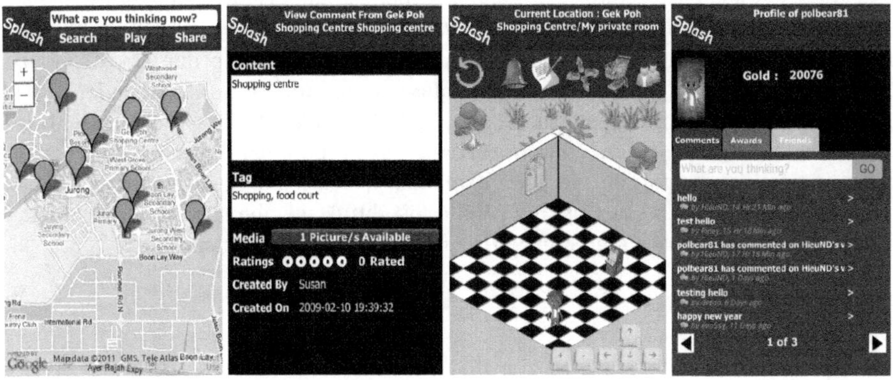

(a) Map view (b) A comment (c) Virtual room (d) User profile page

Fig. 1. SPLASH screen shots

5 Results and Analysis

The study generated a total of 262 entries, with an average of 6.55 per participant.

5.1 Target (T): Who Are the Recipients?

We referred to past work to identify the target audiences of the annotations. (e.g. [2, 12]). First, annotations for "self" were primarily motivated by self-expression, achievement or competition (see Section 5.3). For instance, a participant mentioned that she enjoyed creating annotations on whatever that came to her mind including thoughts, feelings and daily activities. Interestingly, participants motivated by achievement and competition created a larger number of annotations that were incomprehensible. Probing further during interviews, participants mentioned that they were interested in the rewards that came with the creation of these annotations. Many of these annotations thus fell in the "Cheat" category described in Section 5.2.

Annotations for "friends and family" were used to connect with such people, and these were usually used to share personal activities, pictures, or conversations. For instance, a participant mentioned that she liked to create pictures of her friends in class. Lastly, we found that annotations intended for the "public" characteristically contained details of places of attractions, suggestions or recommendations. Examples include *"HAS library, love the decorations here and a nice place to study."*, *"open from 10am till 9pm"*, and *"HSBC discount at swensons"*. In addition, some of these annotations were queries that expected responses from others. Examples include *"anyone living here?"* and *"can I get the yatsuhashi chocolate here?"*

5.2 Annotation (A): What Is Annotated?

The content of all annotations were manually inspected and organized into categories (Figure 2). As shown, "Food" was the top category and annotations typically include reviews and recommendations such as, *"Japanese food here is good"* and *"even though food is not so nice, price is reasonable"*. "Test" was the second most popular category, participants' efforts in trying the application. These annotations include comments such as *"Just splashing"*, *"hello splash"*, or *"testing"*, and were usually the participants' first few annotations created. Interestingly, the third most popular category was termed "Cheat". These annotations were initially classified under "Test" as they consisted of comments such as *"tg"*, *"abc"*, and *"fgo"*. On closer analysis, we realized that the annotations were generated by specific participants within a short span of time. These participants later explained that these annotations were a quick way for them to earn virtual credits. While insightful, the annotations in these latter two categories were useless to others from the content sharing perspective.

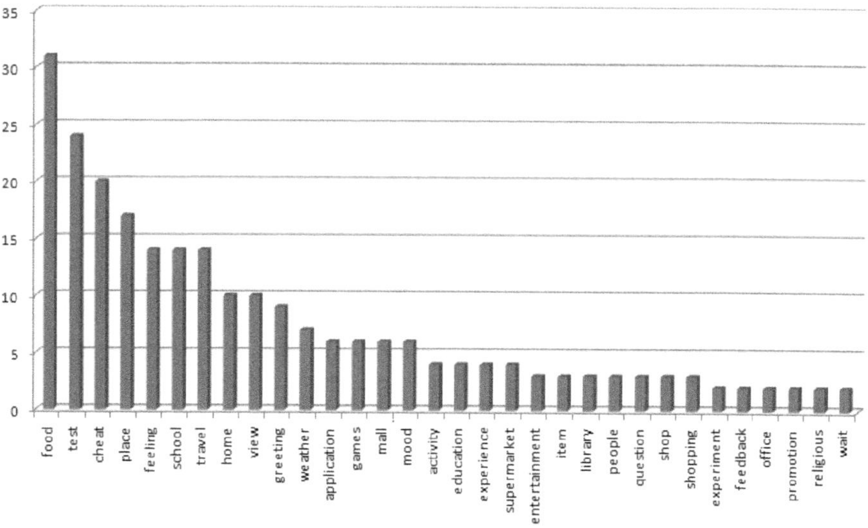

Fig. 2. Annotation categories

"Places" was the next largest category and annotations described a diverse range of places of interest including convenience stores, shopping malls, and attractions. Examples include, *"there is a hdb office on level 2"* and *"nice and romantic place"*. Next, SPLASH served as an outlet for participants express emotions and thoughts and such annotations were classified under the "Feelings" category. Examples include *"feeling sleepy:(:(having loads of wrk to do:(:(god pls help me...."*, or short messages such as *"sad"*, *"surprise"*, and *"emo"*.

"School" and "Travel" annotations highlight two common situations that our participants were frequently in. "School" annotations were explicitly separated from "Places" as they occurred very frequently compared to the other locations. These described a particular place in school or a particular activity relating to school.

We concluded that such annotations to be largely due to the background of the participants (i.e. students) and may not be applicable to other settings. As for "Travel", annotations were created while participants were commuting. Examples included *"the bus is super slow :("*, and *"very crowded train"*. This suggests that participants had pockets of free time which could be filled by using SPLASH.

5.3 Goal: Why Annotate?

We ascertained motivations by relating to earlier work on mobile media sharing as well as creating new categories for annotations that did not fit into existing ones.

Knowledge creation refers to the sharing of information towards a collective benefit [6]. Annotations typically described places of interest. For instance, a participant created an annotation of a bookstore (Figure 3a) and mentioned that by sharing, he was helping others to find it. We also observed that participants offered information beyond facts such as directions to include news and commentary. One participant created an annotation at a sports stadium informing others that it once held 2700 people in a football match. Another shared historical information of a church *"it has grown from 30 members in 1950s to 5000 members today"*.

(a) An annotation for knowledge creation

(b) An annotation for creation/maintenance of social relationship

Fig. 3. Example of photos of annotations

Self-expression refers to the sharing of annotations to express one's views [2]. Annotations were primarily those in the "Feelings" content category. For instance, a participant created an annotation after a long wait for his transportation *"yay, bus is coming!!!!!"*. Apart from being emotional, some participants created annotations from random thoughts such as *"Pain is inevitable, suffering is optional"*. Interestingly, emoticons and symbols were common (e.g. *":)"*, *":("*, "!!!!" , *"~yawns~"*).

Creating and maintaining of relationships refer to annotations that forge or sustain a relationship between the sender and recipient [2]. Such annotations may contain content to provide personal contextual information for others, or messages for friends and family. For example, a participant created an annotation to wish all his friends *"Happy Easter"* (Figure 3b). Another used SPLASH as a chat application by taking turns to annotate at a particular spot to establish a meeting place. Other participants spent time exploring what others were creating and replied to their annotations.

Self-presentation refers to the sharing of annotations to influence how others would view oneself [12]. For example, some participants created annotations of their possessions or pictures of friends. Another bought an expensive ring and shared this purchase via an annotation. Additionally, some participants mentioned that they created additional annotations to earn credits in order to customize their avatar. Indirectly, this was deemed as another form of self-presentation.

Competition refers to a contest of status between participants [10]. In SPLASH, this refers to the scoreboard where participants are able to see the top annotation contributors, individuals with the best annotation ratings, and the wealthiest participants. Those with this motivation were particular about their rankings and would spend time to beat others. One participant mentioned that he would not buy any items as he wanted to be on the top of the scoreboard for being the richest player (*"I want to be on top of everyone"*). On the other hand, other participants mentioned that they kept using SPLASH to "*not be*" at the bottom of the scoreboard.

Achievement involves sharing annotations to accomplish a particular goal [10]. Interestingly, those with this motivation used SPLASH's badges as goals to attain, and would stop playing only after achieving the desired badge. While it was common for them to create "Cheat" annotations, these participants would also announce their triumphs or progress through annotations. Examples included *"I just won balloon shooter"* and *"I need more credits"*. In the interviews, participants mentioned that they would resort to creating "Cheat" annotations especially when they wanted quick wins, such as on the cusp of a particular badge. Further, such participants spent the most time using SPLASH, often playing for hours at a stretch.

6 Discussion and Conclusion

The TAG framework allows application designers to better understand their users and implement features to suit their needs. We summarize our key findings and discuss implications. First, three major categories formed the Target of the annotations (T) and their sharing patterns suggest different features are needed. For the "self" category, helpful features are those that allow individuals to better express themselves. For example, users could select from a predefined set of emoticons or colors to represent their moods. Additionally, features to organize and retrieve such annotations would be desirable due to the potentially large number of annotations created. For annotations intended for "friends and family" or "public", notification tools will be valuable. For instance, users could select the audience that they would like to share and receive annotations.

Findings for annotation content (A) suggest features that could be implemented in response. For example, with "Food" being the top category, developers may consider exploring advertising partnerships with relevant businesses. We also observed a large number of annotations in the "Cheat" category which were not useful to others. Having too many of such annotations may annoy users and negate sustained usage. Thus, developers should be mindful of the pitfalls of external rewards and consider implementing preventive measures. These could include awarding virtual credits only for well-rated annotations, or to users that flag poor quality annotations. Finally, our findings suggest the potential for collecting data pertaining to the interests of users for individual profiling to better facilitate personalization for content creation and access.

For Goal (G), the varied motivations uncovered have important implications for application design. For example, knowledge creation can be afforded by prompting users to create annotations in scarcely populated locations. Further, competition and achievement motivations could be supported by incorporating challenging game mechanisms to foster creation of mobile content. This includes leveling of users, more varied game genres, or missions in which multiple tasks need to be accomplished.

Importantly, we found that games enhanced SPLASH's usage experience in two ways. One, games increased awareness of SPLASH as participants were willing to introduce it to their friends. Two, games encouraged users to spend more time with SPLASH. This increase translated to more exploration and creation of content. Thus, our findings suggest the potential of blending gaming elements with content sharing. Examples of features include scoreboards, badges or ratings on how others value a user's content contributions. Further, SPLASH's games helped resolve the "cold-start" problem that exists in new applications that lack content or users [20]. Hence, designers may consider how to leverage games to encourage usage, for instance, allowing users to unlock "hidden" features when certain objectives are achieved.

Put differently, games and content sharing though seemingly disparate, can be mutually reinforcing. However, this also suggests that developers have to address the twin challenges of usability and game design since users may seamlessly switch between the two activities. Further, such a combination is not without its pitfalls. The occurrences of the "Cheat" category of annotations to earn virtual credits suggest that developers have to be mindful of users who abuse reward systems.

Although this study has yielded valuable insights, there are some limitations in our work. First, many of our participants had IT-related backgrounds. Outcomes could differ with users from non-technical backgrounds. Further, the types of games may also affect the perceptions of our participants. SPLASH gameplay is open-ended and contains action mini-games which may not suit the taste of all participants who may prefer other genres such as puzzles, strategy, and role-playing variants. To obtain a better understanding of motivations, future studies need to be carried out on a diverse audience with different backgrounds. Finally, we would like to integrate more advanced gaming elements including missions, different levels of users, and "unlocking" features as players become more experienced with the application.

References

1. Smith, A.: Mobile Access 2010. Pew Research Center (2010)
2. Goh, D.H., Ang, R.P., Chua, A.Y.K., Lee, C.S.: Why we share: A study of motivations for mobile media sharing. In: Liu, J., Wu, J., Yao, Y., Nishida, T. (eds.) AMT 2009. LNCS, vol. 5820, pp. 195–206. Springer, Heidelberg (2009)
3. Java, A., Song, X., Finin, T., Tseng, B.: Why we Twitter: Understanding microblogging usage and communities. In: 9th WebKDD and 1st SNA-KDD 2007 Workshop on Web Mining and Social Network Analysis, pp. 56–65. ACM, New York (2007)
4. Taylor, C.A., Anicello, O., Somohano, S., Samuels, N., Whitaker, L., Ramey, J.A.: A framework for understanding mobile internet motivations and behaviors. In: CHI 2008 Extended Abstracts, pp. 2679–2684. ACM, New York (2008)
5. Jung, Y., Perez-Mira, B., Wiley-Patton, S.: Consumer adoption of mobile TV: Examining psychological flow and media content. Computers in Human Behavior 25, 123–129 (2009)

6. Lee, C.S., Goh, D.H., Chua, A.Y.K., Ang, R.P.: Indagator: Investigating perceived gratifications of an application that blends mobile content sharing with gameplay. Journal of the American Society for Information Science and Technology 61, 1244–1257 (2010)
7. Casey, S., Kirman, B., Rowland, D.: The gopher game: A social, mobile, locative game with user generated content and peer review. In: 2007 International Conference on Advances in Computer Entertainment Technology, pp. 9–16. ACM, New York (2007)
8. Matyas, S., Matyas, C., Schlieder, C., Kiefer, P., Mitarai, H., Kamata, M.: Designing location-based mobile games with a purpose: Collecting geospatial data with CityExplorer. In: Proceedings of the 2008 International Conference on Advances in Computer Entertainment Technology, pp. 244–247. ACM, New York (2008)
9. Tuite, K., Snavely, N., Hsiao, D.-Y., Smith, A.M., Popovic, Z.: Reconstructing the world in 3D: bringing games with a purpose outdoors. In: Fifth International Conference on the Foundations of Digital Games, pp. 232–239. ACM, New York (2010)
10. Sherry, J.L., Lucas, K., Greenberg, B.S., Lachlan, K.: Video game uses and gratifications as predictors of use and game preference. In: Vorderer, P., Bryant, J. (eds.) Playing Video Games. Motives, Responses, and Consequences, pp. 213–224. Lawrence Erlbaum Associates, Mahwah (2006)
11. Goh, D.H., Ang, R.P., Lee, C.S., Chua, A.Y.K.: Fight or unite: Investigating game genres for image tagging. Journal of the American Society for Information Science and Technology 62, 1311–1324 (2011)
12. Low, G., Goh, D.H., Lee, C.S.: A multifaceted approach to exploring mobile annotations. In: Chowdhury, G., Koo, C., Hunter, J. (eds.) ICADL 2010. LNCS, vol. 6102, pp. 71–80. Springer, Heidelberg (2010)
13. Grant, L., Daanen, H., Benford, S., Hampshire, A., Drozd, A., Greenhalgh, C.: MobiMissions: the game of missions for mobile phones. In: ACM SIGGRAPH 2007 Educators Program, p. 12. ACM, New York (2007)
14. House, N.V., Davis, M., Ames, M., Finn, M., Viswanathan, V.: The uses of personal networked digital imaging: An empirical study of cameraphone photos and sharing. In: CHI 2005 Extended Abstracts, pp. 1853–1856. ACM, New York (2005)
15. Espinoza, F., Persson, P., Sandin, A., Nyström, H., Cacciatore, E., Bylund, M.: GeoNotes: social and navigational aspects of location-based information systems. In: Ubicomp 2001: Ubiquitous Computing, pp. 2–17. ACM, New York (2001)
16. Ames, M., Naaman, M.: Why we tag: Motivations for annotation in mobile and online media. In: Proceedings of the SIGCHI Conference on Human Factors in Computing Systems, pp. 971–980. ACM, New York (2007)
17. Kindberg, T., Spasojevic, M., Fleck, R., Sellen, A.: I saw this and thought of you: Some social uses of camera phones. In: CHI 2005 Extended Abstracts, pp. 1545–1548. ACM, New York (2005)
18. Razikin, K., Tan, K.T., Goh, D.H., Chua, A., Lee, C.S.: Perspectives on mobile socializing, playing and content sharing. In: Proceedings of 8th International Conference on Information Technology: New Generations, pp. 873–878. IEEE Press, Los Alamitos (2011)
19. Kelly, D.: Methods for evaluating interactive information retrieval systems with users. Foundations and Trends in Information Retrieval 3(1-2), 1–224 (2009)
20. Shapira, B., Shoval, P., Tractinsky, N., Meyer, J.: ePaper: A personalized mobile newspaper. Journal of the American Society for Information Science and Technology 60, 2333–2346 (2009)

Flexible Publication Workflows Using Dynamic Dispatch

Sebastian Schick, Holger Meyer, and Andreas Heuer

Database Research Group,
University of Rostock
{schick,hme,heuer}@informatik.uni-rostock.de

Abstract. Publication processes within Digital Libraries are seldom supported by a workflow management system (WFMS). Publication workflows are often described within the application logic due to its data dependency — publication processes are data-driven. Though, documents and the processes should not be treated independently of each other. Rather, processes should dynamically react to changes of document structure and content.

We present an approach for flexible, data-centric publication workflows. The approach extends the control-flow perspective of a WFMS with concepts for handling process adaption at run-time, depending on a document's structure and its content.

Keywords: Digital Library, Publication Process, Flexible Workflow Modeling.

1 Introduction

Nowadays, Digital Libraries (DLs) are used to manage all kind of document types, e.g. academic articles, handwritings or course materials. A multimedia document encompasses different kinds of media types, e.g. text, video and audio. These media types are associated with lots of specific operations, which have to be described and implemented in the DL system. For example, supporting content based retrieval on each media type requires a complex feature extraction and indexing process. That's why the integration of multimedia documents within DLs is a challenge.

A common solution to control publication processes is the integration of a WFMS within the DL. Usually, the publication process of a certain document type is described by a process model. This model consists of activities which define atomic processing steps within the publication process. Besides this, execution constraints, resources (e.g. authors or IT services) and workflow relevant data have to be defined within the scope of a process model. The effort to maintain these process models increases with each document type added to the document model. Thus, documents and the processes should not be treated independently of each other. Processes have to react to document changes in a flexible way, i.e. often the control flow has to be adapted.

C. Xing, F. Crestani, and A. Rauber (Eds.): ICADL 2011, LNCS 7008, pp. 257–266, 2011.

In this paper we begin with the control flow perspective of the workflow and extend it with operations to handle dynamic process changes depending on document structure and content. We introduce a model to define relationships between a document's structure and content and dynamically adapted process structures.

The rest of the paper is organized as follows. In Section 2, we provide a motivating example together with requirements for publication process modeling and discuss related work. In section 3, we give a model for dynamic publication process specification and describe the composition of flexible process parts. The implementation of the approach using the YAWL [1] workflow engine is shown afterwards. Finally, in Section 4 we conclude and give an outlook on further activities.

2 Publication Processes and Digital Libraries

2.1 Running Example

The Publication Process. describes the creation of complex structured documents in digital libraries. In DLs authors often lack support during the creation of complex multimedia documents, e.g. when they structure and assemble different media types into one logical document. These systems often assume that this work was done before. Nevertheless, some processes within the DL depend on the content type like indexing, which runs differently for media types like text, image or video. In addition, content creation should not be handled independently of its usage.

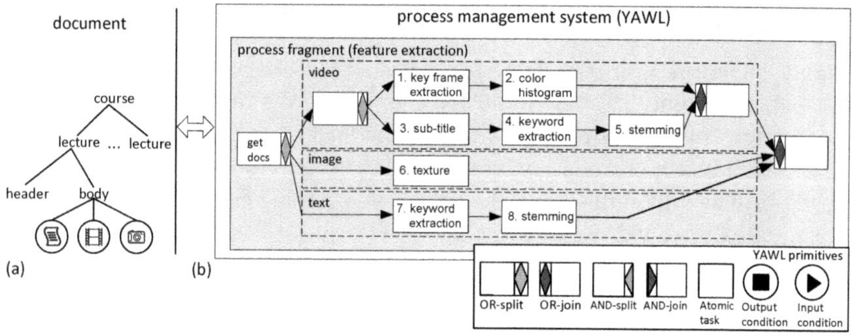

Fig. 1. Workflow for a course management

Fig. 1 (a) shows a simple document for a course management system. A course document consists of several lectures. Each individual lecture can make use of different media types, e.g. a slideshow consisting of a sequence of images, a video shown and discussed in the lecture and the full text of the lecturer's notes. In sub-figure (b) the feature extraction part of the document indexing in the DL is depicted. When the document for a specific lecture contains a video, key-frame extraction takes place first. Then methods for feature extraction for still image are applied to the sequence of key-frames, e.g. the calculation of the color histogram for each image. When sub-titles exist, they are extracted and text indexing takes place afterwards. This can be done in parallel to key-frame extraction and indexing but only if a video is present in the document.

So, some of the activities can be carried out in parallel, e.g. key-frame extraction in parallel to sub-title analysis, but others can take place only in a certain order, e.g. sub-title extraction before full text indexing. Some activities can only take place if the related media type is present in the document. There are other sub-tasks omitted here which must be executed in dependence of certain document parts. For example different metadata extractors are used for different file formats.

2.2 Requirements and Publication Process Modeling

There is a strong need for the description of both the content specific and the application-specific parts of the publication process. This is reflected by dependencies between documents or certain document parts (based on their structure and/or content) and instances of workflow tasks. As shown in the example above, different document parts need different indexing techniques, which are performed by sophisticated sub-workflow sometimes. This poses different requirements on the overall process modeling part of a DL. We list some of them which can be tackled by our approach:

Req. 1 *Avoid complex process specifications.* Moreover, the process specification should concentrate on the application-specific parts of the publication process in a comprehensive manner. **Req. 2** *Decouple the application-specific from the content specific part of the process specification.* Ideally, a high level description of the publication process concentrates on the individual, application-specific phases. **Req. 3** *Support the reuse of process parts.* **Req. 4** *Allow for run-time adaption of document specific process parts.* Each process instance should only contain process parts which are directly related to the active documents it processes. **Req. 5** *Document changes have to be associated with process parts.* Editing operations (e.g. insert, delete, ...) on documents influence process specifications at run-time.

2.3 State of the Art

Within the Digital Library Community different approaches are discussed. Supporting authors by defining flexible interfaces is proposed in [3] and [7]. We propose the use of a WFMS that can control all processes within a digital library, because of supporting authors through flexible user interfaces does not consider technical services such as indexing of different document types.

SCOPE [6] is a framework for generic publication processes. It uses publication components as part of the publication chain which transforms documents from one state to another state. However, the described publication process refers only to publication of documents e.g. MS Word or Latex. Flexible building of collections or complex document structures is not considered.

Greenstone [2] is a Framework for building digital library collections. The architecture is divided into seven parts which represents phases within the publication process. Within these parts different plug-ins are managed to process documents in the right way. Nevertheless, the publication process is restricted to seven phases and a dynamical adaption of publication process is also not possible. Flexibility is supported only by the choice of various plug-ins.

Current WFMS offer different approaches to support flexibility. A common solution to manage different publication processes are process variants [4,5]. Applied to DLs, process variants are produced by using a reference process model (e.g. publication process) which is configured according to different requirements, e.g. different document types. The intention is to avoid *choices* within process models where decisions can be predicted. Since the publication process is highly dynamic during runtime, the model's configuration has only a small effect. Content-related process parts must still be completely modeled (Req. 2,4,5 are not met).

Pockets of flexibility [8] uses the concept of *open instances*. Within the process model pockets of flexibility are defined within a core process. A pocket is a special build activity which composes activities depending on different constructs (e.g. fork sequence, etc.). This approach is very similar to ours regarding the choice of the late modeling concept [10]. However, the composition is left to the user and only restricted afterwards by conditions. Constraints related to external environment are not considered at all (Req. 5 is violated).

3 Flexible Document Aware Workflows

To illustrate our approach for dynamic publication processes Fig. 2 outlines the newly introduced concepts. On the left side of Fig. 2 the dynamic process model is depicted together with the related logical document model underneath. The two models will be discussed in more detail in Sec. 3.1. The dynamic parts of the approach, the process composition and execution, are shown in the center and right part of Figure 2. The details will be discussed in Sec. 3.2.

3.1 Dynamic Process Specification

The dynamic publication process model (cf. Fig. 2) will be adapted in dependency of the current document instances during runtime using a *dynamic dispatch* of activities. Therefore, our approach observes document changes and specifies where corresponding changes in the workflow should take effect, i.e. where dynamically generated sub-processes are executed. We use a notation of a core process which is extended with special *observer* and *generator* tasks. In order to achieve flexibility within the publication process the full specification of the process model is completed at runtime.

Bricklets b_i are the building blocks for re-using specific activities which are bundled into sub-processes (Req. 3). They are not directly part of the process specification but will be executed at well-defined points. The bricklets are a mean

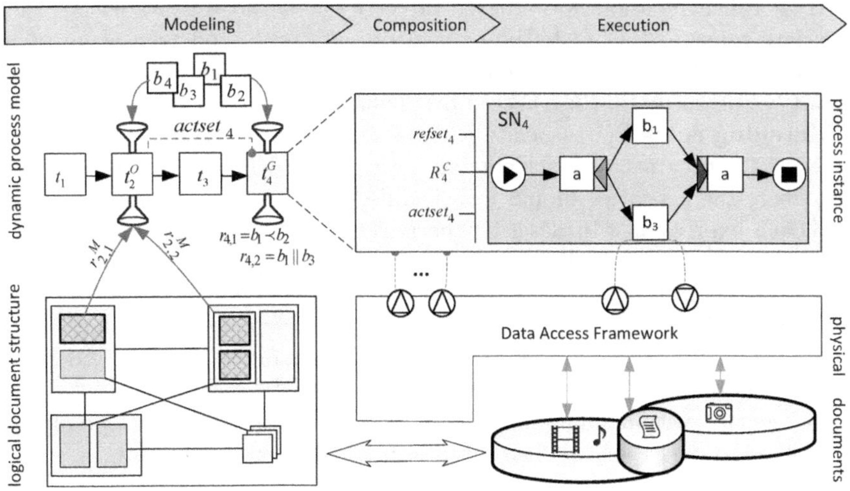

Fig. 2. Supporting Dynamic Publication Processes

for separating the application-specific process from data specific parts which rely on current document structures (Req. 2).

Definition 1 (Bricklet). *Let* $B = \{b_1, b_2, \cdots, b_i\}$ *with* $i \in N$ *be the set of* bricklets. *A bricklet* b_i *is a valid process model, which contains at least one task definition. Bricklets will be assembled into subnets* SN_i.

Observer tasks t_i^O define points within the dynamic process model where the status of documents to be published should be investigated. Usually, they are inserted after activities which cause major changes of the document. The task t_2^O in Fig. 2 denotes an observer task which makes use of the *matching rule set* (R_2^M). The rules ($r_{2,1}^M, r_{2,2}^M$) specify which activities should be added or removed from the process if there are certain parts in document or if they are absent.

Definition 2 (Observer Task). *Let* T^O *denote the set of* observer tasks $t_i^O = (R_i^M, match)$ *with* $i \in N$. *Then:*

- $R_i^M = \{r_{i,1}^M, r_{i,2}^M, \cdots, r_{i,j}^M\}$ *with* $j \in N$ *is a set of matching rules.*
- $r_{i,j}^M : match(pexpr_j) \mapsto (op, t_m^G, B_j, pfrag_j)$ *is a matching rule.*
- $pexpr_j$ *is used to specify data parts expected within the data. It is basically a XPath expression.*
- $op = \{add, delete, merge, undo\}$ *is the set of change operations.*
- $t_m^G \in T^G$ *is a generator task.*
- $B_j \subseteq refset_m$ *is a set of predefined bricklets.*
- $pfrag_j$ *contains the resulting XPath 1.0 node-set using* $pexpr_j$.

Whether a document fragment exists or not is determined using the path expression $pexpr_j$ and the function $match(pexpr_j)$. Using our *Data Access Framework*

(DAF), each expression is evaluated directly on the documents within the appropriate repositories [9]. If the expression gets evaluated true corresponding bricklets are added to or removed from the construction set $actset_4$ of *generator tasks* t_4^G. Furthermore, the selected fragments are returned as $pfrag_j$.

Matching rules R_i^M associate parts of a document (based on content and/or structure) with a set of bricklets (activities) and define points in the control flow where the activities should be scheduled (Req. 2 & 4). As an consequence, an XPath expression will describe the parts within document instances which should (not) match and trigger activities in the subsequent workflow. Sometimes the existence or absence of a document fragment will not only add but remove also scheduled activities depending whatever the default behavior may be. If for example sub-titles are added to a video then their fulltext can be indexed and content based analysis of key-frames can be omitted.

Each $pexpr_j$ within a matching rule $r_{i,j}^M$ is closely associated with an operation *op*. Where an operation *op* may only use bricklets from set $refset_i$.

Definition 3 (Change Operation). *Let* $op = \{add, delete, merge, undo\}$ *denote the set of possible operations to manipulate the construction set* $actset_i$ *(hereinafter* $actset$*) of generator tasks* t_i^G*.*

The add *operation appends the set of activated bricklets* B_j *to the activation set* $actset$*. The* merge *operation appends a set of predefined bricklets* B_j *only into* $actset$ *if they are not member of it. And the* delete *operation removes bricklets from* $actset$*.* B_j^{-1} *is the compensation of* B_j *and the* undo *operation appends a set of compensating bricklets* B_j^{-1} *to the activation set* $actset$ *to rollback operations* B_j *after a data fragment was removed from the document.*

Generator tasks[1] t_i^G specify points within the flow of control where bricklets are combined at run-time to build up a subnet of activities, e.g. SN_4 in Fig. 2. The resulting subnet is then deployed and executed. Essentially, the generator tasks are responsible for the dynamic dispatching of the activities/bricklets like selecting and executing method calls in object-oriented systems (Req. 4 & 5). To build up the subnets a set of composition rules (R_i^C) and set of scheduled activities ($actset_4$) is used. The scheduled activities must belong to a set of allowed activities ($refset_4$) per distinct generator task. If a generator task is executed within the process, it has to compose a valid execution order for the activated bricklets. The generator tasks provide flexibility into the process instance using a late modeling, descriptive approach and under-specification [10].

Definition 4 (Generator Task). *Let* T^G *be the set of* generator tasks $t_i^G : (R_i^C, actset_i, refset_i) \mapsto SN_i$ *with* $i \in N$*. Then:*

- SN_i *is a valid subnet executed if* t_i^G *is processed within the control flow.*
- $R_i^C = \{r_{i,1}^C, r_{i,2}^C, \cdots, r_{i,j}^C\}$ *with* $j \in N$ *defines a set of construction rules.*
- $actset_i$ *is a set of active bricklets* (b_k) *and corresponding number of data fragments* ($f(b_k)$) *chosen by different* observer tasks t_m^O.
- $refset_i$ *defines all bricklets allowed for* t_i^G.

[1] It resembles the idea of *pockets of flexibility* introduced in [8].

Construction rules R_i^C define relationships between bricklets and how they are combined into a resulting control flow. If a bricklet is a pre-requisite for another, a sub-sequent order can be specified. Further, a bricklet can be executed sequential or parallel n-times. If not stated otherwise, bricklets can be executed arbitrarily and in parallel.

Definition 5 (Construction Rules). *Let R_i^C be the set of construction rules $r_{i,j}^C$. A construction rule $r_{i,j}^C \in \{b_k \prec b_l, b_k \overset{n}{\prec}, b_k \overset{n}{||}\}$ defines how a* bricklet b_k *is inserted into the subnet SN_i iff $b_k \in actset_i$.*

- $b_k \prec b_l$*: Iff bricklet $b_l \in actset_i$, b_l is immediately executed after b_k.*
- $b_k \overset{n}{\prec}$*: The bricklet b_k will be inserted sequentially n times.*
- $b_k \overset{n}{||}$*: The bricklet b_k will be inserted n times in parallel.*

By using these concepts, we avoid complex process structures (Req. 1). The primary process specification is a model of the application's point of view. Wherever message-specific activities have to be carried out, they are hidden by generator tasks and descriptive matching and composition rules. These rules determine the dynamic execution of a re-usable set of message-specific activities.

3.2 Composing Sub-processes

After the activation of bricklets, which is done by the observer tasks, the composition of a valid sub-process has to be controlled by the generator task (cf. center and right part of Fig. 2) . Therefore, we provide an algorithm for combining bricklets b_k into a sub-process using the construction rule set R_i^C. The composition is done during runtime to offer a flexible generation of sub-processes.

Since $actset_i$ changes during runtime, SN_i only needs to be generated when t_i^G is activated. This is a two-step procedure. First a directed acyclic graph is created with all activated bricklets $b_n \in actset_i$. In the second step we transform the graph into a valid sub-process (YAWL subnet).

Definition 6. *Let $G_i^C = (V, E)$ be a digraph. V is the set of vertices and E is the set of directed edges. The graph G_i^C contains only one starting vertex "start" $\in V$ and one ending vertex "end" $\in V$.*

Definition 7. *The indegree $deg^-(b_k)$ is the number of head endpoints for bricklet b_k. The outdegree $deg^+(b_k)$ is the number of tail endpoints for b_k.*

Listing 1.1 shows a simple algorithm to calculate SN_i. Since the subnet is composed during runtime no deferred choice is needed and no composition rule is mapped to OR-splits, either. We avoid cycles in the constructed subnet graph by enforcing acyclicity of the construction rule set R_i^C at modeling time[2].

[2] This can be done by applying R_i^C on $refset_i$

Listing 1.1. Algorithm to calculate subnets

initialize G with $G.V = \{start, end\} \cup actset_i$ and $G.E = \{\}$
foreach $r^C_{i,j} \in R^C_i$ {
  if $r^C_{i,j}$ equals $b_k \prec b_l$ and $\{b_k, b_l\} \subseteq actset_i$ {
    add directed edge (b_k, b_l) }}
foreach $r^C_{i,j} \in R^C_i$ {
  //expand replaces b_k *by* n *nodes* $b_{k.m}$ *of type* $b_k \wedge m \in \{1, \ldots, n\}$
  if $r^C_{i,j}$ equals $b_k \overset{n}{\prec}$ and $b_k \in actset_i$ {
    n = $actset_i.f(b_k)$; expand(b_k, n);
    add directed edge between successive $b_{k.m}$ }
  if $r^C_{i,j}$ equals $b_k \overset{n}{\|}$ and $b_k \in actset_i$ {
    n = $actset_i.f(b_k)$; expand(b_k, n)
    if $\exists r^C_{i,m} \in R^C_i \wedge r_{i,m}$ equals $b_k \prec b_l \wedge b_l \in actset_i$ {
      foreach $b_{k.o} \in \{b_{k.1}, \ldots, b_{k.n}\}$ { add directed edge $(b_{k.o}, b_l)$ }}
    if $\exists r^C_{i,m} \in R^C_i \wedge r_{i,m}$ equals $b_l \prec b_k \wedge b_l \in actset_i$ {
      foreach $b_{k.o} \in \{b_{k.1}, \ldots, b_{k.n}\}$ { add directed edge $(b_l, b_{k.o})$ }}}}
foreach $b_k \in G.V$ {
  if $deg^-(b_k) = 0$ { add directed edge $(start, b_k)$ }
  if $deg^+(b_k) = 0$ { add directed edge (b_k, end) }}

The digraph for subnet SN_i is not executable within YAWL, therefore it is transformed into a valid YAWL net. The transformation is based on rules $R_{1\ldots6}$:

R_1 maps the *start* node to a YAWL *Input Condition* where the process starts.
R_2 maps the *end* note to an *Output Condition* where the process ends.
R_3 maps a sequential path from node b_i and b_j to corresponding tasks b_i, b_j.
R_4 maps the split of node b_k to a *And-Split task* b_k. Nodes b_i, b_j are mapped to corresponding tasks b_i, b_j. Node b_n is mapped to a *And-Join task* b_n.
R_5 if the start node *start* is part of a parallel execution, *start* is mapped to an *Input Condition* together with an *And-Split task* a. The dummy task a will be connected with the following tasks $b_i \ldots b_j$.
R_6 if the end node *end* is part of a parallel execution, *end* is mapped to a *Output Condition* together with a dummy *And-Join task* a. The dummy task a will be connected with the incoming tasks $b_i \ldots b_j$.

Example 1. In Fig. 3 (a) the process from Fig. 1 is realized using our approach. Task t_1 (*get docs*) is modeled as an observer task which controls the dynamic dispatch of activities in generator task t_2 (*feature extraction*). Task t_2 provides the set of selectable bricklets $refset_2 = \{b_1, b_2, b_3, b_4, b_5, b_6\}$, which correspond to the tasks (1.…6.) in Fig. 1. The observer task t_1 uses a set of matching rules $R^M_1 = \{r^M_{1,1}, r^M_{1,2}, r^M_{1,3}\}$ with $r^M_{1,1} = add(t_2, b_6,$"`//course[../doctype='image']`"), $r^M_{1,2} = add(t_2, \{b_1, b_2, b_3, b_4, b_5\},$"`//course[../doctype='video']`") and $r^M_{1,3} = add(t_2, \{b_4, b_5\},$"`//course[../doctype='text']`")[3]. For the sub-process construction in t_2, the rule set $R^C_2 = \{b_1 \prec b_2, b_3 \prec b_4, b_4 \prec b_5, b_1 \overset{2}{\prec}, b_4 \overset{2}{\|}, b_6 \overset{2}{\|}\}$ is used, which reflects the order of the activities in Fig. 1. The observer task t_1

[3] Tasks for keyword extraction(4.) and stemming (5.) are reusable for tasks 7. and 8.

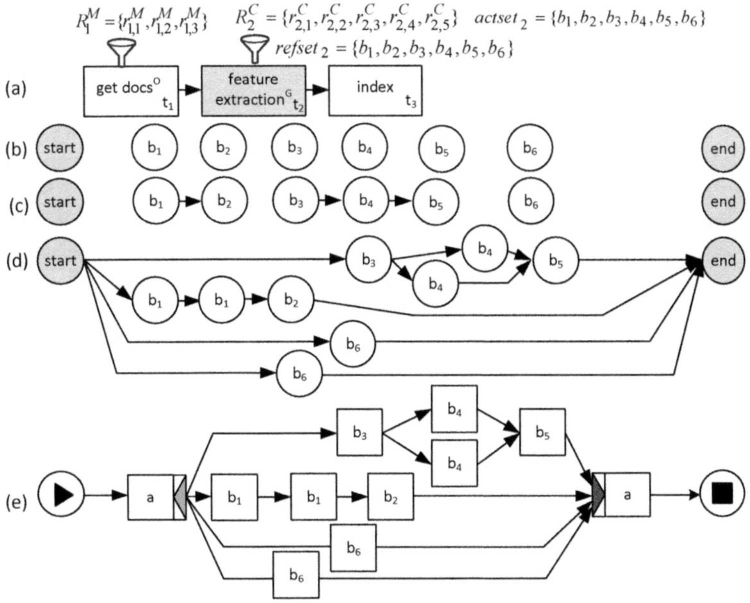

Fig. 3. Digraph construction and transformation into a YAWL subnet

activates the bricklets in $actset_2 = \{b_1, b_2, b_3, b_4, b_5, b_6\}$ due to matching rules $r^M_{1,1}$, $r^M_{1,2}$, $r^M_{1,3}$.

Fig. 3 (b)–(d) depict the construction of digraph G. First, all activated bricklets $(b_1, b_2, b_3, b_4, b_5, b_6)$ will be inserted into $G.V$ (Fig. 3 (b)). The application of rules $b_1 \prec b_2$ (edge from b_1 to b_2), $b_3 \prec b_4$ (edge from b_3 to b_4) and $b_4 \prec b_5$ (edge from b_4 to b_5) is shown in (c). The application of the other rules is shown in (d): $b_1 \overset{2}{\prec}$ (edge from first b_1 to second b_1), $b_4 \overset{2}{\parallel}$ and $b_6 \overset{2}{\parallel}$ (both b_4 and b_6 are duplicated). The final step (d) connects all vertices with $deg^-(b_i) = 0$ or $deg^+(b_i) = 0$ with the "start" and "end" nodes. The YAWL subnet (e) results from transforming digraph G into YAWL. This subnet gets deployed and executed at runtime by generator task t_2.

The approach presented above was exemplified using YAWL [1] and the corresponding WFMS. Two *Custom Component Services* for the observer and generator task types were implemented. We have extended the YAWL editor to describe the matching and construction rule sets. This allows for modeling everything within the standard YAWL environment. The bricklets are implemented as YAWL nets which contain always a start and an end condition.

4 Conclusion and Future Work

The publication of multimedia documents in DLs is a field that bears much practical relevance. We provided an approach for a flexible publication process of multimedia documents in DLs. In contrast to other approaches, we achieve the flexibility by monitoring document changes and specifying at which part in the control flow corresponding changes should take effect. This is done by extending a WFMS with observer tasks to monitor the document changes. Generator tasks allow for flexible construction and execution of process instances. Our technique presented here can be described as a late modeling, descriptive approach using under-specification [10]. The main contribution of our approach is that both context conditions on external data and the resulting changes of the process instances are described within one process model. Additionally, all components of our approach are provided and executed by extensions of the WFMS YAWL. The prototype implementation finally allows for a detailed case study and evaluation.

Future work will extend the construction rule mechanism to provide more freedom in combining bricklets (process fragments) into sub-processes. The construction of sub-processes should take data dependencies between bricklets into account, e.g. based on data flow analysis. Additionally a library of bricklets is built, which offers a set of generic operations for different media types, e.g. extracting metadata, for converting file formats, thumbnail generation, and others.

References

1. van der Aalst, W.M.P., ter Hofstede, A.H.M.: Yawl: yet another workflow language. Information Systems 30(4), 245–275 (2005)
2. Buchanan, G., Bainbridge, D., Don, K.J., Witten, I.H.: A new framework for building digital library collections. In: JCDL, pp. 23–31 (2005)
3. Davis, S., B. II, P.L., Cifuentes, L., Francisco-Revilla, L., Furuta, R., Hubbard, T., Karadkar, U., Pogue, D., S. III, F.M.: Template-based authoring of educational artifacts. In: JCDL, pp. 242–243 (2006)
4. Gottschalk, F., van der Aalst, W.M.P., Jansen-Vullers, M.H., Rosa, M.L.: Configurable workflow models. Int. J. Cooperative Inf. Syst. 17(2), 177–221 (2008)
5. Hallerbach, A., Bauer, T., Reichert, M.: Managing process variants in the process life cycle. In: ICEIS, vol. (3-2), pp. 154–161 (2008)
6. Müller, U., Klatt, M.: Scope — a generic framework for xml based publishing processes. In: Rauber, A., Christodoulakis, S., Tjoa, A.M. (eds.) ECDL 2005. LNCS, vol. 3652, pp. 104–115. Springer, Heidelberg (2005)
7. Park, Y., Karadkar, U., Furuta, R.: Component-based authoring of complex, petri net-based digital library infrastructure. In: Lalmas, M., Jose, J., Rauber, A., Sebastiani, F., Frommholz, I. (eds.) ECDL 2010. LNCS, vol. 6273, pp. 22–29. Springer, Heidelberg (2010)
8. Sadiq, S.W., Orlowska, M.E., Sadiq, W.: Specification and validation of process constraints for flexible workflows. Inf. Syst. 30(5), 349–378 (2005)
9. Schick, S., Meyer, H., Heuer, A.: Enhancing workflow data interaction patterns by a transaction model. In: ADBIS Research Communications (accepted for Publication, 2011)
10. Weber, B., Sadiq, S., Reichert, M.: Beyond rigidity — dynamic process lifecycle support. Computer Science — Research and Development 23, 47–65 (2009)

An RDF-Based Platform for E-Book Publishing

Kornschnok Dittawit and Vilas Wuwongse

Computer Science and Information Management,
Asian Institute of Technology, Thailand
{Kornschnok.Dittawit,vilasw}@ait.ac.th

Abstract. E-books have become an important player in the industry. The market is now flooded with e-readers and tablets designed to optimize e-book reading experience. However, the majority of e-books being published today fail to take full advantage of the electronic form and often appear mostly identical to their printed book counterparts. A platform for e-book publishing is proposed to assist authors in creating rich e-books and enhance reading experience. With the underlying data model that makes use of the Resource Description Framework, the platform has been designed to reap the benefits of the large pool of information available on the internet and on the Linked Open Data (LOD) cloud whenever possible to give authors and readers access to relevant information resources. The developed platform has been compared with other similar tools and the results showed that it is able to provide useful services while maintaining the basic functions of e-book authoring and reading.

Keywords: e-books, electronic publishing, authoring tools, Resource Description Framework, Linked Open Data.

1 Introduction

Electronic publishing plays an important role in today's world. All books and publishing materials start their lives in electronic forms [1]. The popularity and need for printed books started to subside with definite proof from the drop in physical book sales as e-books gain more popularity each year [2].

E-books offer several advantages over printed books; the most useful ones being those features that can never be offered in traditional printed books such as embedded videos and audio, search feature, and built-in dictionary. Additionally, newer e-readers and tablets start to incorporate features that require internet access such as sharing passages and posting comments which could pose a problem some years ago. However, today a large population has wide access to the internet [3] and some people even have access all day. Thus, e-books have the potential to be more useful and become a greater source of information and knowledge if they are to connect to the internet and harvest the vast pool of information available in the cloud.

This paper proposes a platform for e-book publishing that aims to enable e-book authors and readers to obtain additional information readily available on

C. Xing, F. Crestani, and A. Rauber (Eds.): ICADL 2011, LNCS 7008, pp. 267–276, 2011.

the World Wide Web (WWW). The platform is driven by the Resource Description Framework (RDF), a semantic web language that has been chosen as the underlying data modeling language, which allows flexible metadata insertion to all components in e-books and enables linking e-books to relevant web resources. With additional metadata coupled by external information resources, the platform is capable of providing useful services such as searching for information on the same subjects, linking to referenced authors' webpages, and searching for similar images and videos.

2 Reviews of E-Book Authoring and Reading Software

The proposed platform does not intend to change the way people write and read e-books. Instead, the platform contains the same set of features common in existing e-book authoring and reading software supplemented with extra beneficial features. Understanding the essential features for e-book authoring tools and e-book reading software is the key to building the platform. It also helps in designing the data model that serves as the basis for the platform.

2.1 Reviews of E-Book Authoring Tools

Six e-book authoring tools have been selected and studied to derive the set of common features necessary for e-book authoring. The six tools include Adobe InDesign (`http://www.adobe.com/products/indesign.html`), Microsoft Word (`http://office.microsoft.com/en-us/word`), Sigil (`http://code.google.com/p/sigil`), eCub (`http://www.juliansmart.com/ecub`), Jutoh (`http://www.jutoh.com`), and DesktopAuthor (`http://www.desktopauthor.com`). Note that Microsoft Word is not a dedicated e-book authoring tool but is used by some non-professional authors due to its rich features and familiarity. The summary of a list of common features is presented in Table 1.

Table 1. Summary of features in e-book authoring tools

Function Group	Features	Supported Tools**
Content Editing	- What-You-See-Is-What-You-Get (WYSIWYG) editor - can easily insert images and/or multimedia objects	AWSJD AWSJD
Content Type	- images - videos/audio files	* AWD
Content Source	- allows inserting web links - allows multimedia objects from external websites	* AWD
Navigation	- provides function for creating table of contents - allows jumping to different parts of the book	* *
Others	- allows specifying e-book metadata	*

** A = Adobe InDesign, W = Microsoft Word, S = Sigil, E = eCub, J = Jutoh, D = DesktopAuthor, * = all

2.2 Reviews of E-Book Reading Software/Devices

Five e-book reading software/devices have been selected and studied to derive the set of common features necessary for e-book reading. The five pieces of software/devices include Adobe Digital Edition (`http://www.adobe.com/products/digitaleditions`), Amazon Kindle (`https://kindle.amazon.com`), iBooks (`http://www.apple.com/ipad/built-in-apps/ibooks.html`), Adobe Reader (`http://www.adobe.com/products/reader.html`), and Bookworm (`http://bookworm.oreilly.com`). The summary of a list of common features is presented in Table 2.

Table 2. Summary of features in e-book reader software/devices

Function Group	Features	Supported Tools**
Content Viewing	- automatic reflow of contents	AKIB
	- supports image and/or video rendering	*
	- allows decreasing/increasing font size or zooming	*
Navigation	- allows jumping to different parts of the book	*
	- supports bookmarking	AKI
Others	- supports annotations	AKIR
	- supports text highlighting	AKIR

** A = Adobe Digital Edition, K = Amazon Kindle, I = iBooks, R = Adobe Reader, B = Bookworm, * = all

3 E-Book Data Model

Any tool or platform is often built on top of a data model that serves as the abstraction [4] and helps defining important functions and the relationship between each component. Because there exist a number of modeling languages, it is necessary to analyze the nature of the information to model and choose the appropriate language that best describes the information.

3.1 RDF as a Modeling Language

An e-book is made up of text and multimedia content. The content could be grouped into chapters or sections. An e-book often has one table of contents and a list of references as well as an index. From the description, a simple e-book structure could be modeled as a graph as in Fig. 1. E-book is considered a source of information so e-book metadata is important and the ability to specify e-book metadata is implemented in all major e-book authoring tools. As a result, RDF was chosen as the modeling language due to its graph nature as well as the simplicity and flexibility it offers [5]. Using RDF as the data model allows adding metadata to e-books easily and allows defining e-book structure in a meaningful way. Apart from the simplicity and flexibility, RDF has the advantage of being a semantic web language that serves as the basis for the Linked Open Data (LOD)

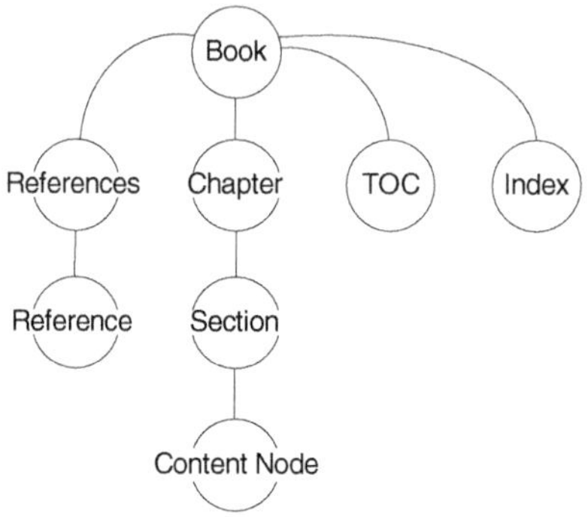

Fig. 1. Representing e-book structure as a graph

cloud [6] which is a vast pool of interconnected RDF resources from various domains such as geography, medicine, famous places and people, and definitions for general vocabularies in different languages. Establishing links from e-books to these resources will help expand the pool of information and offer authors and readers addtional information resources.

Figure 2 shows the example for an e-book modeled with RDF. The e-book consists of a table of contents representing the hierarchical structure of the book, a list of references, and a book index listing all topics that occur in the book. Because e-books are publishing documents, some elements from the Dublin Core (DC) [7] metadata set are used mainly for major components and properties along with a newly defined metadata set annotated by rdfb namespace prefix for some e-book components and properties that are not defined by DC. As a future enhancement, the data model can be made more flexible with the use of application profiles where a set of required elements are specified along with other elements specific to the systems. This makes it possible for local usage of the platform when certain elements are needed to store system specific data.

3.2 Content Nodes and Reading Paths

This section introduces two concepts called "content nodes" and "reading paths". E-book contents can be broken down into atomic elements called "content nodes". Each content node corresponds to a set of consecutive paragraphs that are related to the same subjects. The way in which content nodes are broken down is left to the author to decide. This concept makes it possible to specify metadata to only parts of the content as in Fig. 3 which will be useful when readers read the partial content and would like to query for more information about a specific subject associated with that part. When an author cites the work of another person, he can add the referenced work as the reference for that content node and the reference will also appear in the list of references for the e-book. This gives a stronger tie between the part of the content and the reference than only

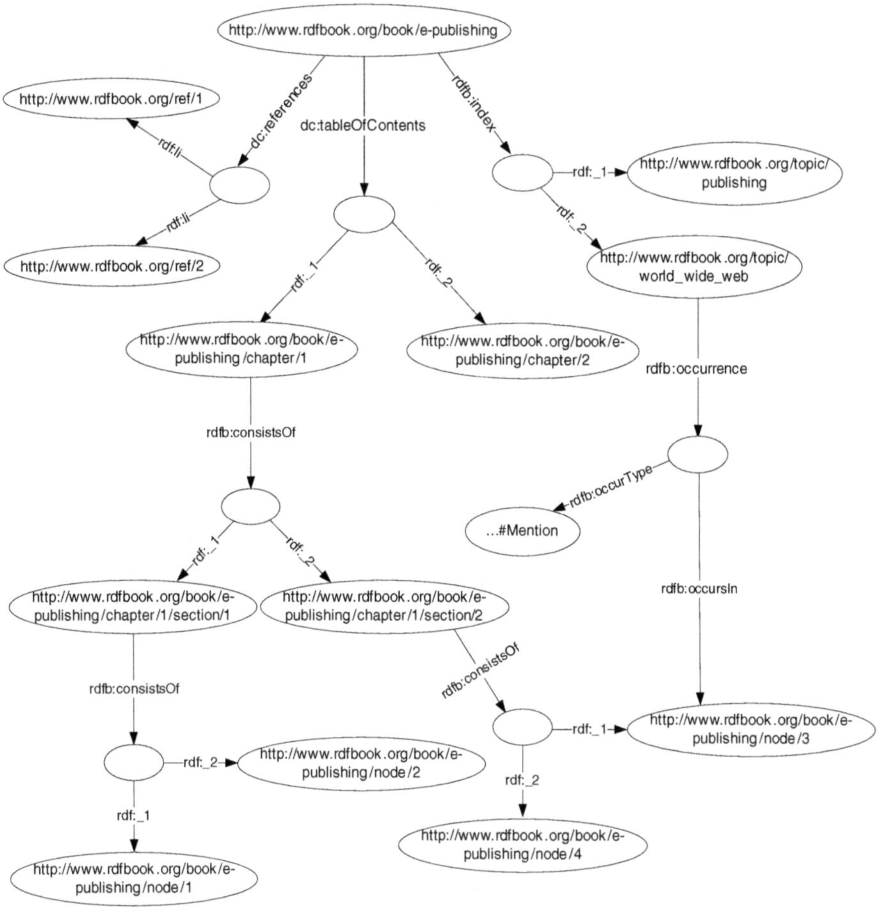

Fig. 2. An example e-book data model using RDF

using the usual way of adding citation that points to a certain reference. A more interesting aspect of the proposed e-book publishing platform is the feature that allows either authors or readers to create "reading paths". A reading path is merely a partial table of contents that serve as a guidance on which chapters the readers should read in order to grasp the essence of certain areas of focuses. This can be most useful for text books whereas readers only read some chapters of interest and leave out the rest. Reading paths can assist readers by telling them which chapters to read if they would like to focus on certain aspects of the book. The data model for a reading path can be seen in Fig. 4. From the figure, the reading path is a sequence of chapters. The sections and content nodes under each chapter are omitted for readability but they are in fact attached to the chapter resources. Because the reading path resource points to the same chapter resources existed in the graph, there is no duplication of resources.

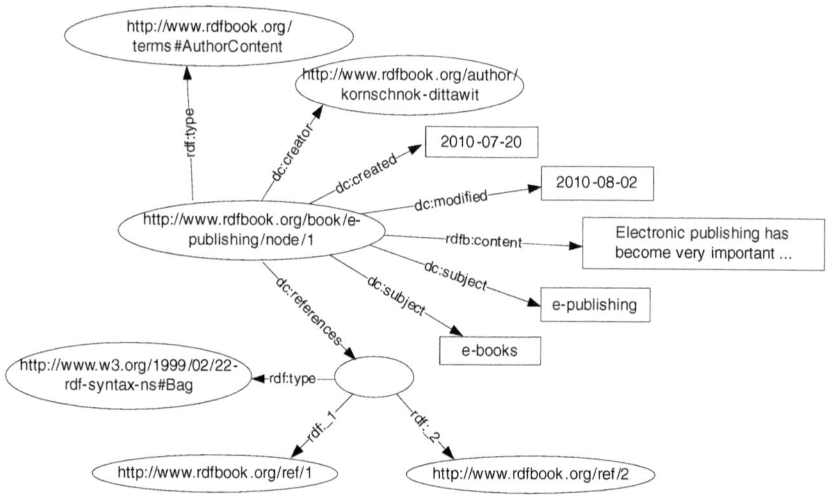

Fig. 3. Modeling e-book content nodes with RDF

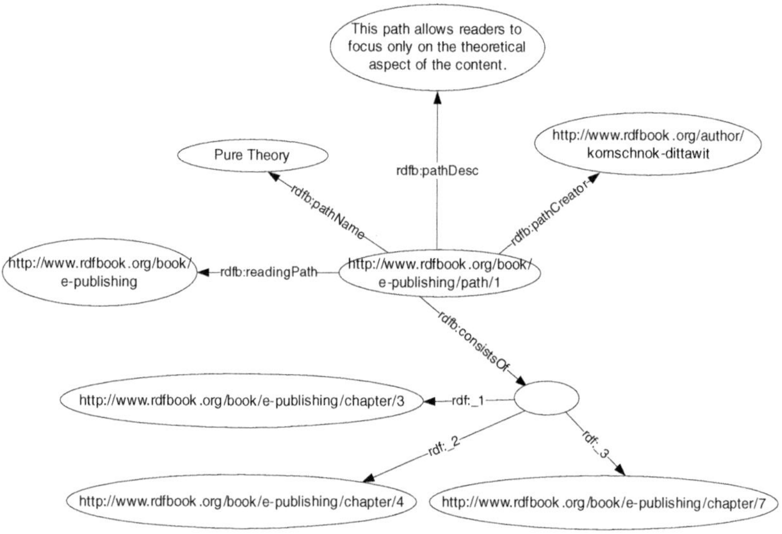

Fig. 4. RDF for reading path data model

4 E-Book Publishing Platform

A platform for e-book publishing has been developed on top of the RDF data model explained in Chap. 3. The platform aims to enhance authoring and reading experience by providing useful services to both authors and readers.

4.1 Platform Architecture

The platform has been developed as a web application using PHP programming language running on Apache HTTP Server. It makes use of several APIs for searching and retrieving additional information from the web. Figure 5 shows the architecture for the developed platform.

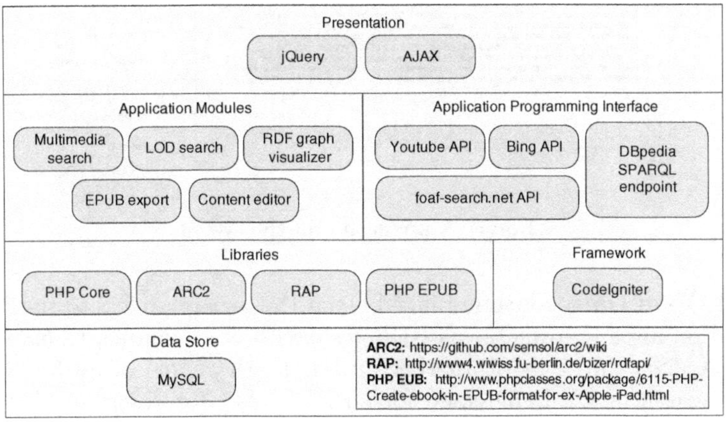

Fig. 5. Platform architecture

4.2 Features

The platform provides features necessary for authoring and reading e-books as well as some extra features as listed in Table 3. Figure 6 shows the system screenshot.

Table 3. List of features in the developed platform

WYWIWYG editor	Managing table of contents
Insert multimedia contents	Edit/view metadata
Save as EPUB	Multimedia/author search
Search for similar resources	Reading path creation

4.3 Platform Components

This section discusses the major components in the developed publishing platform. Only the unique components to the proposed platform will be explained.

Multimedia search. Multimedia search is the built-in feature that allows e-book authors to search for web resources of various types including websites, images, PDF files, PowerPoint slides, and video clips. When the author chooses web resources of types websites, PDF files, or PowerPoint slides to include, the URLs of the selected web resources will be inserted as new content nodes where each node corresponds to each web resource. In case of images and video clips, they will be embedded as new content nodes and can be viewed without visiting the URLs first.

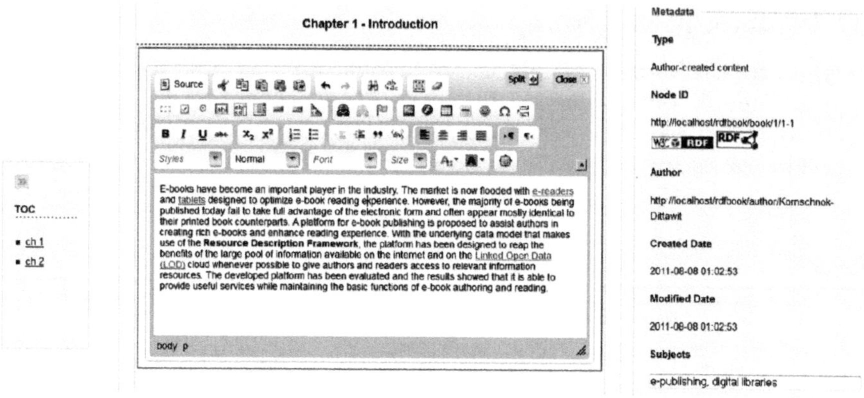

Fig. 6. A screenshot of the system

Linked Open Data search. Linked Open Data search refers to the search for information already available as RDF resources. For example, when an e-book author A cites the work of another author B, the name of author B will be used to search in the LOD cloud for an RDF resource describing his personal information. This will be used by the platform to provide information about the referenced author to the readers.

RDF graph visualizer. The RDF graph visualizer presents an e-book's metadata in the form of a graph which is easier to understand by human. Because the graph for the whole e-book can be very large, the platform allows viewing a subgraph representing a specific component in the e-book such as one content node or one reference.

5 Comparison with Existing E-Book Authoring and Reading Software

The developed platform has been compared with existing e-book authoring tools and e-book reader software/devices. The purpose of the comparison is to assert and confirm that the developed platform offers all features necessary for e-book authoring and reading and at the same time provides additional services fruitful to the users. The results are presented separately in the following two sections.

5.1 Comparison of E-Book Authoring Tools

Six e-book authoring tools introduced in Sect. 2.1 were used in comparison with the developed platform. The main criteria for the comparison include "content editing", "content type", "navigation", and "others". Features in each criterion were listed and each tool was rated as "provided" represented by an "O" or "not provided" represented by an "X". The results are shown in Table 4.

Table 4. Comparison of e-book authoring tools

Criteria	Features	A	W	S	E	J	D	P
Content Editing	provides WYSIWYG editor	O	O	O	X	O	O	O
	can insert images and/or videos	O	O	O	O	O	O	O
Content Type	allows web links	O	O	O	O	O	O	O
	allows multimedia objects from websites	O	O	X	X	X	O	O
Navigation	supports table of contents creation	O	O	O	O	O	O	O
	allows jumping to parts of the book	O	O	O	O	O	O	O
Others	allows specifying e-book metadata	O	O	O	O	O	O	O
	allows specifying e-book content metadata	X	X	X	X	X	X	O
	multimedia search integration	X	X	X	X	X	X	O

A = Adobe InDesign, W = Microsoft Word, S = Sigil, E = eCub, J = Jutoh, D = DesktopAuthor, P = Proposed Platform

5.2 Comparison of E-Book Reader Software/Devices

Five e-book reader software/devices introduced in Sect. 2.2 were used in comparison with the developed platform. Several features under "content viewing", "navigation", and "others" criteria were listed and each tool was given a rating as either "provided" or "not provided". "O" characters represent "provided" while "X" characters represent "not provided". The results are shown in Table 5.

Table 5. Comparison of e-book reader software/devices

Criteria	Features	A	K	I	R	B	P
Content Viewing	automatic reflow of contents	O	O	O	X	O	X
	supports image and video rendering	O	O	O	O	O	O
	allows adjusting font size or zooming	O	O	O	O	O	O
Navigation	supports table of contents	O	O	O	O	O	O
	supports bookmarking	O	O	O	X	X	X
	allows jumping to parts of the book	O	O	O	O	O	O
Others	supports annotations and text highlighting	O	O	O	O	X	X
	search for web resources of the same subject	X	X	X	X	X	O
	search for similar images and videos	X	X	X	X	X	O
	automatically links to authors' web pages	X	X	X	X	X	O
	supports reading paths	X	X	X	X	X	O

A = Adobe Digital Edition, K = Amazon Kindle, I = iBooks, R = Adobe Reader, B = Bookworm, P = Proposed Platform

5.3 Results

The results from the two comparison sessions showed that the developed platform is able to provide additional useful features to users while maintaining the basic features necessary for e-book authoring and reading.

6 Conclusion

The world is slowly embracing e-books. The internet is growing. Computers and tablets are getting cheaper. These elements are tied together and they should not develop separately. A good platform for e-book publishing needs to take authoring and reading experience to another level with the utilization of the internet. This paper proposes an RDF-based e-book publishing platform that makes use of the resources on the internet for providing relevant information with the help of author-provided metadata. While it might be true that the information on the internet at present might appear scattered and largely unorganized, there will come a time when all those pieces are connected to each other in an organized and semantically meaningful manner. When that day comes, electronic publishing platforms have to be ready to take full advantage of the change.

References

1. Harrison, B.L.: E-Books and the Future of Readings. IEEE Computer Graphics and Applications 20(3), 32–39 (2000)
2. Jiang, Y., Katsamakas, E.: The Impact of e-Book Technology on Book Retailing. In: 2010 43rd Hawaii International Conference System Sciences (HICSS), pp. 1–8 (2010)
3. Internet World Stats: Internet Usage Statistics (2011), `http://www.internetworldstats.com/stats.htm`
4. Boone, J.: A Survey of Data Models for Hypermedia. IBM Corporation (1992)
5. Nair, S.S., Jeevan, V.K.J.: A Brief Overview of Metadata Formats. DESIDOC Bulletin of Information Technology 24(4), 3–11 (2004)
6. Bizer, C., Health, T., Berners-Lee, T.: Linked Data - The Story So Far. International Journal on Semantic Web and Information Systems 5(3), 1–22 (2009)
7. Dublin Core Metadata initiative: DCMI Metadata Terms (2010), `http://dublincore.org/documents/dcmi-terms/`

Visual Sentiment Summarization of Movie Reviews

Jin-Cheon Na, Tun Thura Thet, Christopher S.G. Khoo, and Wai Yan Min Kyaing

Wee Kim Wee School of Communication and Information,
Nanyang Technological University,
31 Nanyang Link, Singapore 637718
{tjcna,ut0001et,assgkhoo,kymwai}@ntu.edu.sg

Abstract. A prototype digital library of social media content was developed to present a summarized view of public opinion in a visual interface. The domain of the study was movie reviews of multiple genres harvested from weblogs, discussion boards, user and critic review Web sites, and Twitter. The system performs fine-grained analysis to determine both the sentiment orientation and sentiment strength of the reviewer towards various aspects of a movie, such as overall opinion, director, cast, story, scene, and music. Various visual interface components were developed to present an overview of public opinion on multiple aspects of each movie, and a usability evaluation was conducted to observe their effectiveness. Aspect-based sentiment summarization interface has the highest score for usefulness while a sentiment link analysis graph visualizing how positive and negative sentiment terms are associated with review aspects has the highest score for overall rating.

Keywords: Sentiment Summarization, Information Visualization, Movie Reviews, Digital Libraries.

1 Introduction

With the rapid growth of user-generated content on the Internet, many researchers have been studying effective algorithms for sentiment analysis [8]. Sentiment analysis is a type of subjectivity analysis which analyzes sentiment in a given textual unit with the objective of understanding the sentiment polarities (i.e. positive, negative, or neutral) of the opinions toward various aspects of a subject [1, 4]. For instance, when looking at the reviews of a folk song, the opinions in the review text cover not only the overall sentiment but also many specific aspects such as vocals, lyric, recording quality, creativity, and so on.

This study uses an aspect-based sentiment analysis approach and focuses on movie reviews. It determines the sentiment intensity (i.e. sentiment scores) and polarity towards various aspects of a movie, such as overall opinion, cast, director, story, and music. Our approach performs sentiment analysis at the clause-level using a linguistic approach, and details of the sentiment analysis method have been reported in [14]. In this study, a prototype digital library of movie reviews was developed to assess the usefulness of presenting the sentiment summary in a visual interface. This paper provides a brief summary of the sentiment analysis method, and then mainly focuses on

C. Xing, F. Crestani, and A. Rauber (Eds.): ICADL 2011, LNCS 7008, pp. 277–287, 2011.

the visualization methods used to present the summary and overview of the sentiment information extracted as well as the user evaluation.

2 Related Work

Sentiment analysis has been studied by many researchers in recent years. Sentiment analysis approaches often require resources such as sentiment lexicons to determine which words or phrases are positive or negative in a general context. General Inquirer [13] is a manually compiled resource often used in sentiment analysis. Many techniques have been proposed for learning the polarity of sentiment words [3, 10]. Most of the early studies were focused on document-level analysis for assigning the sentiment orientation of a document [7, 9]. However, these document-level sentiment analysis approaches are less effective when review texts are rather short, and in-depth sentiment analysis of review texts is required. More recently researchers have carried out sentence-level sentiment analysis to examine and extract opinions toward various aspects of a reviewed subject [1, 11]. In contrast to most studies which focused on document-level or sentence-level sentiment analysis, our approach uses clause-level sentiment analysis so that different opinions on multiple aspects expressed in a sentence can be processed separately in each clause. Some researchers have studied phrase-level contextual sentiment analysis, but phrases are often not long enough to contain both sentiment and feature terms together for detailed analysis [15].

Some of the studies on sentiment analysis used graphical visualization techniques to provide sentiment summarization. However, evaluation of the user-interface aspects has not been studied much [8]. For instance, Yi and Niblack [16] used a stacked-bar chart where the sizes of positive and negative bars represent the number of documents. Liu et al. [5] used a sentiment bar chart where the portions of the bar above and below the horizontal line represent how many documents are positive and how many documents are negative. Sentiments of the reviewers on three different electronic products are summarized for four reviewed aspects: general, display, speed and battery. Morinaga et al. [6] used the principal component analysis technique to visualize the association of products with positive and negative sentiment terms. For instance, it can be observed that cellular phone A is associated with positive terms, whereas cellular phone B is associated with negative terms. In our study, various visual interface components were explored to present an overview of public opinion on multiple aspects of each movie. Especially, a sentiment link analysis graph, called SentiGraph, was developed to visualize how sentiment terms are associated with movies. Compared to Morinaga et al.'s work [6], SentiGraph provides more detailed analysis results. For instance, it shows what kinds of sentiment terms are used for discussing each aspect of a target object (i.e. a movie).

3 Summary of Sentiment Analysis Method

Figure 1 gives an overview of the algorithm used for clause-level sentiment analysis of movie review texts. Firstly, for each sentence of the review text, semantic annotation (such as movie, director, cast, and character names) is performed, and a prior sentiment score is assigned to each word. The prior sentiment scores of about 32,000 individual words are derived from SentiWordNet [2] with the help of a subjectivity lexicon of [15]. Then, the grammatical dependencies are determined, and the sentence is broken into independent clauses. For each clause, the contextual sentiment score is calculated by traversing the dependency tree based on its clause structure. The review aspect of the clause is determined by locating an occurrence of a movie aspect feature word. After processing all the clauses of the sentence, the sentiment score for each review aspect is calculated by taking the average of the clauses addressing the same aspect. The sentiment score for the whole sentence is determined from the calculated scores of multiple aspects. The output sentiment scores can be used to identify the most positive and negative clauses or sentences with respect to particular movie aspects.

```
FOR each sentence
    Perform semantic annotation
    Assign prior sentiment scores to words
    Generate grammatical dependencies
    Break a sentence into clauses
    FOR each clause
        Calculate clause-level sentiment score
        Determine Aspect
    END FOR
    Calculate sentence-level sentiment score
END FOR
```

Fig. 1. Algorithm for clause-level sentiment analysis

The algorithm is applied to informal review postings on a movie discussion board. We conducted experiments with a dataset of 1000 sentences: 500 positive and 500 negative. The movie review sentences were collected from the discussion board of a movie review site, IMDB (www.imdb.com). The dataset covered 34 movies (17 positive and 17 negative). Experimental results show the effectiveness of the proposed approach. The accuracies of clause-level sentiment classification for overall movie, director, cast, story, scene, and music aspects are 75%, 86%, 83%, 80%, 90%, and 81% respectively. Full details of the sentiment analysis method and its experimental results have been reported in [14].

4 A Prototype for Sentiment Summarization of Movie Reviews

4.1 System Design

A prototype system was implemented using Fedora (the *Flexible Extensible Digital Object Repository Architecture*), which is open-source digital repository software [12] and capable of serving as a digital content repository. The front-end web application was developed using J2EE technology, and calls the web services exposed by Fedora for accessing and managing digital objects stored in Fedora. We use the Dublin Core (dublincore.org) to put metadata related to three main types of digital objects in the system: Movie, Director, and Cast objects (A Movie object contains its movie reviews). The digital objects stored in Fedora are making use of the object relationship capability in Fedora. Thus, a digital object of Movie type has relations with Director and Cast objects. This is useful, for example, when we need to search the list of movies directed by a certain director or movies where a particular cast member plays in. Additionally, a customized crawler was developed to automatically harvest various movie review documents from critic review sites, IMDB user reviews, IMDB discussion board, and Twitter. The back-end Fedora repository was ingested by the collected movie review data.

4.2 Dataset

For this study, online review documents of 10 recent movies with various review ratings were selected. Around 5 critic reviews, 25 user reviews, 25 discussion threads, and 100 tweets (Twitter posts) were collected for each movie. For discussion threads and tweets, a simple filtering method was used to remove irrelevant posts. Irrelevant discussion threads were filtered by analyzing the title and the first post of each thread. Only the threads containing both movie-related terms and sentiment words were selected for further analysis. For Twitter posts, relevant sentences were automatically identified by looking for the presence of both sentiment term and the movie name. This simple method could be improved to remove various types of irrelevant sentences, such as opinion spam. In addition to the 10 movies, the critic reviews of 800 movies were also collected. After data collection, the clause-level sentiment analysis method was applied to the dataset.

4.3 User Interface Components

Basic user interface components, i.e. sentiment bar, sentiment meter, sentiment term cloud, sentiment treemap, and sentiment time series chart, were implemented using Google Visualization API (code.google.com/apis/visualization/documentation /gallery.html) to visualize sentiment summarization results. The composite user interface components, i.e. aspect-based sentiment summarization, sentiment link analysis graph, and top most positive and negative sentences, were also implemented. Figure 2 shows the main page of the movie *Avatar* that displays most of user interface components except for sentiment time series and top most positive and negative sentences.

Sentiment bar displays the distribution of positive and negative reviews. It has two sub-bars: positive bar on the right is displayed in green color, and negative bar on the left is displayed in red color. Sentiment meter is a gauge that points to a sentiment score which is a real number within the range of -1 (most negative) to +1 (most positive). It is similar to speedometer or fuel gauge in a car. Sentiment term cloud is similar to tag cloud visualization. This visual component shows sentiment terms used in review documents, and the hyperlinks point to the sentences in the original review documents. Rotating Flash interface is used so that all the terms can be accessed by a mouse click in order to follow the hyperlink. Positive sentiment terms are displayed in green color while negative sentiment terms are displayed in red color. The sentiment terms with higher term frequencies are displayed with larger font sizes than those with lower term frequencies. Sentiment treemap is similar to sentiment term cloud. The size of the box containing a sentiment term varies according to the term frequency.

The boxes containing sentiment terms with higher term frequencies are larger than those with lower term frequencies. It has positive terms on the left-hand side and negative terms on the right-hand side. If there are more positive terms in the reviews, the positive area will be larger than the negative area or vice versa. Sentiment time series chart provides the trend of reviewers' sentiment over time. This chart is suitable to visualize trends in reviewers' sentiments towards a movie over a period of time.

Aspect-based sentiment summarization displays overall sentiment summarization and sentiment analysis results for each review aspect. Firstly, the total number of reviews, the ratio of positive and negative reviews, and the average sentiment score of the movie are displayed. Next, it provides sentiment summarization for each aspect by displaying the number of clauses, the ratio of positive and negative clauses, and the average sentiment score of the review aspect. This in-depth sentiment summarization allows the user to understand the overview of the sentiment towards each review aspect of the movie at a glance without needing to read through many review documents. The user interface component of top most positive and negative sentences displays top 10 most positive and negative sentences with their sentiment scores. The following section describes a sentiment link analysis graph in detail since we believe that it is a novel approach for visualizing sentiment analysis results compared with the other user interface components.

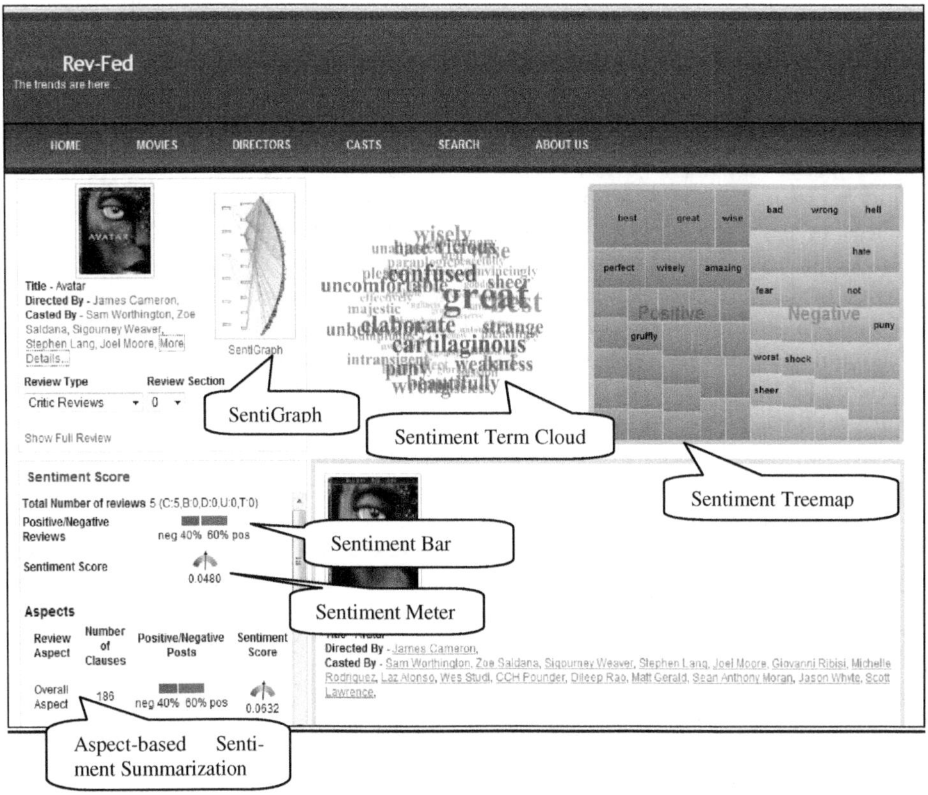

Fig. 2. Movie page: Avatar

4.4 Sentiment Link Analysis Graph

SentiGraph is a sentiment link analysis graph which visualizes how sentiment terms are associated with review aspects through feature terms for a given set of review documents. For instance, as shown in Figure 3, the feature term *"soundtrack"* is used to associate the sentiment terms *"good"* and *"amazing"* with *music* aspect; the important terms (e.g., *"soundtrack"* and *"sound effect"*) indicating a specific movie aspect (e.g., *"music"*) are called feature terms. The association of sentiment terms to each review aspect is identified by detecting the co-occurrence of feature terms and sentiment terms together in the same clause. SentiGraph has been developed as a Java applet using Picolo Toolkit (www.cs.umd.edu/hcil/jazz/), and is visualized on a single canvas. The zooming and panning of the entire canvas and its components can be performed to explore the details.

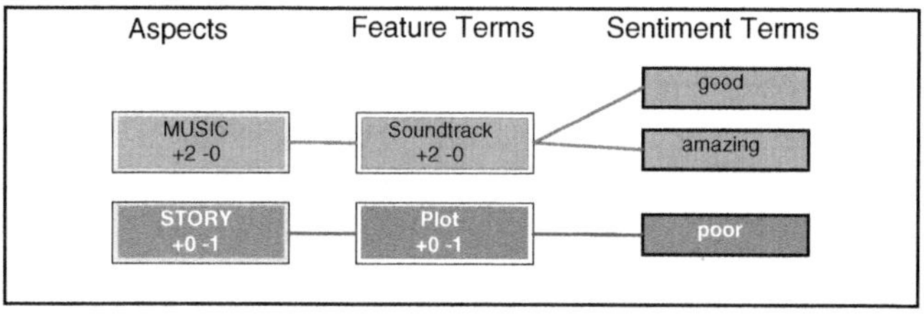

Fig. 3. Sentiment link analysis graph

In SentiGraph, sentiment terms, feature terms, and review aspects are displayed as thin-bordered-boxes, gray-bordered-boxes and cyan-bordered-boxes respectively. The line between a sentiment term and a feature term represents the co-occurrence of the terms in the same clause, while the line width represents the frequency of the co-occurrences. The color of the box containing a sentiment term represents the orientation of the term based on its prior sentiment score. Similarly, the color of the line between a sentiment term and a feature term is decided based on the sentiment orientation of the term. The ratio of the positive and negative clauses containing a feature determines the color of the box containing the feature term. The line between a feature term and a review aspect represents the relationship of the feature term and the review aspect. The color of the line also depends on the sentiment orientation of the feature term connecting to it. The width of the line represents the total number of clauses containing the feature term. Finally, the color of the box of a review aspect is determined by the ratio of positive and negative clauses for all related feature terms covering a particular review aspect. For instance, in Figure 3, the boxes containing the terms *"music"*, *"soundtrack"*, *"good"*, and *"amazing"* are in green color representing positive sentiment while the boxes containing the terms *"story*, *"plot*, and *"poor"* are all in red color representing negative sentiment. The sentiment terms *"good"* and *"amazing"* co-occur with the feature term *"soundtrack"*, which belongs to the music aspect. Since the sentiment terms *"good"* and *"amazing"* are positive, the feature term *"soundtrack"* becomes positive since it is associated with 2 positive clauses, which makes the music aspect positive. The sentiment term *"poor"* co-occurs with the feature term *"plot"*, which belongs to the story aspect. Similarly, since the sentiment term *"poor"* is negative, the feature term *"plot"* associated with 1 negative clause becomes negative, which makes the story aspect negative.

Figure 4 shows a sentiment link analysis graph of a movie, Avatar. From the figure, we can tell that the feature term *"Avatar"* (movie name) co-occurs with both positive (e.g., *"nice"*, *"grace"*, *"lucky"*, and *"perfect"*) and negative (e.g., *"bad"*, *"wrong"*, and *"disability"*) terms. Since it co-occurs with more number of positive terms (+28 vs. -15 clauses), it becomes positive. Similarly, we can easily observe that the overall aspect is positive and linked with the positive feature terms *"Avatar"* and *"movie"*; the director aspect is also positive and linked with the positive feature terms *"James Cameron"* (director name), and *"directing"*, and *"filming"*. As shown in Figure 4, the overall color distributions of the review aspects, feature terms, sentiment terms, and their connected links on the graph help the user to quickly get an overview of sentiment orientations and opinions on various movie aspects.

However, a canvas with too many components could be visually complex for the users. Therefore, various options are provided to fine-tune the visualization of the components. The user can adjust the number of sentiment terms to be displayed by setting a threshold value for the prior sentiment score or term frequency. The option for displaying either a specific review aspect or all the review aspects is also provided. The user can specify all or specific POS terms to be included in the display. Stop-word option can also be used to filter out stop words in the display. Additionally, the user can choose all or a specific review genre to be displayed on a canvas. For example, if the user is only interested in opinions from Twitter posts, review genre filter can be set to Twitter. This option allows the user to focus on the sentiment analysis results of a specific online genre. SentiGraph can be also used for comparing opinions from different online genres by displaying two or more genres' sentiment link analysis graphs side by side on a single canvas. It helps the user to compare sentiment orientations and opinions in one online genre to those of other genres. It also allows displaying two movies side by side on a single canvas for comparing various opinions towards the two movies.

4.5 User Evaluation

A user evaluation was carried out with 12 participants (8 PhD students, 3 research project officers, and 1 professor) who have relevant experience in user interface design or a good knowledge in cinema studies. At the beginning of each evaluation session, an overview of sentiment analysis and summarization was introduced to each subject, and each subject was given an hour to have hands-on experience with the prototype. The subject was also asked to compare the sentiment summaries of two movies (*"avatar"* and *"clash of the titans"*) which have opposite ratings from the reviewers (high-rated vs. low-rated movies). The task was to evaluate if the difference in sentiment orientations towards two movies is reflected well in the results of the automatic sentiment analysis and summarization. At the end of each session, the subject was asked to answer the questionnaire.

The evaluation results are summarized in Table 1. Aspect-based sentiment summary has the highest score for usefulness while sentiment treemap has the lowest.

As for usability, sentiment bar has the highest score while sentiment term cloud has the lowest. As for novelty, aspect-based sentiment summarization has the highest score while sentiment bar has the lowest. For overall rating, SentiGraph has the highest score while sentiment treemap has the lowest.

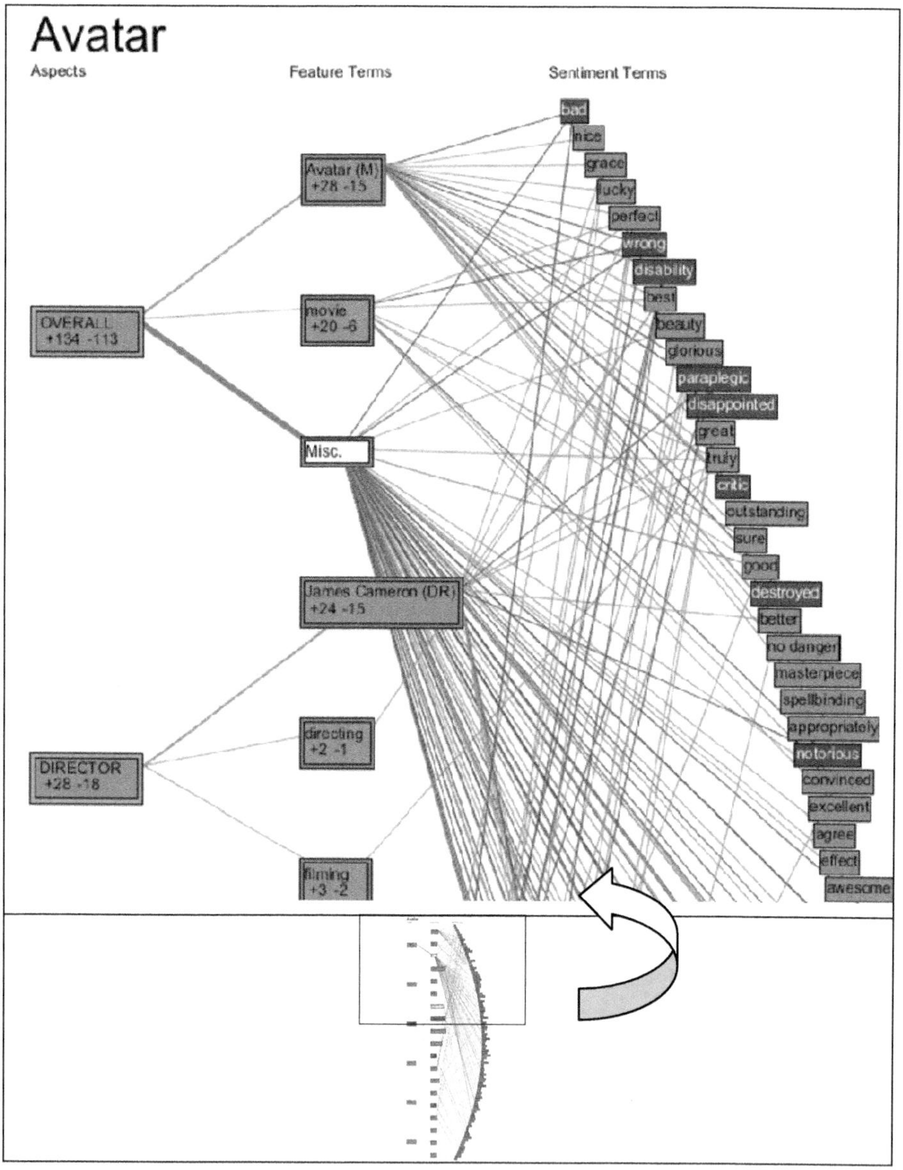

Fig. 4. Sentiment link analysis graph of a movie, Avatar

The valuable feedback given by the subjects are summarized as follows:

- **Sentiment Meter:** The subjects gave positive feedbacks on sentiment meter as useful, intuitive, and easy to understand. Its resemblance to motorcars' fuel gauge makes the users more comfortable with the interface.
- **Sentiment Bar:** Most of subjects agreed that it is useful and intuitive, but the subjects thought that this is the least novel since bar charts are commonly used in many applications. One subject suggested aligning the bars to the centre so that the lengths of positive and negative bars could be compared more easily.
- **Aspect-based Sentiment Summarization:** This user interface component received the most positive feedbacks from the subjects. The subjects commented that this is useful as it provides a holistic overview of the sentiment analysis. The subjects also stated that it provides detailed accounts of aspects and allows the user to skip or concentrate on certain aspects.
- **Sentiment Term Cloud:** The subjects found this user interface visually appealing, but they also found it hard to use. One subject pointed out that the sizes of the sentiment terms are difficult to compare as they change as the cloud rotates. Another subject pointed out the difficulty of tracking and clicking the term to drill down further details.
- **Sentiment Treemap:** Most of subjects were not sure about the usefulness of this user interface. Some mentioned that it is hard to understand and confusing. It may be partially because Google Visualization API does not allow us to customize this user interface, such as font colors of positive and negative words.
- **SentiGraph:** Most of subjects found this user interface useful and novel. They believed that this is useful for advanced users who are interested in the details of sentiment analysis. They also agreed that the integrated drilldown feature is useful as they can read and cross-check with the original texts for verification. However, they raised their concerns that this could be visually complicated for ordinary users with simple needs.

In addition to feedback on the user interface components, general suggestions were also given. One meaningful suggestion made by some subjects is to allow the users to comment, adjust, or correct system-generated sentiment scores or orientations accordingly. One subject also suggested incorporating sentiment scores for ranking of search results, which has been incorporated in the system already.

Table 1. The evaluation results of user interface components

	Usefulness (Out of 5)	**Usability (Out of 5)**	**Novelty (Out of 5)**	**Overall (Out of 5)**
Sentiment Meter	3.83	4.08	3.83	4.00
Sentiment Bar	4.25	**4.17**	<u>3.50</u>	4.08
Aspect-based Sentiment Summarization	**4.45***	4.00	**4.08**	3.92
Sentiment Term Cloud	3.73	<u>3.25</u>	3.83	3.92
Sentiment Treemap	<u>3.42</u>*	3.42	3.83	<u>3.50</u>
SentiGraph	3.98	3.48	3.98	**4.11**

*: Numbers in **Bold** for the highest score and <u>underlined</u> for the lowest score

5 Conclusion

With the rapid growth of user-generated content on the Internet, visualization techniques for sentiment summarization will become important features in digital libraries. In our study, especially, a sentiment link analysis graph, called SentiGraph, was developed to help the user to analyze sentiment orientations and opinions on various movie aspects by visualizing how sentiment terms are associated with review aspects through feature terms. Most of subjects in the evaluation found it useful and novel. We believe that SentiGraph is a useful information visualization tool for the domains which require in-depth sentiment analysis. Currently, we are applying the proposed approach to health and medical domains, particularly focusing on public opinions on drugs with various aspects, such as side effects, effectiveness, dosage, cost, etc.

References

[1] Ding, X., Liu, B., Yu, P.S.: A holistic lexicon-based approach to opinion mining. In: The International Conference on Web Search and Web Data Mining, pp. 231–240. ACM, New York (2008)
[2] Esuli, A., Sebastiani, F.: Determining term subjectivity and term orientation for opinion mining. In: The European Chapter of the Association for Computational Linguistics, pp. 193–200 (2006)
[3] Hatzivassiloglou, V., McKeown, K.R.: Predicting the semantic orientation of adjectives. In: The 35th Annual Meeting of the ACL and the 8th Conference of the European Chapter of the ACL, pp. 174–181 (2007)
[4] Hu, M., Liu, B.: Mining and summarizing customer reviews. In: The 10th ACM SIGKDD International Conference on Knowledge Discovery and Data Mining, pp. 168–177 (2004)
[5] Liu, B., Hu, M., Cheng, J.: Opinion observer: Analyzing and comparing opinions on the web. In: The 14th International Conference on World Wide Web (2005)
[6] Morinaga, S., Yamanishi, K., Tateishi, K., Fukushima, T.: Mining product reputations on the Web. In: The ACM SIGKDD Conference on Knowledge Discovery and Data Mining (KDD), pp. 341–349 (2002)
[7] Pang, B., Lee, L.: Seeing stars: Exploiting class relationships for sentiment categorization with respect to rating scales. In: The Association for Computational Linguistics, pp. 115–124 (2005)
[8] Pang, B., Lee, L.: Opinion Mining and Sentiment Analysis. Foundations and Trends in Information Retrieval 2(1-2), 1–135 (2008)
[9] Pang, B., Lee, L., Vaithyanathan, S.: Thumbs up? Sentiment classification using machine-learning techniques. In: The 2002 Conference on Empirical Methods in Natural Language Processing, pp. 79–86 (2002)
[10] Qiu, G., Liu, B., Bu, J., Chen, C.: Expanding domain sentiment lexicon through double propagation. In: The 21st International Joint Conference on Artificial Intelligence, pp. 1199–1204. Morgan Kaufmann, San Francisco (2009)
[11] Shaikh, M.A.M., Prendinger, H., Ishizuka, M.: Sentiment assessment of text by analyzing linguistic features and contextual valence assignment. Applied Artificial Intelligence 22(6), 558–601 (2008)

[12] Staples, T., Wayland, R., Payette, S.: The Fedora Project: An Open-source Digital Object Repository Management System. D-LIb Magazine 9(4) (April 2003)
[13] Stone, P.J., Dunphy, D.C., Smith, M.S., Ogilvie, D.M.: General Inquirer: a computer approach to content analysis. The MIT Press, Cambridge (1966)
[14] Thet, T.T., Na, J.-C., Khoo, C.: Aspect-Based Sentiment Analysis of Movie Reviews on Discussion Boards. Journal of Information Science 36(6), 823–848 (2010)
[15] Wilson, T., Wiebe, J., Hoffmann, P.: Recognizing contextual polarity in phrase-level sentiment analysis. In: The Conference on Human Language Technology and Empirical Methods in Natural Language Processing, pp. 347–354 (2005)
[16] Yi, J., Niblack, W.: Sentiment mining in WebFountain. In: The 21st International Conference on Data Engineering, ICDE (2005)

Towards Ontology-Based Knowledge Visualization

Yigang Zhou

Peking University Library, Beijing 100871, P.R. China
zhouyg@lib.pku.edu.cn

Abstract. This paper introduces the successful implement of two ontology visualization tools in ontology-based knowledge retrieval model, which supplies solutions for knowledge visualization.

Keywords: Knowledge Visualization, Ontology, Visualization Tool.

1 Introduction

As a critical technology of Semantic Web, ontology solved the problem of knowledge description, knowledge sharing and other issues[3].But many ontologies contain very complex structures to be clearly understood. A lot of researchers make efforts to tide over this problem. One of the challenges, ontology visualization obtained considerable results. It helps us to understand structures of ontology that has concepts, hierarchy of concepts and relationship between concepts. There are many tools for visualization. Now, we try to explain two tools of ontology visualization used in our projects, which supplies solutions for knowledge visualization.

2 Visualization of Ontology-Based Knowledge Retrieval Model

TouchGraph[1]and Prefuse[2]are used in our two projects based on ontology. The principle of visualization in these two projects are almost the same, the technical implementation differs a little. The first project mainly explains the principle and user experiences of ontology visualization. And the paper tries to interpret ontology visualization at a technical perspective from the second project.

2.1 Visual Knowledge Retrieval for Digital Library Based on Ontology

This project developed a history field retrieval system to implement basic function of ontology, including ontology construction, ontology retrieval, ontology reasoning, ontology visualization, among which, ontology visualization is a highlight point.

For visualization part of this retrieval system, We have carried out second development to the TouchGraph toolkit, TouchGraph doesn't support reading or writing semantic data, we extract data from OWL file using ontology processing engine "Jena", then convert them into several triples. According to subject and object of the triples, we create nodes, and the predicate defines the relation(Property) between subject and object which is represented into directed edge. The component "GraphLayout" in TouchGraph realizes effects such like Zoom, Pan, Focus/Context and Incremental Navigation, as shown in Fig.1.

C. Xing, F. Crestani, and A. Rauber (Eds.): ICADL 2011, LNCS 7008, pp. 288–291, 2011.

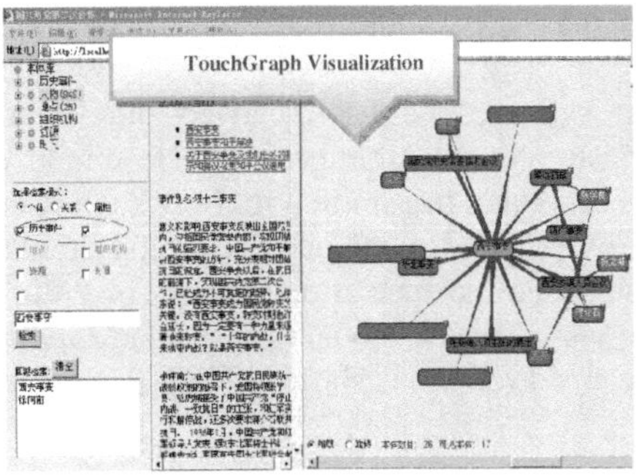

Fig. 1. Visualization Development by Touch Graph

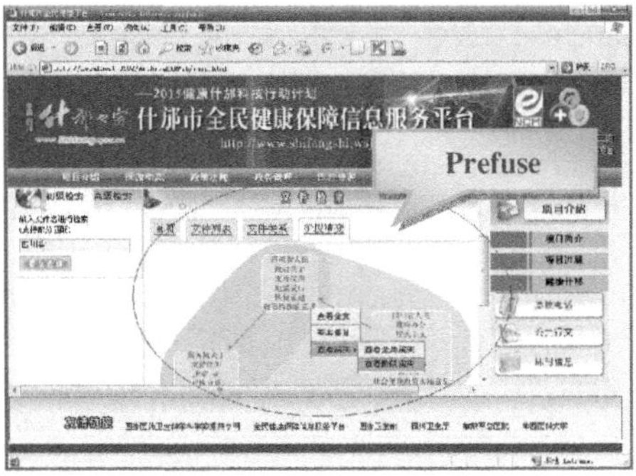

Fig. 2. Visualization Development by Prefuse

The graphical display of ontology is implemented by a Java Applet. When users click a result entry to look for details, one or more connected networks will appear in the display area, this instance and its related ontology instances composed the network's nodes, whose relationship composed the network's edges. In the network, every node has a text label, which display the ontology instance name. Nodes are connected by a directed edge, the thicker end is the start of this edge, the thinner end is the end of this edge, this direction represents the relation between subject and object in property.

For example, we search a history event "Xian event", then, related event, people, location, time etc are displayed, and their relation was represented by different edges, there is a bidirectional edge connecting "Xian event"(A famous military event happened in Dec.12 , 1936, Shanxi Province, China.) and "Xueliang Zhang", which means Mr. Zhang is a pivotal figure in "Xian event", and there is another bidirectional edge connecting "Xian event" and "The eight allegations", which means "Xian event" reduced "The eight allegations", and "The eight allegations" is the production of "Xian event". These significations can't be delivered by traditional search engine. It embodies ontology's excellent character to reveal knowledge link and visualization's intuition to express knowledge.

When ontology is colossal, the relationship between ontology is complex, the related instances of an ontology may be a big amount, it is unnecessary to represent these instances on the screen all at once, so we only display instances which are directly related with the focus ontology, at the same time, we classify these related instances with different colors according to ontology type which we have defined when we build the knowledge base. Users can right-click focus node, select command from the pop-up menu to look up one type or several types of related instances. The graph provides "Zoom", "Rotate" toolbar, which were convenient for users to adjust view for better observation. Users can also drag the graph directly.

2.2 Visual Document Retrieval Model Based on Ontology

In another project, we developed an Ontology-based document retrieval system. The retrieval system can be used to query medical and health policy information stored in database, and visualize the results to reveal the connectivity between policy documents, as shown in Fig.2.

For visualization part of this retrieval system, We have carried out second development to the Prefuse toolkit, the basic system architecture is resemble to Prefuse, the difference is the data model layer. What's transmitted to the visualization system is Ontology object which is created by popular open-source Ontology toolkit Jena, such as OntModel and Individual, rather than the special data object of Prefuse ("table", "graph" and "tree"). Therefore, the visualization model has a Data Mapping Layer, which switches Ontology to data object of Prefuse, as shown in Fig. 3.

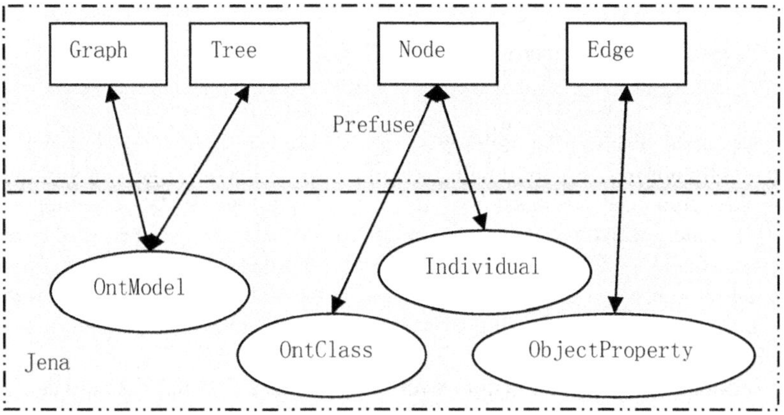

Fig. 3. Data Mapping

The subsystem of retrieval module transmit the OntModel to data mapping layer of visualization module, OntModel contains a great deal of classes, relationships and instances. According to the mapping rule in Fig.3, OntClass and individual are mapped to corresponding nodes, Object Property is mapped to edges, and all Data Property such as names, URL, description of ontology object are stored as Prefuse Object's property. The following code sample presents how does data mapping layer of visualization module convert OntClass in Jena to Node in Prefuse.

```
  Node currNode = m_graph.addNode ();
  currNode.setString ("URI", p_currentClass.getURI ());
  currNode.setString ("name", p_currentClass.getLocalName
());
  currNode.setString ("type", "class");
```

p_currentClass is an "OntClass" instance to be mapped into "Node" instance, m_graph is a instance of "Graph" data type in Prefuse. m_graph create a new node "currNode" in Prefuse, which is processed as the mapping copy of p_currentClass, then, data property like ontology names, URI, class are added as "name", "URI", "type" property into currNode. In this way, a simple mapping from ontology object to Prefuse object was accomplished.

The data returning from retrieval module first flow to data mapping layer to finish data type Conversion. After data mapping, couples of Prefuse data object are produced in system, these object will be sent to upper layer of the system.

In the 'abstract visual data' layer, appearance data such as colors, shapes are bind with specific Prefuse data object, in succession, renderer, interaction, flash are bind with specific structure one after the other. The following code sample presents binding gray to font color.

```
  ColorAction aText = new ColorAction (NODES,
VisualItem.TEXTCOLOR);
  aStroke.setDefaultColor (ColorLib.gray (100));
  aStroke.add ("_hover", ColorLib.rgb (255, 100, 100));
```

In the 'interaction' layer, renderer presents all visual object in the display area in foreground. The following code sample presents managing renderer.

```
  DefaultRendererFactory drf = new DefaultRendererFactory
();
  drf.setDefaultRenderer (labelR);
  drf.add ("ingroup ('aggregates')", polyR);
  drf.add (new InGroupPredicate (EDGES), edgeR);
```

After interaction between user and graph, controller sent a series of operations to 'data mapping' layer, 'abstract visual data' layer, 'interaction' layer to trigger a new round of rendering. The sample code is omitted.

3 Conclusion

The essence of ontology is to reveal relationship between knowledge, the benefit of visualization is to show the natural characteristics of ontology in a vivid graphic way. The combination of them will make it easier to understand knowledge.

We have successfully used two ontology-based visualization tools in our projects , However, there are several issues that need to be considered in future works, and our efforts are ongoing. Our tools should be refined and updated further as more is learned about visualization and the specific needs of ontology-based application. To do this, we will find and test more mature tools of ontology visualization. And we will also do more secondary development based on these tools.

Acknowledgments. This work is supported by NSFC(National Nature Science Foundation of China) project(70373047). It is supported by the project (2006-x-29) sponsored by the State Archives Administration of China as well.

References

1. TouchGraph, `http://www.touchgraph.com/navigator.html`
2. prefuse: A Toolkit for Interactive Information Visualization, `http://prefuse.org/doc/manual/`
3. 董慧: Ontology and Digital Library.武汉大学出版社 (2008)

A Feedback Enabled Multimedia WebQuest Model for College Public English Learning

Zheng Zhang[1], Yan Zhang[2], and Yiyu Jia[2]

[1] Department of Language and Communication, Canvard College,
Beijing Technology and Business University,
1 South Road, Song Zhuang, Tongzhou District 101118, Beijing, China
[2] Vermont Information Processing, 402 Watertower Circle,
Colchester, Vermont 05446, USA
zhang-zheng@sina.com, {zxlm80,yiyu.jia}@gmail.com

Abstract. With the rapid development of information technologies, the college English reform of China has arrived at a spot where the mature technology platform is able to support it to make every possible novel attempt. Of many experimental teaching practices, one representative Web based instructional model is WebQuest. In this article, we outline a WebQuest model designed for college public English learning. The novel aspect of this model includes a student feedback module with the WebQuest's *Resource* component. The students are able to give score ratings to the existing resource items, or recommend new resource items. These information will then be taken into account when refining the design of the same WebQuest project. The contribution of this proposed model is two-fold: (1) it promotes the resource reuse by applying advanced data mining methods, so as to improve the efficiency and effectiveness of the WebQuest project maintenance; (2) it opens a channel to collect the thoughts and ideas from the students, so as to achieve the intelligence integration from multiple sources.

Keywords: College English, WebQuest, Data mining, Feedback.

1 Introduction

College public English or college English, is a required subject for any non-English major college student in China. Currently, the college English learning is facing many challenges. First, along with the increase enrollment of college students in recent years, the lack of English teachers has become a serious problem. According to the statistics from authoritative institution, the teacher student ratio has reached 1:130 [1]. Second, the teaching philosophy and pedagogy is in need of injecting the advanced technologies for better learning effect. According to the *College English Curriculum Requirements* published by the ministry of education, "The new teaching and instruction model should be developed with the support of the modern information advances, especially the network technologies. The essential goal is to push the development of English learning towards the direction of personalized learning, learning without the restriction of time and place, and self-motivated learning." [2]. Following this call, many educators start to apply advanced information technologies into the classroom of college English teaching. Researchers from English education, information systems, and computer science have contributed their effort in related areas [3,4,5].

C. Xing, F. Crestani, and A. Rauber (Eds.): ICADL 2011, LNCS 7008, pp. 292–295, 2011.

As a representative of reformative teaching practices, the WebQuest instruction model developed by Bernie Dodge and Tom March has been drawn a lot of attention [6]. It is a learning activity based on the exploration of the resources from the World Wide Web. Its main feature is to construct the bridge from receptive learning to independent learning, so as to help the students achieve the learning habit transition. When applying WebQuest model into the college English learning in China, we should realize that the English learning involves unique characteristics that differ from other subjects. In addition to understanding a particular concept, the other major learning objectives include vocabulary memorization, phrases and sentence patterns usage, English listening and speaking practice. Therefore, an ideal situation is to incorporate some multimedia elements such as audio and video resources, as well as the speech recolonization system, into the WebQuest project. The implanted audio and video elements serve the function of gaining students listening abilities through news or story telling and situational dialogues. Compared to regular WebQuest projects, we therefore face a new challenge in designing a WebQuest project for English learning. Because of the timeliness of some multimedia resources, it requires additional work load to maintain and refine the resources within the WebQuest project, in order to reuse it in the next semester or next year's class.

In this article, we present a design of a novel feedback enabled multimedia WebQuest model for college public English learning. The main contribution of this design is two-fold. First, it promotes the resource reuse by applying advanced data mining methods, so as to improve the efficiency and effectiveness of the WebQuest project maintenance. Second, it opens a channel to collect the thoughts and ideas from the students, so as to achieve the intelligence integration from multiple sources. We organize the rest of this article as follows. In Section 2, we illustrate the design of the proposed feedback enabled multimedia WebQuest model. In Section 3, we present the usage of this model into a case study in college English. In Section 4, we present our future work and concluding remarks.

2 Design

A WebQuest project consists of six key components: *Introduction*, *Task*, *Process*, *Resource*, *Evaluation*, and *Conclusion*. Of these six components, *Resource* includes a list of materials that are provided by the teacher. The students explore and collect information from these materials, so as to find their useful information to complete the "Task". In our design of a WebQuest project for college English learning, we further improve the functionality of the *Resources* component. Specifically, we include the multimedia resources such as audio and video materials into the resource list. Furthermore, we maintain a student feedback system that keep track of the score ratings on each existing and new recommended resource items. By analyzing this feedback system, the teacher will make the decision to remove existing resources with low ratings, and/or to include some new recommended materials into the *Resource* list for the next group of students who need to take the same WebQuest project. We illustrate this idea in Table 1 and Table2.

3 Case Study

When we give the lecture 8 "Legal and Moral Implications of Cloning" using the textbook *New Horizon College English: Reading and Writing Tutorial 3*, we find it is a very abstract topic [7]. If adopting the conventional classroom teaching style, the stu-

Table 1. An Example List of Resources

R_ID	Category	Source	Address
1	article	Internet	http://· · ·
2	article	Internet	http://· · ·
3	article	Lib A	article ID
4	audio	Internet	http://· · ·
5	audio	Lib A	audio ID
6	video	Lib A	video ID

This table shows an example list of resources that are provided by the teacher for the corresponding WebQuest project. The column "R_ID" denotes the ID of the corresponding resource item, the column "Category" briefly describes what type of material this item belongs to, the column "Source" and "Address" show where this resource item is located. The resources can be either reading, audio, or video materials. They can be found either from the Internet, or from available digital libraries.

Table 2. Resource Feedback

R_ID	Score	Comments
1	5	...
2	4	...
3	5	...
4	4	...
5	5	...
6	5	...

During the process of exploring these on-list resources, the students will give scores to rate every piece of resources. The score can be in the range between 1 and 5, where a higher score indicates a higher degree of satisfaction. Comments to the corresponding resources can be provided too. When a student carries out the WebQuest project, he/she may go seek information resources other than the already provided resources. He/she could provide the information of these new resources during the process of exploring this project.

dents may feel this topic is very dry to learn, and lose the learning interest. As a new rising technology, the application of cloning can bring significant impact on people's daily life. We consider it appropriate to construct a WebQuest project on this topic, so as to improve the learning effect through the independent and explorative study. In the following paragraphs, we briefly present how the proposed feedback module could be applied into the *Resource* component.

In the *Introduction* component, we propose following introduction: "Cloning is a new technology, and it will bring a great change in human's life. Suppose you are a writer and you are going to write a book about cloning technology, so you interview some 'scientists' to help you." In the *Task* component, we require every four students joining into one group, in which they each play the role as medical expert, agriculturalist, environmental protection expert, and the parent who wants to clone his/her child, respectively. Every student is required to find out the related information of cloning technology in the field represented by their corresponding role. They need to take notes on these information in the form of audio, video, picture, or text, and organize them into a pamphlet. In the *Process* component, we design a series of questions that guide the students' thinking process. For example, the students who play the medical expert role need to answer questions like "If my patient need organ transplant, would it be possible to adopt cloning technology?"; "What the advantages are for doing organ cloning?"; "Is the cloning technology able to relieve the patients who suffer from genetic diseases?". The students who play the role of agriculturalist need to explore questions like "Is the cloning technology able to help increase the yield of crops?"; "How to cultivate new vegetable varieties with high quality?". Through answering these questions, the students may understand the advantages and disadvantages the cloning technology brings to the human beings. During the explorative study, as the resources provided are all English reading or audio materials, the students practice their reading and listening ability, as well as new learnt vocabularies and the corresponding usage.

Table 3. Case Study Resources and Score Rating

R_ID	Category	Source	Address	Score	Comments
1	article	Internet	http://www.ornl.gov/sci/techresources/ Human_Genome/elsi/cloning.shtml	5	...
2	audio	Internet	http://www.tingroom.com/lesson/crazy/ 2/4365.html	5	...
3	article	Internet	Ad Hoc Committee on an International Convention against the Reproductive Cloning of Human Beings	4	...
4	other	Internet	http://un-interpreters.org/glossaries.html	2	...

In the *Resource* component, we provide a resource list, and the students need to make use of these resources to finish the assigned task. The students will give score rating to each resource item. They could also nominate new resource items to penitentially improve this WebQuest project. We show a part of the resource list in Table 3. Here we assume the scores are in the range 1 to 5, where a bigger score indicates the corresponding resource item is more satisfactory.

Finally, in the *Evaluation* component and the *Conclusion* component, we construct the evaluation rules to score each student's WebQuest project, and guide the students to conclude their study by giving a presentation.

4 Future Work and Conclusion

In this article, we have outlined a framework to include a student feedback system into a regular WebQuest project. By introducing a specific case study on College public English, we have shown where this feedback module could take part in. In the next stage of study, we will focus on how to efficiently and effectively analyze these feedback messages. We will apply machine learning and data mining methods into this feedback module. Specifically, we need to (1) refine resource list, and keep the information up-to-date; (2) design extra credit scheme for students who recommend good resources; (3) try to understand what topics the students consider interesting through their feedback.

References

1. Zhang, Y.: Rethinking about college english reform. China University Teaching 12, 19–21 (2003) (in Chinese)
2. Ministry of Education. College English Curriculum Requirements. Shanghai Foreign Language Education (2007)
3. Li, T., Zhou, H.: Design and research of self-adaptive learning diagnosis agent. Journal of Convergence Information Technology 5(1), 54–59 (2010)
4. Sox, A., Rubinstein-Ávila, E.: Webquests for english-language learners: Essential elements for design. Journal of Adolescent & Adult Literacy 53(1), 38–48 (2009)
5. Lu, Y.: The instructional design of webquest based viewing, listening, and speaking of college english learning. Journal of Yangtze University (Social Sciences) 33(2) (2010)
6. Dodge, B.: Some thoughts about webquests (1997), http://webquest.sdsu.edu/about_webquests.html
7. Hu, Q.: New Horizon College English: Reading and Writing Tutorial, vol. 3. Foreign Language Teaching and Research Press, China (2009)

Retrieval Effectiveness of Cross Language Information Retrieval Search Engines

Schubert Foo

Division of Information Studies,
Wee Kim Wee School of Communication and Information,
Nanyang Technological University, Singapore 637718
sfoo@pmail.ntu.edu,sg

Abstract. This study evaluates the retrieval effectiveness of English-Chinese (EC) cross-language information retrieval (CLIR) on four common search engines along the dimensions of recall and precision. We formulated a set of simple and complex queries on different topics including queries with translation ambiguity. Three independent bilingual proficient evaluators reviewed a total of 960 returned web pages each to assess document relevance. Findings showed that CLIR effectiveness is poor with average recall and precision values of 0.165 and 0.539 for monolingual EE/CC searches, and 0.078 and 0.282 for cross lingual CE/EC searches. Google outperformed Yahoo! in the experiments, and EC and EE searches returned better results than CE and CC results respectively. As this is the first set CLIR retrieval effectiveness measurements reported in literature, these findings can serve as a benchmark and provide a better understanding of the current CLIR capabilities of Web search engines.

Keywords: cross language information retrieval, search engines, Google, Yahoo!, English-Chinese translation, Chinese-English translation, evaluation studies, recall, precision.

1 Introduction

The ability to search and retrieve information in multiple languages is becoming increasingly important and challenging in today's environment. Consequently, multi-lingual and cross-lingual (language) information retrieval (MLIR and CLIR) search engines have received more research attention and are increasingly being used to retrieve information on the Internet. CLIR refers to searching, translating and retrieving information in different languages, but mainly between a source language and a target language [1].

Few studies and methodologies have been developed to study the retrieval effectiveness (recall and precision) of Internet search engines, but particularly more so for cross-lingual or multi-lingual search engines due to the added barrier of language among different groups of language users and the problem of estimating data recall on documents retrieved from the Web [2]. As such, we are interested to assess both aspects of mono-lingual and cross-lingual retrieval to yield a more

C. Xing, F. Crestani, and A. Rauber (Eds.): ICADL 2011, LNCS 7008, pp. 296–306, 2011.

comprehensive evaluation and assessment of the effectiveness of such search engines which is basically lacking in information retrieval (IR) research literature. As part of the study, we propose a simplified methodology to estimate and compute the measures of recall and precision based on the original work of Clarke and Willet [3].

2 Related Work

Evaluation studies on search engines to date are largely focused on reporting features, search capability, collection crawled, indexing policies, user interface, usability, and so on. They have largely omitted quantitative evaluation of the retrieval effectiveness of the system. This is even more so for cross lingual search engines [4]. Studies that attempted to do so deal mainly with monolingual search engines, for example, Kumar and Prakash [5] on Google and Yahoo!. While there is a rich body or work done on evaluating retrieval effectiveness, these mainly used a laboratory-based test collection approach through initiatives such as TREC, NTCIR and CLEF. These tests rely on a test collection, a set of queries and relevancy judgment between the retrieved documents and queries to support research, benchmarking and reporting of results. The collection size was generally manageable so that relevance judgement information to compute data precision and recall was possible. But when we move away from a fixed document collection to the Internet where search engines prevail, we run into the challenge of computing data recall since neither exhaustive relevant documents nor absolute relevancy judgments exists. Some form of estimation is therefore required to determine the relevant documents for specific queries. One way is assigned scores to documents retrieved and to pool together results of different search engines to estimate the pool of relevant documents [3,5].

While there are numerous search engines that are currently in existence, few support truly cross-language retrieval. Many search engines are monolingual but have the added functionality to carry out translation of the retrieved pages from one language to another, for example, Google, EZ2Find and AltaVista [6]. The development of web-based CLIR systems have been reported by Ogen et al. [7], Ogden and Davis [8], Capstick et al. [9] and Penas et al.[10]. These focus mainly on operation and interaction of the systems with little or no users tests on performance. Work on query translation for different language pairs for CLIR found that the coverage and quality of the translation dictionaries used played an important role in the quality of answers retrieved [11]. CLIR researchers posits that current approaches are deficient due to the fact that monolingual retrieval has become language dependent, relying on specialized stemmers and stopwords, which means that retrieval is conducted independently to each language without consideration of the other languages the search engine has being requested to search [12]. Additionally, after documents from different languages are retrieved independently, and merged without information from the translation and retrieval giving rise to a lack of collection cohesion in delivering the search results.

While CLIR research and developments continues to advance, CLIR search engines continue to be used by users to seek multi lingual information from their queries. With almost no literature being reported on a systematic evaluation of commercial CLIR search engines, this study aims to fill the gap to evaluate two major search engines that is used widely in Asia to search for English and Chinese documents. In our study, we chose the popular Google and Yahoo! in Taiwan and Hong Kong (Google.tw, Google.hk, Yahoo.tw and Yahoo.hk) that indexes a good mix of English and Chinese documents. Monolingual dominant search engines like Baidu in China and Google in Singapore were omitted since they fall out of scope for evaluation. We originally included Google China for the study but had to abandon it due to its cessation of operation. We aim to study the retrieval effectiveness of EE, EC, CC and CE pairs using a set of queries on these four search engines.

3 Evaluation Methodology

3.1 Query Selection

As all four search engines being investigated are general search engines without specific subject emphasis, we developed a set of characteristics to aid query formulation and selection, and to ensure a good coverage of the query topics being evaluated. These include: people, event, places, simple to complex queries/general to specific queries, queries with potential language ambiguity, official government documents (possibly translated in more than one language), domain specific (Hong Kong and Taiwan) topics, time-based and/or recent topics, controversial and/or special topics. An original set of 18 queries were developed but were subsequently reduced to 6 queries for the evaluation (Table 1) to make the experiment feasible in terms of the time required to carry out the work.

Table 1. Queries for Experiments

	English Query	Chinese Query	People	Event	Place	Complex	Domain specific	Official documents	Time-based	Translation ambiguity
Q1	Amazon in America	汉堡的历史			X					X
Q2	Education reform highlights	教育改革大纲		X				X		
Q3	Brother Sharp	兄弟尖锐 (Google translation)	X				X			X
Q4	Sharp Brother (Google translation)	犀利哥	X				X			X
Q5	Disneyland 2000-2010	迪士尼乐园2000到2010			X	X	X		X	
Q6	New projects of Disneyland after 2000	2000年以后迪士尼乐园的新项目		X	X	X		X		

Each query has to be tested four times (EE, EC, CC, CE) on four search engines resulting in 16 sets of results per query. In the retrieval process, both simple and advanced search mode were used. For EE and CE, the option "Results in English" is selected. For EC and CC, the option "Results in both simplified Chinese and traditional Chinese" is selected. If an option to choose the target language is unavailable, then "Web page all over the world" is selected. Subsequently, the top 10 valid results for each query (i.e. 160 documents) were identified to allow relevance judgements to be carried out by a panel of evaluators.

3.2 Web Page Result Selection, Relevance Criteria and Evaluation Panel

We chose the established method to select the top 10 results in the target language retrieved by each search engine for the evaluation [3]. The searches were carried out on all four search engines on the same day to avoid any influence of indexing latency. In addition:

- Only web pages written in Chinese and English languages were considered. All pages in other languages were ignored.
- The web page content was used to determine the language of the web page. For instance, if the webpage came from an English domain website or only the heading was in English, but the major content was written in Chinese, then the page was regarded as a Chinese web page. If the content was in both languages in almost the same proportion, then the category of web page will be determined as the language that appeared first.
- Web pages containing mainly pictures or videos were ignored.
- Duplicate web pages retrieved by the same search engines were discarded and replaced by the next valid web page sequentially.
- If the result pointed to a dead link, "the file is not found" or "the server is not responding" message was encountered, a re-attempt was made subsequently to access the page later on the same day, failing which, the web page was discarded and replaced by the next valid web page sequentially down the list of results.

We used Clarke and Willett's [3] approach using a 3-scale score to determine the relevance of the results. Scores of 1, 0.5 and 0 were given to relevant web pages, partially relevant web pages and irrelevant web pages respectively. The following specific criteria were used for the assessment:

- A relevant web page was given a score of 1, and an irrelevant web page a score of 0. Partially relevant web pages, for example, a forum with certain content related to the query topic, was given a score of 0.5.
- If the result web page contain a series of links rather than the web page with required information, and one or some of these links were expected to be relevant to the query, then a score of 0.5 is given.
- If one web page was retrieved more than once by different search engines, the same score is applied to all occurrences.
- Duplicate web pages were judged based on content of the web page rather than the URL, headings or other criteria (as the same content may be hosted by different web servers and assigned different URLs).
- Web pages on sales advertisement (which occurred frequently on Yahoo!) was assigned a score based on the relevance of its content. If it was relevant to the query topic, it was given a score of 0.5, otherwise 0 was given.

Three post-graduate students who were enrolled in the Master of Science in Information Studies program at Nanyang Technological University, Singapore, acted as evaluators to assess the retrieved document relevance against the query topics. All of them were familiar with using common search engines including Google and Yahoo!. They were all fluent English-Chinese bilingual speakers. Their ages range from 20 to 30. The evaluation was carried out independently by each evaluator without consulting each other. Each evaluated a total of 960 web pages for this study.

3.3 Data Recall and Precision Computation

Recall is calculated as the sum of the scores of document retrieved by a search engine against the total score of relevant documents in the pool retrieved by all the search engines in the experiment.

Precision is calculated as the sum of the scores of documents retrieved by a search engine against the total number of retrieved documents for evaluation.

In order to obtain the relevant documents in the pool, any overlapping documents retrieved by the search engines must be discarded. In our case, if 4 search engines A,B,C and D retrieve *a,b,c* and *d* relevant documents, and there is an overlap of e documents, then the recall for A is $a/(a+b+c+d-e)$, and so on.

This formula is a simplified version of that proposed by Clarke & Willet [3] which included a second round of search to identify further potential documents using samples of text from relevant pages which had not been retrieved by all search engines to obtain a more accurate value for recall. We felt that this was unnecessary for our experiments as we are dealing with more than one language and following the same process could potentially yield language ambiguity problems and add more noise to the results. As such, this extra effort is not justifiable for CLIR evaluations when our main interest is to do a relative comparison among the search engines.

As we are using a group of 3 evaluators with different subjective assessments of relevance, we take the average of the scores for retrieval effectiveness computation. Denoting the individual scores as $x1$, $x2$, and $x3$, we obtain an Average X $(=(x1+x2+x3)/3)$ and with a merged pooled of relevant documents Y $(Y=a+b+c+d-e)$, we compute Recall=Average X/Y and Precision=Average X/10.

4 Experimental Results

Table 2 shows the results of the recall evaluation. The individual recall of EC and CE for cross lingual search is shown in columns C2 and C3, and EE and CC for monolingual search is shown in columns C5 and C6 for the four search engines (Rows R2 to R5). The table also shows the results for the combined average recall for cross lingual search (C4), monolingual search (C7) and the average recall of all search engines in the last row (R6).

Similarly, Table 3 shows the individual precision of EC and CE for cross lingual search (columns C2 and C3), and EE and CC (columns C5 and C6) for monolingual search for the four search engines is given. The results for the combined average precision for cross lingual search (C4), monolingual search (C7) and the average precision of all search engines in the final row (R6). It should be noted that the average precision is computed as the simple average of the precision values. This is different from the Mean Average Precision in traditional IR literature which takes into account the precision-recall levels at 11 points [13,14]. For ease of comparison, Figure 1 shows the average recall and precision for the cross lingual and monolingual search (C4 and C7) for the evaluated search engines.

Table 2. Recall for Four Search Engines

	C1	C2	C3	C4	C5	C6	C7
R1	Search Engine	Search Type		Average Recall	Search Type		Average Recall
		EC	CE	EC&CE	EE	CC	EE&CC
R2	Google.tw	0.134	0.049	0.092	0.231	0.167	0.199
R3	Google.hk	0.154	0.057	0.105	0.217	0.147	0.182
R4	Yahoo.tw	0.052	0.040	0.046	0.182	0.075	0.128
R5	Yahoo.hk	0.105	0.029	0.067	0.193	0.111	0.152
R6	Average	0.111	0.044	0.078	0.206	0.125	0.165

Table 3. Precision for Four Search Engines

	C1	C2	C3	C4	C5	C6	C7
R1	Search Engine	Search Type		Average Precision	Search Type		Average Precision
		EC	CE	EC&CE	EE	CC	EE&CC
R2	Google.tw	0.447	0.217	0.332	0.761	0.539	0.650
R3	Google.hk	0.478	0.261	0.370	0.653	0.503	0.578
R4	Yahoo.tw	0.175	0.203	0.189	0.556	0.319	0.437
R5	Yahoo.hk	0.339	0.131	0.235	0.581	0.403	0.492
R6	Average	0.360	0.203	0.282	0.638	0.441	0.539

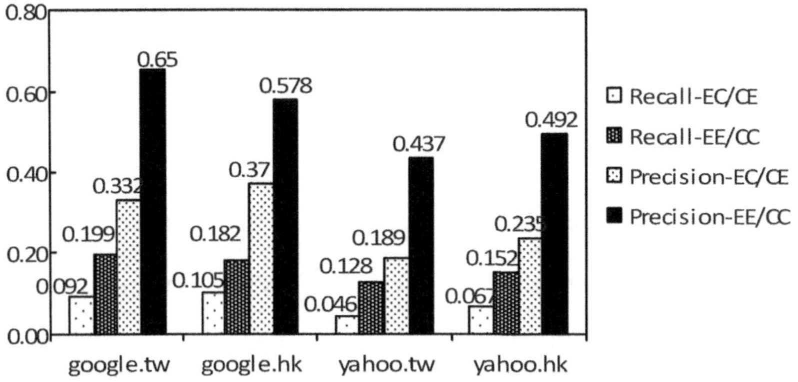

Fig. 1. Retrieval Effectiveness of Google and Yahoo Search Engines

From Tables 2 and 3, and Figure 1, we draw the following observations:

Recall and precision of monolingual (EE and CC)(columns C5-C7) searches are better than cross lingual search for all individual search engines, Google and Yahoo! as a group (Google.tw and Google.hk, and Yahoo.tw and Yahoo.hk), and collectively for all 4 search engines as a whole. Thus, we conclude that the effectiveness of cross lingual search is poorer than monolingual search.

Google outperforms Yahoo! in both recall and precision for both monolingual and cross lingual search in both the sub domains in Taiwan and Hong Kong, and as a combined group (columns R2-R3 vs. R4-R5).

The recall performance of EC is higher than CE for all 4 search engines (Table 2, columns C2 and C3). In other words, using an English query to search information in Chinese in these search engines will draw more relevant documents as opposed to using a Chinese query to search information written in English. On average, the recall performance of EC is about 2.5 times better as that of CE (0.111 vs. 0.044).

In terms of precision performance, EC is higher than CE for 3 of the 4 search engines (Table 3, columns C2 and C3). On average, the precision performance of EC is about 1.8 times better as that of CE (0.36 vs. 0.203).

When we examine the effectiveness of individual search engines for cross lingual search, Google.hk gives the best recall performance for both EC and CE with values of 0.154 and 0.057 respectively. Yahoo.tw yielded the poorest recall result for EC search (0.052) while Yahoo.hk yielded the poorest result for CE search (0.029). A similar observation is made for the precision performance, with Google.hk providing the best results for both EC and CE cross lingual search with precision values of 0.478 and 0.261 respectively. Yahoo.hk yielded the poorest precision for CE search (0.131) and Yahoo.tw yielded the poorest precision for EC search (0.175).

When we examine the effectiveness of individual search engines for monolingual search, Google.tw gives the best recall performance for both EE and CC with values of 0.231 and 0.167 respectively. Yahoo.tw yielded the poorest recall result for both EE search (0.182) and CC search (0.075). The same observation is made for the precision performance, with Google.tw providing the best results for both EE and CC with precision values of 0.761 and 0.539, and Yahoo.tw yielding the poorest precision for both EE search (0.556) and CC search (0.319).

In summary, Google.hk came out tops in cross lingual search while Google.tw came out tops in monolingual search. Nonetheless, the retrieval effectiveness of EC and CE searches is far from satisfactory. Even the best search engine, Google.hk, bears a low average recall of 0.105 and a low average precision of 0.370.

5 Discussion

The findings show that the retrieval effectiveness of EC and CE cross lingual search is much lower than that of EE and CC monolingual search. For EC and CE search, an average recall of 0.078 and an average precision of 0.282 are obtained, while those of EE and CC search are 0.165 and 0.539, respectively. Such differences may be attributed to various reasons. First, the majority of documents indexed by the search engines are written in English than in Chinese. As a result, more English documents are expected to be retrieved for random queries. English words have clear delimiters (space and punctuations) between words while Chinese words do not. The additional process of segmenting Chinese words into distinct word entities to support search and retrieval can increase the chance of errors in matching query terms to retrieved documents, thereby affecting both recall and precision [15]. The process of cross lingual retrieval is more complex, requiring an additional step of translation which again adds potential inaccuracies and noise in the retrieval process.

In our experiment, we also included two queries with translation ambiguity. Query 3, "Brother Sharp" is automatically but incorrectly translated into "兄弟尖锐" by the Google Translator as its equivalent Chinese query. The correct translation should be "犀利哥" (Query 4) which is in turn translated to 'Sharp Brother" for CE by the same Google Translator. When "兄弟尖锐" is used for CE search, no English documents were found. On the other hand, using the correct translation of "犀利哥" yielded relevant documents, with recall values of 0.110 for Google.tw, 0.152 for Google.hk, 0.124 for Yahoo.tw, and 0.058 for Yahoo.hk. From this pair of corresponding queries, it is evident that query translation quality plays a critical role in cross lingual search.

When we attempted to compare out results with previous works, we were unable to find any similar literature for cross lingual evaluation. Nonetheless, we found a related piece of work by Shafi and Rather [16] for monolingual retrieval that showed that Google with a recall of 0.20 and precision of 0.29 in their experiments. Our monolingual experiments (EE and CC) yielded an average recall of 0.19 and precision of 0.61 for Google (hk and tw). This appears to support Google's improvement in retrieval effectiveness over this period from 2004 to 2010. With the exponential rise of web documents, these two experiments suggest that recall was maintained (0.20 to 0.19) and precision improved (from 0.29 to 0.61). While this experiment is small and cannot lead to any firm conclusion, the result is nevertheless interesting to demonstrate the improvement in retrieval effectiveness of commercial search engines. With the rising importance of Chinese language, we can expect more Chinese documents to be increasingly authored and indexed and more effective and accurate translation methods be further developed to support improved cross lingual searching.

In our experiment, we also found that a large portion of Chinese documents were indexed both in English and Chinese. While the contents of Chinese documents were not fully translated or authored in English, many Chinese documents had accompanying English titles and headers. Hence, when we carry out EC searches using English queries, these corresponding Chinese documents were retrieved. However, the reverse is not true for English documents. This relatively smaller number of English documents that is indexed in both languages will therefore result in less chance for them to be retrieved when we use a Chinese query to search for English information (CE search). This difference of indexing exhaustivity is probably the main cause of difference in retrieval effectiveness for EC and CE searches using the same search engine, with EC producing better results than CE for all four search engines (R6, Tables 1 and 2). This observation seems to suggest that if we can provide basic metadata (such as document title, keywords, headings) in multiple languages, this would help enhance the quality of cross lingual searches if such metadata is indexed by the search engines.

Finally, when we probed deeper into Yahoo.tw and Yahoo.hk relatively poorer performance, we found that Yahoo! results contain significant amounts of commercial advertisements. For example, in the "Brother Sharp' query, the first 6 out of 10 results were advertisements for Brother printers and photocopiers, printing paper, Panasonic and other companies. As another example, the EE search of "Amazon in America" yielded all 10 ten results about the Amazon online shop, including six pure advertisements recommending books and other products. In contrast, the same query on Google.tw retrieved 2 results on the Amazon online shop advertisements, but included other relevant results about the Amazon rainforest, Amazon Basin and Amazon River. This should somewhat not be surprising since the business model of commercial search engines dictates and impacts on the retrieval characteristics of the search engine.

6 Limitations of Study

Due to the limited number of queries, we acknowledge that the findings are at best indicative and therefore provide a preliminary understanding of the effectiveness of the search engines being evaluated. Average values for recall and precision were selected and presented in preference over individual results for each query to cater to the requirement of manuscript length, although the latter form of presentation could give a better indication of reliability of the findings. Other forms of measures for retrieval effectiveness for ranked search engine evaluations such as the Normalized Discounted Cumulative Gain (NDCG) [14,17] or error measures such as the MUC ERR measure [18], or Slot Error rate (SER) [19] based on work done on speech recognition research, R-precision[13.14] or marginal relevance [14] were not considered in this study.

7 Conclusions

The Internet has paved opportunities for increasing multi-lingual information exchange and retrieval in future. While this future will be mitigated by the quality of cross lingual search engines to create the connectivity among multi-lingual documents, our study suggest that the current quality of CLIR search engines are poor with scope for much improvement.

Our experiment on four search types (EC, CE, EE and CC) on four commonly used search engines (Google.tw, Google.hk, Yahoo.tw and Yahoo.hk) demonstrated the overall higher retrieval effectiveness of monolingual searches (EE and CC) over multilingual searches (EC and CE). EC searches returned superior results than CE as a result of a larger pool of English-enabled documents for indexing. More complex queries yielded lower recall and precision in comparison to simple queries. Clear expression of the information needs using simple and unambiguous queries are possible ways to improve the retrieval effectiveness. Due to the poor recall of cross lingual searches, searching across multiple CLIR search engines or utilizing meta-CLIR search engines becomes even more essential to support comprehensive searches for multi lingual documents.

Creating accurate metadata in different languages in documents or good translation of key information in documents can help improve the quality of the index and retrieval. Searching in context which is affected by language and translation ambiguity remains a challenge that can benefit from methods such as domain specific corpus, dictionary or ontology-based, or other approaches.

This study highlights the need for further research efforts for CLIR search engines to seek improvements in translation methods, index quality, multilingual retrieval techniques, and user interface for results presentation in order for a new generation of CLIR search engines to emerge and be widely and successfully used as their monolingual counterparts. Other evaluation techniques mentioned in the previous section are logical extensions of this work to provide additional evidence to assess these CLIR engine retrieval effectiveness.

Acknowledgement. The author gratefully acknowledges the contributions of Gao Xiuqing, Katherine Chia and Tang Jiao who carried out this study as part of their Critical Inquiry Project in their M.Sc. (Information Studies) program.

References

1. Hansen, P., Petrelli, D., Beaulieu, M., Sanderson, M.: User-centred interface design for cross-language information retrieval. In: 25th Annual International ACM SIGIR Conference on Research and Development in Information Retrieval, pp. 383–384. IEEE Press, New York (2002)
2. Kralisch, A., Mandl, T.: Barriers to information access across languages on the internet: network and language effects. In: 39th Annual Hawaii International Conference on Systems Sciences, vol. 3, pp. 4–7 (2006)
3. Clarke, J.S., Willett, P.: Estimating the recall performance of web search engines. Aslib Proceedings 49(7), 184–189 (1997)
4. Youssef, M.: Cross language information retrieval: Universal usability in practice. Department of Computer Science, University of Maryland (2001), `http://otal.umd.edu/uupractice/clir/#r11` (retrieved May 30, 2011)
5. Kumar, B.T., Prakash, J.N.: Precision and relative recall of search engines: A comparative study of Google and Yahoo. Singapore Journal of Library & Information Management 38, 124–137 (2009)
6. Zhang, J., Lin, S.Y.: Multiple language supports in search engines. Online Information Review 31(4), 516–532 (2007)
7. Ogden, W.C., Cowie, J., Davis, M., Ludovik, E., Nirenburg, S., Molina-Salgado, H., et al.: Keizai: An interactive cross-language text retrieval system. In: Ananiadou, S., Hayashi, Y., Jacquemin, C., Leong, M.K., Chen, H.H. (eds.) Workshop on Machine Translation for Cross Language Information (1999), `http://citeseerx.ist.psu.edu/viewdoc/download?doi=10.1.1.37.5775&rep=rep1&type=pdf` (retrieved May 30, 2011)
8. Ogden, W.C., Davis, M.W.: Improving cross-language text retrieval with human interactions. In: 33rd Annual Hawaii International Conference on System Sciences, vol. (3), p. 3044 (2000)
9. Capstick, J., Diagne, A.K., Erbach, G., Uszkoreit, H., Leisenberg, A., Leisenberg, M.: A system for supporting cross-lingual information retrieval. Information Processing & Management 36(2), 275–289 (2000)

10. Peñas, A., Gonzalo, J., Verdejo, F., Lenguajes, D., Informáticos, S.: Cross-language information access through phrase browsing. In: Applications of Natural Language to Information Systems. LNI, pp. 121–130 (2001)
11. Airio, E.: Who benefits from CLIR in web retrieval. Journal of Documentation 64(5), 760–778 (2007)
12. Gey, F., Kando, N., Peters, C.: Cross-language information retrieval: the way ahead. Information Processing and Management 41, 415–431 (2005)
13. Baeza-Yates, R., Ribeiro-Neto, B.: Modern Information Retrieval, 2nd edn. Addison-Wesley, Reading (2010)
14. Manning, C.D., Raghavan, P., Schutze, H.: Introduction to Information retrieval. Cambridge University Press, Cambridge (2008)
15. Foo, S., Li, H.: Chinese word segmentation and its effect on information retrieval. Information Processing and Management 40(1), 161–190 (2004)
16. Shafi, S., Rather, R.: Precision and recall of five search engines for retrieval of scholarly information in the field of biotechnology. Webology 2(2), Article 15 (2005), `http://www.webology.org/2005/v2n2/a12.html` (retrieved May 30, 2011)
17. Croft, B., Metzler, D., Strohman, T.: Search engines: Information retrieval in practice. Addison Wesley, Reading (2009)
18. Chinchor, N., Dungca, G.: Four scores and seven years ago: The scoring method for MUC-6. In: Proc. MUC-6 Conference, Columbia, MD (1995)
19. Makhoul, J., Kubala, F., Schwartz, R., Weischedel, R.: Performance measures for information extractors. In: Proc. DARPA Broadcast News Workshop, Herndon, VA, USA (1999)

Term Familiarity to Indicate Perceived and Actual Difficulty of Text in Medical Digital Libraries

Gondy Leroy and James E. Endicott

School of Information Systems and Technology, Claremont Graduate University
{Gondy.Leroy,EndicotJ}@cgu.edu

Abstract. With increasing text digitization, digital libraries can personalize materials for individuals with different education levels and language skills. To this end, documents need meta-information describing their difficulty level. Previous attempts at such labeling used readability formulas but the formulas have not been validated with modern texts and their outcome is seldom associated with actual difficulty. We focus on medical texts and are developing new, evidence-based meta-tags that are associated with perceived and actual text difficulty. This work describes a first tag, 'term familiarity', which is based on term frequency in the Google corpus. We evaluated its feasibility to serve as a tag by looking at a document corpus (N=1,073) and found that terms in blogs or journal articles displayed unexpected but significantly different scores. Term familiarity was then applied to texts and results from a previous user study (N=86) and could better explain differences for perceived and actual difficulty.

Keywords: Natural Language Processing, Health Informatics, Perceived Difficulty, Actual Difficulty, Meta Information, Lexical Tags.

1 Introduction

In healthcare, pamphlets and brochures are often used to explain medical procedures, effects and guidelines. The text needs to be written in a manner that consumers, who may have low literacy, can understand. Today's readability formulas could form an excellent starting point to facilitate writing easier text. Unfortunately, they suffer from several shortcomings. They have not been validated for modern texts, topics or sources. Moreover, they have not been clearly demonstrated to be related to text understanding. Many studies rely on expert opinions instead of measured user understanding. We are working on an evidence-based difficulty checker that indicates where text may be too difficult for consumers to understand.

Difficulty levels of text are often measured with readability formulas. While many different formulas exist, the Flesch-Kincaid readability grade level is among the most popular in the U.S. It is based on syllable and word counts and its outcome represents the education needed to understand the text (e.g., 12.0 = finished 11th grade and started 12th grade). Most writing guidelines, e.g., by the American Medical Association [1], advise writing at 6-8th grade level. However, based on these formulas most English-language Internet sites require a 10th grade or even college level education [2]. There are readability formulas that leverage lists of common words (e.g., Dale-Chall) but they used fixed word lists that are evenly weighted.

C. Xing, F. Crestani, and A. Rauber (Eds.): ICADL 2011, LNCS 7008, pp. 307–310, 2011.

We distinguish between perceived and actual difficulty because both are useful, but for different reasons. It is important for text to have an easy perceived difficulty so it is not intimidating. Actual difficulty level is important to facilitate understanding of the material. The distinction between perceived and actual difficulty is important but seldom acknowledged. However, evidence for the distinction can be found in different models. For example, the Health Belief Model contains 4 dimensions: perceived susceptibility, perceived severity, perceived benefits, and perceived barriers. In a large review study [3], perceived barriers was found to be the most significant dimension in explaining health behaviour.

2 Term Familiarity

We are working on defining text features, to label text difficulty levels, in an evidence-based and computationally efficient manner. We focus on 3 groups: grammatical, lexical, and composition features. In earlier work [4], we found several grammatical metrics (noun phrase complexity, function word density, and grammatical structures) that are strongly related to perceived text difficulty. However, these were only slightly related to actual difficulty as measured by understanding and retention of information.

This work focuses on a lexical measure: term familiarity. We use the frequency of terms in the Google corpus as an approximation of term familiarity. The Google web corpus contains n-gram models calculated based on a corpus of a trillion words from the Google collection of public Web pages. It contains 1,024,908,267,229 tokens and provides frequency counts for unigrams, bigrams, trigrams, 4-grams and 5-grams. For example, the corpus shows that the word "apnea" is less common than "obesity". We believe such frequencies may help explain difficulty where readability formulas would fail, e.g., most formulas would judge "obesity" as the more difficult term based on syllable or letter counts.

2.1 Corpus Analysis - Feature Detection

Study Design. A corpus was developed containing patient blogs from Blogspot (up to 6 posts per blogger) representing 'easy text' and full-text journal articles from Wiley Online Library (top 100 matches) representing 'difficult' text. For each, we selected documents on 7 topics: alcoholism, autism, cancer, depression, diabetes, obesity and post-traumatic stress disorder.

We calculated the ratio of terms in each document that were nouns, verbs, adjectives, and adverbs as these are the main content bearing terms. Each term was submitted to the Google corpus and its frequency was noted. We then calculated our term familiarity measure for each document as the weighted average of term frequencies adjusted for their relative proportion in the document.

Analysis and Results. In total, 372 blogs and 697 journal articles were found. Difficulty in finding patient blogs without over-representing a single author is the reason that fewer blogs were selected for each topic. Naturally, journal articles are much longer than blogs and therefore all measures are normalized by word length.

Table 1 provides an overview of the results. It is interesting that the proportion of nouns is significantly higher in the articles, while the proportion of verbs is significantly higher in the patient blogs. The term frequencies also differ, with patient blogs using nouns and adjectives that have significantly higher frequencies. The frequency of the verbs did not differ between the two groups. Our familiarity measure, which takes the ratio of each word group into account when combining frequencies per document, is significantly higher for patient blogs than for journal articles.

Table 1. Corpus Descriptive Statistics (*: significant at $p < .001$, independent samples two-tailed t-test, +with Bonferroni correction $\alpha/4$)

Feature	Patient blog	Journal Article
Average Total Word Count:	726	5158
Average Percentage Nouns*+:	25	39
Average Percentage Verbs*+:	19	14
Average Percentage Adjectives*+:	7	10
Average Percentage Adverbs*+:	6	4
Average Noun Frequency*+:	131,114,039	77,111,994
Average Verb Frequency:	718,591,781	702,892,877
Average Adjective Frequency*+:	203,763,085	104,070,055
Average Adverb Frequency*+:	508,593,450	588,474,771
Term Familiarity*:	215,181,482	159,667,621

2.2 User Study - Feature Application

Study Design. In a previous user study [4], we evaluated 8 sentences (Part 1) and 2 paragraphs (Part 2) for 3 features: noun phrase complexity (the simple version had fewer compound nouns), sentence structure (the simple version had shorter and active sentences), and function word density (the simpler version had more function words).

- **Part 1:** For each sentence, all 8 versions were shown and participants chose the sentence that 'looked' the easiest.

- **Part 2:** We worked with 1 paragraph on depression and 1 on heart disease and constructed an easy and difficult version for each. Each participant received an easy and a difficult paragraph with the topics and order counterbalanced. To measure perceived difficulty we asked the participants to indicate from which paragraph they thought they remembered most. To measure actual difficulty, we posed 3 multiple-choice questions for each paragraph.

Analysis and Results (N=86). As reported earlier [4], we found very clear effects of our grammatical metrics on perceived difficulty for the individual sentences (Part 1). The results for the paragraphs (Part 2) could not all be explained by the differences in grammatical features. However, Table 2 shows how these results also are in line with term familiarity. The first unexplained result was that 62% of all participants chose the depression document as the easiest (perceived difficulty) of the two documents they read, even though half of the participants received its difficult version. However, term familiarity was higher in both version of the depression document. The second unexplained result was that the percentage correct answers (actual difficulty) was only slightly influenced by surface metrics. Again, these results are in line with term familiarity for the different documents. For both results, higher familiarity is in line with documents perceived as easier and more correct answers.

Follow-up studies will be conducted to collect more data to verify or refute these relationships statistically.

Table 2. Actual difficulty in relation to grammar and lexical features

Topic:	**Depression**		**Heart Disease**	
Difficulty Level:	Easy	Difficult	Easy	Difficult
Actual Difficulty (% correct):	67	57	57	54
Term Familiarity:	303,670,073	296,258,700	229,500,975	158,856,104

3 Conclusion

We are working toward the development of meta-information that indicates text difficulty. This work introduced a composite feature, term familiarity, which can be assigned to text by computational means. From a corpus analysis, we can conclude it distinguishes between easy and difficult text. It also served to clarified hereto unexplained results of a user study for perceived and actual difficulty of text.

Later work will use larger samples sizes to statistically verify or refute these findings. Term familiarity represents only one aspect of text and we plan to combine it with others for a more comprehensive measurement of text difficulty.

Acknowledgments. This work was supported by the U.S. National Library of Medicine, NIH/NLM 1R03LM010902-01.

References

[1] Weis, B.D.: Health Literacy and Patient Safety: Help Patients Understand. Manual for Clinicians. In: AMA and AMA Foundation, 2nd edn. (2007)

[2] Berland, G.K., Elliott, M.N., Morales, L.S., Algazy, J.I., Kravitz, R.L., Broder, M.S., Kanouse, D.E., Muñoz, J.A., Puyol, J.-A., Lara, M., Watkins, K.E., Yang, H., McGlynn, E.A.: Health Information on the Internet: Accessibility, Quality, and Readability in English and Spanish. JAMA 285, 2612–2621 (2001)

[3] Janz, N.K., Becker, M.H.: The Health Belief Model: A Decade Later. Health Education Quarterly 11, 1–47 (1984)

[4] Leroy, G., Helmreich, S., Cowie, J.: The Influence of Text Characteristics on Perceived and Actual Difficulty of Health Information. International Journal of Medical Informatics 79, 438–449 (2010)

An Entailment-Based Question Answering System over Semantic Web Data

Shiyan Ou and Zhenyuan Zhu

Department of Information Management, Nanjing University, China
{oushiyan,MG1014017}@nju.edu.cn

Abstract. This paper reports a novel knowledge-based Question Answering (QA) method with the use of Semantic Web technologies and textual entailment recognition. Different from most of ontology-driven QA methods, this method does not perform deep question analysis to transform a natural language question into an ontology-compliant query for answer retrieval. Instead, it performs textual entailment recognition to discover the question template entailed by a user question from the whole machine-generated set and then takes the associated SPARQL query template to produce the complete query for retrieving the answers from the Semantic Web data that subscribe to the same ontology. An evaluation was carried out to assess the accuracy of the QA method, and the results revealed that the generated question templates can cover almost all the user questions and 65.6% of the user questions can be correctly answered with the support of a semantic entailment engine.

Keywords: Question Answering, Semantic Web, Textual Entailment, Ontology.

1 Introduction

In Question Answering (QA), structured knowledge-based QA has developed in parallel with free text-based QA. It extracts answers from structured data rather than free texts and thus is well adapted to the applications requiring complex queries and accurate answers [1]. Early knowledge-based QA systems are mainly database-based, which are not flexible and powerful enough due to the dependency on the database schema and the lack of semantics and reasoning. The emergence of Semantic Web technologies offers a new opportunity to develop novel, sophisticated forms of knowledge-based QA, where ontologies play a crucial role. In the Semantic Web, ontologies are used to conceptually model the knowledge in various domains and semantically mark up unstructured texts to produce Semantic Web data represented in the format of RDF. To extract the answers from the semantic RDF data that subscribe to an ontology, a natural language question needs to be precisely transformed into an structured query that is expressed in a RDF query language (e.g. SPARQL) and also compliant with the same ontology. However, this is a complex task involving deep linguistic analysis to reconstruct a logical form for the question and even requiring extensive manual work. In addition, because of language variability, the same or similar questions can be asked in various ways. Thus, it is not worth processing each expression individually to produce the same or similar query. To address this problem,

C. Xing, F. Crestani, and A. Rauber (Eds.): ICADL 2011, LNCS 7008, pp. 311–320, 2011.

textual entailment recognition, an inference task of determining whether the meaning of a text fragment *Text* can be inferred from another text fragment *Hypothesis* [2], was proposed as a solution to identify the different question expressions that entail the common semantic meaning and thus can share the same retrieval procedure [3].

This paper presents a knowledge-based QA method for a restricted domain with the use of Semantic Web technologies and textual entailment recognition. Different from most of ontology-based QA methods, the proposed method does not perform deep linguistic processing to transform a natural language question into an ontology-compliant query. Instead, a set of question templates (called *hypothesis questions*) and their corresponding query templates are automatically produced offline on the basis of the domain ontology. These question templates cover basic types of questions and possible items to be queried in a domain, whereas the corresponding query templates provide an instruction on how to obtain the answers for such type of questions from the underlying RDF data. A user question may be a basic question that can be directly matched to a question template or a complex question that can be decomposed into the combination of several basic questions. In this way, the QA task is reduced to the problem of looking for the question template that is entailed by a user question and taking its corresponding query template to produce a complete query for answer retrieval. In comparison with other knowledge-based QA methods, the reported QA method has three main features: (1) It has some degree of reasoning ability due to the use of the ontology and textual entailment recognition; (2) It uses textual entailment recognition to deal with language variability, which can avoid the problem of processing multiple questions that entail same or similar meaning; and (3) It produces question templates and query templates offline on the basis of the domain ontology, which can avoid deep linguistic processing and thus speed up the QA procedure.

The subsequent sections are organized as follows: Section II reviews some related QA work; Section III describes the acquisition and storage of Semantic Web data; Section IV describes the automatic generation of question templates and their corresponding query templates; Section V describes the core architecture of the QA method; Section VI reports the evaluation of the accuracy of the QA method and its results; and Section VII presents the conclusion and future work.

2 Related Work

The use of natural language questions to access relational databases has long been an area of research in the artificial intelligence and database communities. The key point of such database-based QA is how to represent a natural language question to a database query. In the past decade, database-based QA has somewhat gone out of fashion [4]. The Semantic Web provides a new and potentially important context in which the results from this area can be applied [5].

In the Semantic Web, the availability of domain ontologies make it possible to semantically transform a natural language question into an ontology-compliant representation and retrieve the answers from Semantic Web data with some degree of reasoning. This characteristic makes ontology-based QA more smart and powerful than database-based QA. In MOSES [6], a natural language question is first represented as a Question Quasi-Logical form with syntactic analysis, and then further mapped to an ontology-based concept-relation form with the support of the domain ontology aligned with some general linguistic resources. AquaLog [7] is another ontology-based QA system to query the semantically structured web contents. It makes use of some linguistic tools to annotate a natural language question and then translates it into a linguistics triple form. This intermediate triple is further processed at a higher semantic level with the use of the domain ontology and some lexical resources to produce an ontology-compliant triple for answer retrieval. Afterwards, the single-domain QA system Aqualog was extended to the cross-domain QA system PowerAqua [8], which is able to answer questions in multiple domains that are modeled by a set of heterogeneous ontologies. In addition to the same question processing step performed in Asqualog, PowerAqua has a further step to discover the set of the ontologies that jointly cover the user question and then convert the linguistic tripe to equivalent ontological triples. SemanticQA [9] is a hybrid QA system which can extract answers from linked data sets and unstructured web documents. It first matches all word combinations taken from a natural language question to various entities in linked data sets to produce RDF triples and then combines these triples into a single SPARQL query for answer retrieval. If the answers cannot be found in linked data sets, a search engine will be called to retrieve relevant web documents for answer extraction.

The abovementioned QA systems focus on the transformation of natural language questions to ontology-compliant queries or triples. It requires complex linguistic processing and may take longer time in QA. Our QA method is also ontology-based, but seeks to produce ontology-compliant queries by mapping a natural language question to a pre-generated question template through textual entailment recognition and thus can avoid deep question processing.

3 Acquisition and Storage of Semantic Web Data

To construct a knowledge base for providing potential answers to the questions in a specific domain, an ontology needs to be built for conceptually modeling domain knowledge and semantically annotating domain information. We selected the tourism domain as a sample domain and built a tourism ontology using OWL-DL. It covers several important aspects of this domain, including tourism destinations, tourism sites, tourism events, and transportation [10]. A domain ontology usually does not contain general concepts which are applicable to a wide range of domains. To augment the available domain resources through semantic connections and definitions, the domain ontology could be aligned with some upper ontologies, such as WordNet[1]. We developed a semi-automatic method to implement this alignment, whereby unambiguous matches were aligned automatically, multiple matches were disambiguated by a human, and those concepts without any matches were aligned completely manually [10].

We extracted original data mainly from domain-related web sites, for example, hotel data from Google Maps, using web mining techniques, and then converted them into Semantic Web data with embedded semantic markups. A set of wrappers was

[1] http://wordnet.princeton.edu/

developed with the use of HtmlParser[2] to extract snippets of useful information from HTML pages and label them with XML tags based on their field labels detected from the texts. A mapping tool, which can map XML tags with the ontology classes and properties, was then developed based on similarity matching and used to directly convert the XML data to the semantic RDF data.

For storing the RDF data, various storage systems have been implemented with different methods and targets, e.g., Jena[3] and Sesame[4]. In QA, response time is a crucial issue, especially for the enormous amount of data. Thus the storage system must have good scalability and efficiency, and furthermore should provide sufficient reasoning capability to support the semantic requirements of the application. Considering the above requirements, we selected Jena as the storage system and the SPARQL query language as the data retrieval method. Jena now has two persistent storage models: SDB (hash/index layout) and TDB (incremental/parallel index). SDB is a SQL database-supported RDF store and suitable for transactions, while TDB is a high-performance, non-transactional RDF store using custom indexing and storage. According to the experiment results for testing SPARQL query execution time, we found that these two storage models have no clear difference for simple queries but SDB with the index layout is more efficient for complex queries, and thus finally selected SDB-index as the RDF store.

4 Question Template Generation

Since the domain knowledge is modeled by a domain ontology, it is possible to use the same ontology to model the questions that could be answered in this domain. Thus we sought to produce question templates (also called *hypothesis questions*), which represent the basic types of the questions to be queried in a domain, on the basis of the domain ontology. Subsequently we used the tourism ontology as the example to show how to generate question templates for the tourism domain.

In the tourism ontology, there are main classes (e.g. *Site*) which can exist independently and minor classes (e.g. *Address*) which must be associated to a main class. We focused on the main classes and their associated properties within a specific distance to create question templates. The associated properties of the main class represent the possible items which can be queried for an instance of the class. Fig.1 shows the abstract structure of a main class C0 and the properties directly or indirectly associated to it. The distance from the main class to a property is measured as follows:

- 0 – the datatype properties of C_0;
- 1 – the datatype properties of C_{10} which is directly linked with C_0;
- 2 –the datatype properties of C_{20} which is indirectly linked with C_0 through another class C_{10};

[2] `http://htmlparser.sourceforge.net/javadoc/index.html`
[3] `http://jena.sourceforge.net/`
[4] `http://www.openrdf.org/`

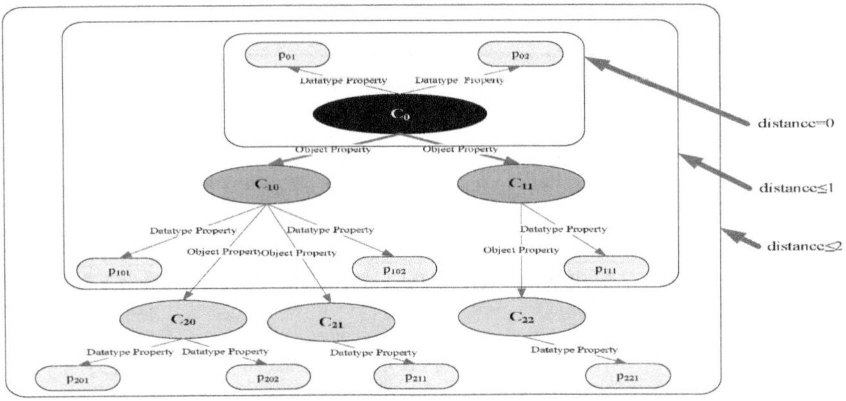

Fig. 1. The abstract structure of a main class and its associated properties

After carrying out a manual analysis to 500 sample user questions, we decided to create two types of question templates:

- Q_{name}: Query the name of a class instance using its several associated properties within a specific distance as the constraints;
- $Q_{attribute}$: Query one of the associated properties of a known class instance.

The user questions that can be covered by these two types of question templates are called basic questions, whereas the rest are called complex questions which can be decomposed into the combination of several basic questions. We furthermore restricted the constraints of the Q_{name} questions and the distance of the associated properties to less than three. The reasons are: (1) 91% of the user questions in the tourism domain contain fewer than three constraints; (2) a user question can entail a question template of the same type containing some of its constraints and thus can obtain the answers that subsume the exact ones; and (3) the link between a main class and an associated property out of the distance three becomes very weak in the tourism ontology. According to the various combinations of the main class and its associated properties (see Fig. 1), a set of question templates was predefined and listed in Table 1.

Once the question templates had been determined, we parsed the OWL ontology using Jena OWL API and produced all the hypothesis questions, which are more specific question templates, on the basis of various predefined templates. For example, if C_0=Hotel, C_{10}=Address, C_{20}=Destination, P_{01}=Hotel.rating, P_{201}=Destiantion.name, the following Q_{name} hypothesis question can be generated:

- H_1: *What is the name of the hotel which has the [Hotel. rating] and is in the [Destination.name]?*

Along the path from the main class C_0 to each of its associated datatype property, the nodes and their relationships can be expressed in a series of RDF triples. Thus the SPARQL query template corresponding to the hypothesis question H_1 can be

Table 1. Predefined Question Templates based on the Tourism Ontology

Type	Dist.	Con.	Question Templates
	0	1	*What is the name of the C_0 which has [$C_0.p_{01}$]?*
		2	*What is the name of the C_0 which has [$C_0.p_{01}$] and [$C_0.p_{02}$]?*
		1	*What is the name of the C_0 which has [$C_{10}.p_{101}$]*
		2	*What is the name of the C_0 which has [$C_{10}.p_{101}$] and [$C_{10}.p_{102}$]*
	≤1	2	*What is the name of the C_0 which has [$C_{10}.p_{101}$] and [$C_{11}.p_{111}$]*
Q_{name}		2	*What is the name of the C_0 which has [$C_0.p_{01}$] and [$C_{10}.p_{101}$]*
		1	*What is the name of the C_0 which has the p_{201} of the C_{20} [$C_{20}.\ p_{201}$]?*
		2	*What is the name of the C_0 which has [$C_{20}.\ p_{201}$] and [$C_{20}.\ p_{202}$]?*
		2	*What is the name of the C_0 which has [$C_{20}.\ p_{201}$] and [$C_{20}.\ p_{202}$]?*
	≤2	2	*What is the name of the C_0 which has [$C_{20}.\ p_{201}$] and [$C_{21}.\ p_{211}$]?*
		2	*What is the name of the C_0 which has [$C_{10}.\ p_{101}$] and [$C_{20}.\ p_{221}$]?*
		2	*What is the name of the C_0 which has [$C_{11}.\ p_{111}$] and [$C_{20}.\ p_{201}$]?*
		2	*What is the name of the C_0 which has [$C_0.\ p_{01}$] and [$C_{20}.\ p_{201}$]?*
	0	1	*What is the p_{01} of the [C_0.name]?*
$Q_{attribute}$	1	1	*What is the p_{101} of the C_{10} of the [C_0. name]?*
	2	1	*What is the P_{201} of the C_{20} of the [C_0. name]?*

produced at the same time based on all the involved RDF triples. In the hypothesis question and its query template, the square-brackets slot represents the value of a class's datatype property (e.g. Hotel.rating), which corresponds to a named entity in a user question. A SPARQL query template can be used to create a real SPARQL query by instantiating its slots with the named entities extracted from a user question.

5 Core QA Architecture

After the generation of possible hypothesis questions, the QA task is reduced to the problem of looking for the hypothesis question that is entailed by a user question and then taking its corresponding query template to create a complete SPARQL query for answer retrieval. As shown in Fig. 2, there are five main steps in this QA procedure.

5.1 Question Annotation

User questions usually contain several named entities, such as person names, places names, numbers, date and time. To better match a user question to a hypothesis question, the named entities were recognized and annotated with the classes and properties defined in the domain ontology, and the temporal entities were annotated with TIMEX2[5]. For example, the user question U_1 is annotated as follows:

- *<U1>Could you tell me the name of a <NE type = "Hotel.rating">3-star</NE> hotel in <NE type = "Destination.name">Beijing</NE>.</U1>*

5 `http://fofoca.mitre.org/`

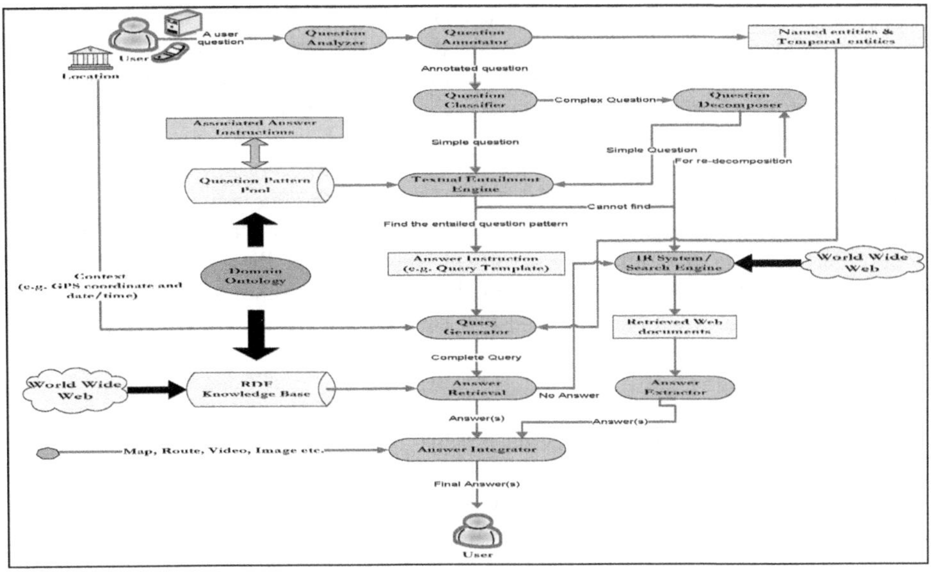

Fig. 2. Core Architecture of the QA Method

5.2 Question Classification and Decomposition

After question annotation, the topic and the type of each user question was identified to narrow the search scope of hypothesis questions to a specific topic and type. The topic of a user question refers to the main class that is involved in it, whereas the type refers to Q_{name}, $Q_{attribute}$ or Complex (i.e. complex questions).

We developed a question-topic classifier and a question-type classifier using machine learning techniques. By experimenting various classification algorithms and question features, we obtained the best question-topic classifier with an accuracy of 95% and the best question-type classifier with an accuracy of 85.5%.

For a basic user question, according to its identified topic and type, the search scope is restricted from the whole machine-generated set of hypothesis questions to the subset for the specific topic and type. For a complex user question, we need to decompose it into basic questions first and then identify the topic and type of these basic questions. For example, the question "*could you give me a contact number for a modern Italian restaurant in Beijing?*" can be decomposed into the combination of

- Q_{name}: *Could you give me the name of a modern Italian restaurant in Beijing?*
- $Q_{attribute}$: *Could you give me a contact number for <the restaurant name answered in Q_{name} >?*

In this paper, we focus on basic questions and leave complex ones for future work.

5.3 Textual Entailment Recognition

A textual entailment engine was used to recognize the entailment relation between a pair of a user question U and a hypothesis question H. In this paper, two entailment engines are presented: a syntactic engine using a bag-of-words method and a semantic engine using semantic information obtained from the domain ontology aligned with WordNet. For a pair of questions, a score ranging from 0 to 1 was used to indicate the confidence with which an engine estimates the entailment relation between them.

For the syntactic engine, the *entailment confidence score* was calculated only as the cosine similarity of the question pair. Besides term co-occurrence, the semantic engine also examined the semantic relations between the questions, including hypernyms/hyponyms defined in the ontology, synonyms in WordNet, and denominalization. Furthermore, the confidence score was weighted according to the number of the common named entities in the question pair so that the hypothesis questions that share more named entities with the user question can obtain a higher score.

5.4 Answer Retrieval

After the entailed hypothesis question is found, its associated query template is taken and used to create a complete SPARQL query by filling in the slots with the named entities and temporal entities extracted from the user question. Then the SPARQL query is used to retrieve the answers from the RDF data through Jena RDF API.

6 Evaluation

In knowledge-based QA, the correct extraction of the answers depends on the generation of the correct query. Thus, for our QA method, we measured to what extent the textual entailment engine was able to select the correct hypothesis question and as a result the correct query template. Furthermore, it was also assessed in this evaluation whether topic and type classification is helpful for textual entailment recognition.

The evaluation was based on 200 user questions. Each user question was marked manually by assigning to it the machine-generated hypothesis questions entailed by it. A user question can entail a hypothesis question containing all the same constraints or some of its constraints. The extent of the entailment was weighted as follows:

$$w = \textit{the number of the constraints of the hypothesis question/ the number of the constraints of the user question} \quad (1)$$

For example, the user question U_1 entails H_1, H_2 and H_3 with different weights:

- H_1: *What is the name of the hotel which has the [Hotel.rating] and is in the [Destination.name]?* (w = 1.0)
- H_2: *What is the name of the hotel which has a [Hotel.rating]?* (w = 0.5)
- H_3: *What is the name of the hotel which is in the [Destination.name]?* (w = 0.5)

While looking for the hypothesis question for each user question, the entailment engine calculated the confidence score with each of the hypothesis questions in the machine-generated set and took the one that obtained the highest score as the final output. The output of the entailment engines was compared against the human-generated gold standard, and the entailment accuracy was calculated as follows:

Entailment accuracy = ∑ *the weight of the entailed hypothesis question for each user question / the total number of the user questions* (2)

In this evaluation, the following three tests were carried out with the two entailment engines presented in the previous section:

- *Test 1:* for each user question, look for the hypothesis question entailed by it from the whole machine-generated set;
- *Test 2:* for each user question, identify its topic first and look for the hypothesis question from the subset for its topic;
- *Test 3:* for each user question, identify its type first and look for the hypothesis questions from the subset for its type;

The experiment results are reported in Table 2. According to the result of Test 1, the semantic engine is able to find the hypothesis questions from the total of 2703 machine-generated ones for 65.0% of the user questions, but the syntactic engine can only handle 42.5% of the questions. Comparing the results between Test 1 & 2 and between Test 1 & 3, the entailment accuracy has a substantial increase for both engines. This indicates that if the topic or type of a user question can be correctly identified, this can help the entailment engine to discover the correct hypothesis question for it. In Test 2, considering the accuracy (95%) of the applied question-topic classifier, the final QA accuracy still has a moderate increase for both engines. However, in Test 3, because of the lower accuracy (85.5%) of the applied question-type classifier, the final QA accuracy has a moderate increase for the syntactic engine but decreases a little bit for the semantic engine. This means that type classification is more helpful for the syntactic engine. It is because the semantic engine has used question type information more or less by detecting questions' expected answer types.

Table 2. Textual Entailment Accuracy and QA Accuracy in the Evaluation

Test	Topic/Type	Num of User Ques.	Num of Hypothesis Ques.	Entailment Accuracy		QA Accuracy	
				Syntactic Engine	Semantic Engine	Syntactic Engine	Semantic Engine
1	All	200	2703	42.5%	65.0%	42.5%	65.0%
2	Site	122	2339	46.6%	69.0%	44.3%	65.6%
	Event	30	223				
	Occurrence	48	141				
3	Q_{name}	97	2458	54.3%	73.2%	46.4%	62.6%
	$Q_{attribute}$	85	245				
	Complex	18	0				

7 Conclusions and Future Work

This paper has reported a novel knowledge-based QA method, which relies on the pre-generation of question templates on the basis of a domain ontology and the textual entailment recognition of discovering the question template entailed by a user question. This method is appropriate for ontology-based QA in a restricted domain and deems to be more efficient than those ontology-based QA systems which require deep questions analysis for producing ontology-compliant queries. The evaluation results revealed that the pre-generated question templates can be used to directly or indirectly answer almost all the user questions and the developed textual entailment engine was able to find the entailed question templates for most of the user questions (65.6%).

In future, we will focus on processing complex questions which involve question decomposition. Furthermore, we will also consider extending the QA method to linked data sets modelled by multiple ontologies.

Acknowledgements. This paper is supported by Humanities and Social Science Foundation of Ministry of Education of China (Project 10YJA870014) and National Social Science Foundation of China (Project 11BT0023).

References

1. Mollá, D., Vicedo, J.: Question answering in restricted domains: An overview. Computational Linguistics 33(1), 41–61 (2007)
2. Dagan, I., Glickman, O., Magnini, B.: The PASCAL Recognising Textual Entailment Challenge. In: Quiñonero-Candela, J., Dagan, I., Magnini, B., d'Alché-Buc, F. (eds.) MLCW 2005. LNCS (LNAI), vol. 3944, pp. 177–190. Springer, Heidelberg (2006)
3. Kouylekov, M., Negri, M., Magnini, B., Coppola, B.: Towards entailment-based question answering: ITC-irst at CLEF 2006. In: Peters, C., Clough, P., Gey, F.C., Karlgren, J., Magnini, B., Oard, D.W., de Rijke, M., Stempfhuber, M. (eds.) CLEF 2006. LNCS, vol. 4730, pp. 526–536. Springer, Heidelberg (2007)
4. Androutsopoulos, I., Ritchie, G.D., Thanisch, P.: Natural language interfaces to databases—an introduction. Natural Language Engineering 1(1), 29–81 (1995)
5. Lopez, V., Uren, V., Sabou, M., Motta, E.: Is Question Answering fit for the Semantic Web? a Survey. Semantic Web–Interoperability, Usability, Applicability (in press, 2011)
6. Atzeni, P., Basili, R., Hansen, D.H., Missier, P., Paggio, P., Pazienza, M.T., Zanzotto, F.M.: Ontology-based Question Answering in a Federation of University Sites: the MOSES Case Study. In: Meziane, F., Métais, E. (eds.) NLDB 2004. LNCS, vol. 3136, pp. 413–420. Springer, Heidelberg (2004)
7. Lopez, V., et al.: AquaLog: An ontology-driven Question Answering System for organizational Semantic intranets. Journal of Web Semantics 5(2), 72–105 (2007)
8. Lopez, V., Nikolov, A., Sabou, M., Uren, V., Motta, E., d'Aquin, M.: Scaling up Question-Answering to Linked Data. In: Cimiano, P., Pinto, H.S. (eds.) EKAW 2010. LNCS, vol. 6317, pp. 193–210. Springer, Heidelberg (2010)
9. Tartir, S., Arpinar, I.B.: Question Answering in Linked Data for Scientific Exploration. In: 2nd Annual Web Science Conference (2010)
10. Ou, S., et al.: Development & Alignment of a Domain-Specific Ontology for Question Answering. In: 6th International Conference on Language Recourses and Evaluation, ELRA (2008)

An Integrated Interactive and Persistent Map-Based Digital Library Interface

Samuel J. McIntosh and David Bainbridge

Department of Computer Science,
University of Waikato,
Hamilton, New Zealand
{sjm84,davidb}@cs.waikato.ac.nz

Abstract. In this paper we explore the enrichment of the interface to a "standard" text-based digital library with interactive digital maps to enhance its capabilities. Central to the design is to maintain two interlocked views, displayed side-by-side: one the traditional text-based digital library view, the other an interactive map view. At any stage of the user's interaction the two views represent the same information, only displayed in the form most fitting to that view. The views are interlocked, in that the user can interact with either view, and as a result both views are simultaneously updated. To achieve this, the underlying digital library infrastructure was extended to include automatic place name detection and disambiguation capabilities, which are used to augment the text view. To assist user navigation within this view, we explore a fisheye and multicolumn representation of the text documents. The work is evaluated through a user study, focusing on the new and extended services in the digital library.

1 Introduction

Digital libraries and digital maps are two fast-growing technologies in the world of computing. In this paper we describe the work undertaken integrating geographic capabilities into an off-the-shelf text-based digital library system. Central to the design approach is to maintain two interlocked views, displayed side-by-side: one the traditional text-based digital library view, the other an interactive map view. At any stage of the user's interaction the two views represent the same information, only displayed in the form most fitting that view. More specifically, the Greenstone 3 digital library system was augmented through the use of the digital mapping system Google Maps to provide the interface described and evaluated here.

The structure of this paper is as follows. First we review prior work, then we give a walkthrough of the developed environment, emphasizing the areas where the existing interface has been enhanced, or is wholly new. This helps provide context to the evaluation of the interface (Section 4) before concluding with a summary of our findings.

C. Xing, F. Crestani, and A. Rauber (Eds.): ICADL 2011, LNCS 7008, pp. 321–330, 2011.

2 Related Work

Interest in the graphical visualization of geographical information within a digital library can be traced back to the very start of the field. Pertinent to the work reported here are the projects GIPSY [11], NewsExplorer [9,8], Perseus [10] and Informedia [2]. The visualization capabilities of these systems varies from system to system based on factors such as resource constraints and the technology available at the time. In terms of visualizing spatial information derived from text documents (the focus of the work reported here), GIPSY is not only the earliest of these systems, it also embodies the most abstract form of visualization. As the Nevada example in [11] shows, the fact that the state is geographically discernible as an elevated area is implicit in the technique used, rather than the result of it being explicitly marked out, as with all the other systems reviewed.

In comparison, the 2004 version of the NewsExplorer system was capable of using three different methods to display maps [9]. The first method used "shape files" which could be used to create GIF, PNG or JPG images. The researchers reported that this method was computationally heavy and slow but had the advantage that it was independent of proprietary software. The second display method produced maps using the DMA (Digital Map Archive) tool from the JRC's ISFEREA project. This program is a commercial tool capable of rendering maps quickly and is also capable of overlaying different layers of information over the map such as roads and population density. The final method used for displaying maps was with SVG. This method provided a relatively simple looking map but had the advantage that it rendered quickly. In the 2006 version of NewsExplorer only the SVG method was retained, with the project moving to Google Earth and a Flash-based system (WorldKit) to deliver their map visualization needs.

The Informedia digital library system used the commercial ESRI MapObjects library to visualize its geographical content. As MapObjects was a library, it meant that they were able to program their own interfaces and functionality to fully utilize its capabilities. In the Perseus project the Bolles collection map browser was implemented as a Java applet, leveraging off vector data from Bartholomew Mapping of contemporary London in addition to the historic rasterized maps. The most recent incarnation of the Atlas feature (plot sites) uses Google Maps.

3 Walkthrough

We now demonstrate the geographically enhanced capabilities to Greenstone 3 through a walkthrough of the interface. We named the resulting system, ATLAS.

Enhanced Text Search. When the user submits a text query, in addition to seeing the standard text-based results set page, the interface brings up a map-based view (Figure 1). Dynamically, ATLAS begins loading each document in the background and scans their content, looking for places that are mentioned within them. Each document in the results set is given a color, with places located

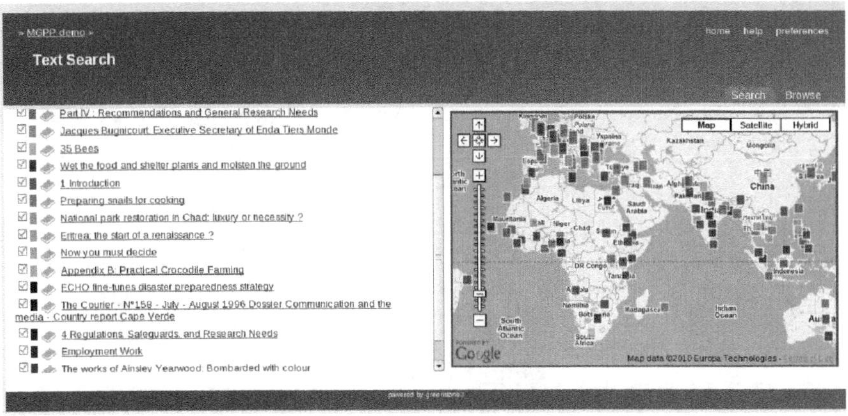

Fig. 1. The text query results page demonstrating the additional ATLAS functionality

in that document marked on the map in the same color. The user may select a document to read from the map view, just as they can from the text-based list of results.

Enhanced document view. ATLAS enhances the standard document view of Greenstone by adding a map to the right-hand side of the screen like on the text results page. This map contains markers showing the locations of the place names that have been found in the document. These places are also highlighted in the text to allow the user to see where they occur in the document. Some of these details can be seen in Figure 2, which also illustrates context menus, discussed next.

Context menus. The user is also able to interact with ATLAS to find out more information. For example, each highlighted place name can be moused over to produce a menu that gives several options specific to that place name. This allows the user to perform tasks such as: centering the map on this specific place, highlighting the place on the map, highlighting the place in the text or correcting the disambiguation done by the system. An example of this menu functionality is shown in Figure 2 where the user has moused over "London" and have chosen to change which London the place name refers to. See [5] for details about the automated name disambiguation technique devised for ATLAS.

Because of the decreased space now available for the document text—due to the map filling half of the screen—two alternative text views were introduced, aimed at making better use of the limited available space: fisheye and multi-column. Their usefulness was evaluated in the user study summarized in Section 4.

Fisheye view. In a fisheye visualization, only the area of the document near to the mouse cursor is made large—with the rest of the document rendered small—and the view dynamically updated as the cursor is moved [3]. The technique is useful in allowing the user to get an overview of the document whilst still being

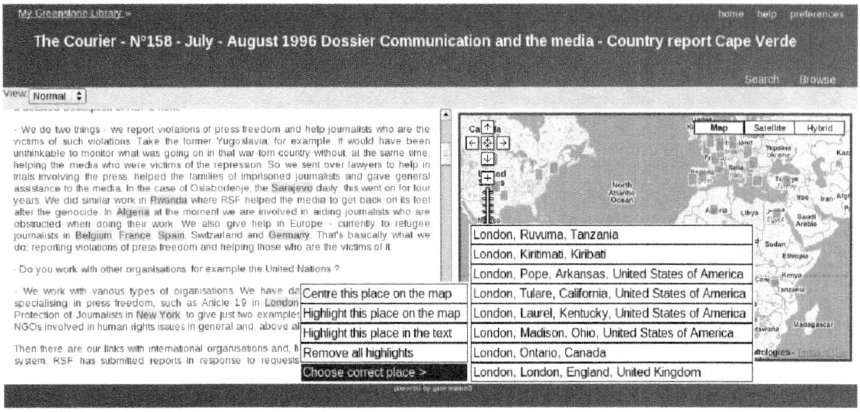

Fig. 2. A menu displaying interaction options for the currently selected place name

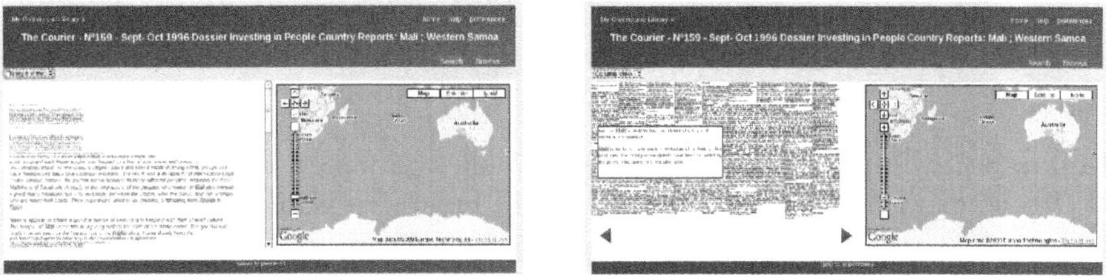

Fig. 3. The map-enhanced document text with companion (a) fisheye, and (b) multi-column view

able to read and navigate it. This is especially useful in the case of ATLAS where places in the text have been highlighted, meaning it is easy to see where they are, even when the text was out of the focus of the fisheye. Figure 3a shows a screenshot of the fisheye view in action.

Several novel techniques were used in the implementation of the fisheye view of HTML web content within a browser. To achieve such a visualization each line in the unmodified text needed to be located, as text resizing was done on a line by line basis. To do this the program initially places HTML *span* tags around each word in the text. Care is taken to exclude any other tags that may be already be present in the text. Placing *spans* around each word can be used to locate the vertical height of each word in the text (through the use of the *getOffsetTop()* DOM method provided by the *span* tag) and in turn can be used to locate the lines in the text. Each line is then wrapped in a *div* tag and stored in an array, in order of their vertical location on the page. All the *span* tags used for line finding are then removed to increase performance as they are no longer needed. This process is demonstrated in Figure 4.

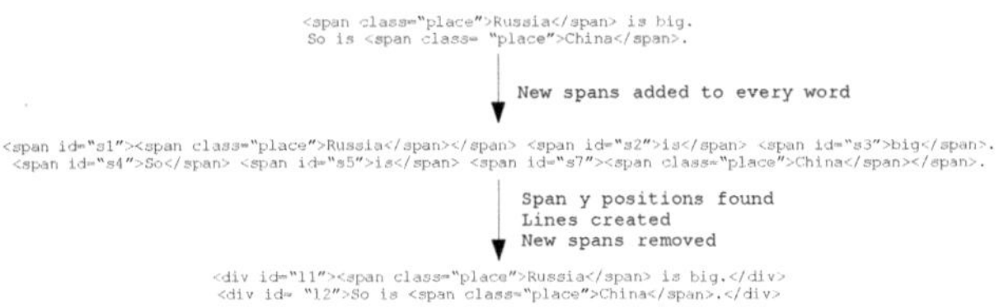

Fig. 4. The line finding algorithm used to set up the fisheye view

A downside to the fisheye view however is that there is much wasted whitespace in the context area (i.e., above and below the focused area) [4]. After experimenting (unsuccessfully) with several parameters used in the fisheye view to see if we could improve the situation, we decided to test out an entirely different type of view based on multiple columns.

Multi-column view. This view was designed to minimize the loss of content to whitespace. It does this by decreasing the font size of the text and arranging it into several columns so that the whole of the document is displayed at once without the need for scroll bars. As the user moves their cursor over the text in the columns, a larger version of the text is displayed in a box centered above the position of the mouse cursor. The difference in use of whitespace, compared with fisheye, can be seen in Figure 3b.

Spatial search. Finally, in addition to augmenting the text searching functionality of Greenstone, ATLAS provides searching functionality of its own in the form of spatial searching. The user selects an area that they wish to locate documents related to and clicks the "Search Area" button to perform the search. The documents that are found are then displayed in order of relevance on the opposite side of the web page.

Selection is performed by clicking on the map, which creates a point at that location. Successive points are joined together to create a polygon which is then used in the spatial query. If the user makes a mistake they can either drag the points to a new location or clear all of the points to start again. Figure 5 shows the spatial search interface with an area around Hamilton, New Zealand selected.

4 Evaluation

To evaluate the usability of the ATLAS system a user study was designed. The focus of the study was to gather information about the experimental aspects of the system, such that we could make informed decisions about their individual and collective effectiveness.

A pilot study was run to identify any teething issues with the interface or evaluation procedure. After some minor adjustments, the full user test was instigated. In line with user evaluation guidelines for testing a prototypes such as the one described here—in the order of 5 to 10 subjects [6]—we formed a relatively small set of participants. For our evaluation we recruited 10 participants

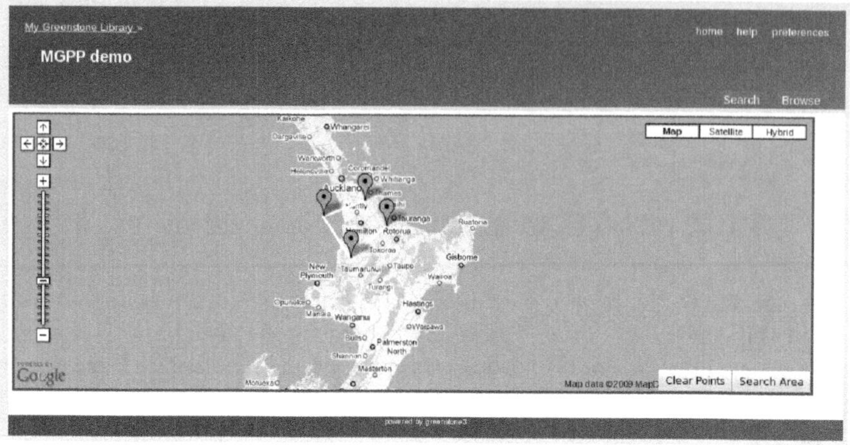

Fig. 5. The ATLAS spatial searching interface

(4 females) with their ages ranging from 20 to 52. Levels of computing experience also varied from frequent computer users to infrequent users with limited computing experience. This diversity in participants is commensurate with user needs of a general purpose digital library, where there is the expectation that the information contained within it should be easily accessible to all who wish to use it.

The Greenstone collection used for the user study was the Humanity Development Library. This collection contains useful information for developing countries, such as how to farm butterflies or snails. This collection was adequate for the user study as most documents contained several place names. Participants were first given a brief introduction to the system, explaining about digital libraries and what ATLAS does to enhance them. They were also given an example of how to browse for a document, how to perform a text search and how to perform a spatial search. After this introduction, participants were asked to complete a series of tasks, each with one or two questions to answer. Tasks were deliberately designed to encourage users to use each of the features that were provided by ATLAS. Once the tasks were completed the participants were then asked to complete a brief questionnaire about their experiences designed to elicit answers to the following questions.

Did participants find ATLAS useful? Almost all participants stated that they believed having a map that shows the places mentioned in the documents they read was something that they would find helpful. Sample responses included: "Yes, because I like to know where in the world places are in what I am reading. It gives more background;" "Yes, often you've heard of the place but don't know where it is;" and "Yes, many people, including myself, don't have a good idea of world geography and therefore the context of information."

Clearly the main reason that participants approve of the idea of marking places in documents on a map is that it gives more background context by allowing them to actually visualize the locations that are mentioned. This in turn gives the user further information such as the distance between places and how they are arranged.

Did participants find the system intuitive and easy to use? There were three main aspects of the software that need to be considered to answer this question as, from the user's point of view, the system is made up of three components: ATLAS, Greenstone and Google Maps. The way these components are combined means that having a weakness in one of these affects the usability of the system as a whole. For example, two users had not used Google Maps before and therefore struggled to complete several of the tasks despite being able to adequately use the rest of the system. Overall however, from both the comments of the participants and the fact that most participants completed the tasks without difficulty, the usability of the system was shown to be acceptable to users with a range of different IT experience.

To test the standard document view, participants were given several tasks that used each of the features available in the place-specific menu. Most users had no trouble using these features and were able to complete these tasks quickly. These participants also made positive comments about the system, for instance: "Very easy - great!" and "Easy, exactly as I'd expect."

Although users found the features ATLAS provided in the standard document view easy enough to use, there were several aspects of its functionality that caused minor problems. Several users expected to be able to use the right mouse button to bring up the menu as is usually possible in other programs. Several other users expected to be able to left click on the place name to be able to bring up the menu. The implementation tested actually required a user to hover the cursor over the place name for small amount of time for this menu to be shown; subsequent to this testing, right-clicking to bring up the menu has been added to the implementation. For the left click functionality two participants suggested that left clicking on a place name should center the map on the corresponding place rather than needing to choose "Center this place on the map" from the menu. As this was the most common menu item selected in the tests it makes sense for it to have a shortcut and using left clicks provides an efficient way of doing this.

The ATLAS-enhanced text search view was evaluated by having the participants search for the term "water" and then looking for places on the displayed map that came from the document *35 Bees*. The goal was that users would look at the color that was given to *35 Bees* and then look for markers of that color on the map. Of all of the features provided by ATLAS, the text search view takes the longest to load. Before this had completed, several users proceeded to click on the link to *35 Bees* document rather than wait for this map to load. The efficiency of this ability in ATLAS could be increased significantly by moving to spatial indexes that are computed when the collection is built. That way, the system will know in advance the places that are present in each document, and therefore will not need to scan each document that is retrieved.

The spatial document search view was tested by having each participant create a query for the island of Tasmania, Australia by drawing a polygon around it. They then clicked the “Search Area” button and were taken to a list of result documents of which they were asked to write down the first 3. All participants completed this task successfully, although those that had not used Google Maps before required a small amount of assistance.

Our user evaluation also highlighted a usability issue in the base Greenstone 3 system, that confused several participants. After a user selects *Browse* on a collection's “about” page, they are taken to a page that gives the user options as to how they wish to browse the collection. Several users were confused at the sight of this page, and most of these confused users did not realize the page was loaded. Indeed, one user thought a table was being loaded and so waited.

There were two major factors contributing to this problem. The first problem was that the page was mostly whitespace with only a small strip of green along the top of the content area, meaning that several users did not notice the options at the top of the page. The other contributing factor was that the browsing options did not appear like clickable links, due to the fact the links were not underlined like regular links. These links should be changed to be consistent with all the other ones in the digital library [7]. Adjustments to address these issues have since been made to the base system.

How helpful did participants find the fisheye and column views? The fisheye view was tested by having the participants navigate to a long document, then activate the fisheye view. Users were then asked to give a rough estimate of how many highlighted place names there were in the document. Although the system had found 49 place names in the text, users' answers varied significantly. The lowest estimate was 12—with the next lowest estimate being 20—and the highest estimate was 60. This large variation is a clear indication that there were some usability problems with the fisheye view, as implemented.

The most obvious reason that there was such a varied array of estimates was that several participants did not realize that there was more to the document than what was initially shown when the fisheye view was presented to them. They counted the places they could currently see and assumed that it was the whole document. This is a fair assumption for these participants to make as it is common for user interfaces that make use of the fisheye concept to attempt to display all of the content at once (such as FishNet [1]). Whether or not to attempt to fit all of the text into the one area was something that was considered when the fisheye view was first developed. It was decided against due to the size of the text needing to be so small for long documents that it was almost invisible due to the smaller font size causing smaller line lengths as the number of characters per line remained the same.

Another reason contributing to the poor usability of the fisheye was that users easily got lost when scrolling the document as the document and the fisheye focus moved at the same time, making it difficult for them to remember where they were up to so that they could continue counting place names. One of the questions asked in the survey was “How easy did you find the fisheye view to use compared to viewing the text normally? What did you like/dislike about it?" The responses to this question included remarks such as: "Made scrolling confusing and ineffective ...;" "I liked it because it was different but disliked it because it confused me;" and "Had trouble with the Fisheye, I couldn't get it to do what I wanted it to do."

Participants getting lost in the text when scrolling the fisheye view was also reported by both Jakobsen and Hornbæk in their evaluation of fisheye for source code [4] and Baudisch et al. with their evaluation of their FishNet system which is a fisheye Internet browser [1]. This finding is particularly problematic for the ATLAS system as, because documents are designed for reading sequentially, having users able to get easily lost in a viewing method that was designed to increase document readability is counter-productive.

The column view was tested by having the participants choose column view in same long document that the users had used the fisheye view. Participants were then asked to locate three different place names in the text and write them down. All participants completed this task easily after a brief explanation of how it worked. This view also received more positive remarks such as: "Easy, its logical for me to use," "... I like to be able to see all the text at the same size so I can scan over it easier," and "Column view was good. You could see how much there was to read, and I found it easy to scroll through ..." The only negative comments were that it was "a bit daunting" and "difficult to skim read."

5 Conclusion

In summary, in this paper we have described how we took the baseline configuration to Greenstone 3, and developed it to support geographical capabilities both in terms of its services and user interface. The work is available, open-source, through Greenstone's extension framework.[1] New digital library services were introduced for supporting a gazetteer and disambiguating place names. Central to the interface design was the idea to display two inter-locked views of the digital library side-by-side: on the left the traditional text-based digital library view; and on the right a map-based view (implemented using the Google Maps API) that provided an alternative view to the same information. In the interface the result of text-based searching and viewing documents were enriched with geographical capabilities. In the case of viewing the document, two document overview techniques were experimented with—fisheye and multicolumn—to see how well they assist users accessing the marked up place names in the document.

A user evaluation of the completed system provided a range of insights. For example, comparing the multicolumn view with the fisheye view, the multicolumn view performed more satisfactorily—in fact issues that have been reported elsewhere for the fisheye view were evident in our implementation also. Within the baseline digital library, we also discovered a common source of user confusion with the browsing page that only a modest change has eliminated. Finally,

[1] http://trac.greenstone.org/browser/gs3-extensions

our evaluation showed that users were able to readily comprehend the side-by-side interlocked viewer interface, and moreover rated highly the enrichment of the digital library with geographical facilities; however, contrary to our expectations, many users indicated they would like to see the map region smaller than the text view—even have the ability to hide one or other of the views. Such adjustments to the interface are straightforward to make.

References

1. Baudisch, P., Lee, B., Hanna, L.: Fishnet, a fisheye web browser with search term popouts: a comparative evaluation with overview and linear view. In: AVI 2004: Proceedings of the Working Conference on Advanced Visual Interfaces, pp. 133–140. ACM, New York (2004)
2. Christel, M.G., Olligschlaeger, A.M., Huang, C.: Interactive maps for a digital video library. In: International Conference on Multimedia Computing and Systems, vol. 1, p. 9381 (1999)
3. Furnas, G.W.: Generalized fisheye views. SIGCHI Bull. 17(4), 16–23 (1986)
4. Jakobsen, M.R., Hornbæk, K.: Evaluating a fisheye view of source code. In: CHI 2006: Proceedings of the SIGCHI Conference on Human Factors in Computing Systems, pp. 377–386. ACM, New York (2006)
5. McIntosh, S.J., Bainbridge, D.: Gazetteer and place name representation and efficiency in a digital library. In: Under review for the International Conference on Asian Digital Libraries (2011)
6. Nielsen, J.: Why you only need to test with 5 users. Alertbox (2000), `http://www.useit.com/alertbox/20000319.html`
7. Nielsen, J., Loranger, H.: Prioritizing Web Usability. New Riders, Berkeley (2006)
8. Pouliquen, B., Kimler, M., Steinberger, R., Ignat, C., Oellinger, T., Blackler, K., Fuart, F., Zaghouani, W., Widiger, A., Forslund, A.-C., Best, C.: Geocoding multilingual texts: Recognition, disambiguation and visualisation. CoRR, abs/cs/0609065 (2006)
9. Pouliquen, B., Steinberger, R., Ignat, C., De Groeve, T.: Geographical information recognition and visualization in texts written in various languages. In: SAC 2004: Proceedings of the 2004 ACM Symposium on Applied Computing, pp. 1051–1058. ACM, New York (2004)
10. Smith, D.A., Crane, G.: Disambiguating geographic names in a historical digital library. In: Constantopoulos, P., Sølvberg, I.T. (eds.) ECDL 2001. LNCS, vol. 2163, pp. 127–136. Springer, Heidelberg (2001)
11. Woodruff, A.G., Plaunt, C.: Gipsy: automated geographic indexing of text documents. J. Am. Soc. Inf. Sci. 45(9), 645–655 (1994)

Towards Very Large Scale Digital Library Building in Greenstone Using Parallel Processing

John Thompson[1], David Bainbridge[1], and Hussein Suleman[2]

[1] Department of Computer Science, University of Waikato, Hamilton, New Zealand
jmt12@students.waikato.ac.nz, davidb@cs.waikato.ac.nz
http://cs.waikato.ac.nz/

[2] Department of Computer Science, University of Cape Town,
Cape Town, South Africa
hussein@cs.uct.ac.za
http://www.cs.uct.ac.za/

Abstract. As very large digital library collections become more commonplace, software tools must adapt appropriately. This paper reports on an evolution of the Greenstone Digital Library software to support parallel processing during the collection building phase. A series of experiments were conducted to first establish a basic speed-up factor, and then deconstruct the parallelisation process to understand the execution profile of the application. Several bottlenecks were identified and resolved to further improve the performance. The adaptation of Greenstone confirms that the build phase is indeed a suitable candidate for parallelisation; and suggests that parallelisation of processing is a new avenue for exploration in emerging digital library architectures.

Keywords: Greenstone, VLDL, Parallel Processing, Open MPI.

1 Introduction

The last few decades have seen an explosion in the amount of digital information being collected, and the issue at hand is matching this with the development of systems for presenting and making this information accessible. Key amongst these systems is digital library software, which provides a structured, browseable, and searchable gateway to a moderated and metadata annotated collection of electronic documents. Digital libraries are now being created that advance into the millions if not tens of millions of documents—for example, the HathiTrust Digital Library (www.hathitrust.org) has, at the time of writing, full text indices for approximately eight million documents.

The creation and practical use of VLDLs has also raised many new issues, such as those discussed during the international VLDL workshop series (e.g., Mangi et al. [5]), particularly related to size (in aspects such as disk space, number of documents, and number of unique terms or words), sustainability, and interoperability. But technology has also advanced and, with multi-processor/hyper-threading

C. Xing, F. Crestani, and A. Rauber (Eds.): ICADL 2011, LNCS 7008, pp. 331–340, 2011.

now the norm in consumer computers, explicitly developing techniques for digital library software that exploits parallel processing capabilities looks promising as a solution to some of these scale concerns. For example, Stanfill [7] proposed a number of ways to leverage the power of parallel processing applicable to the information retrieval component of a digital library system, and acting on indices in the order of terabytes.

Meanwhile, Greenstone [8] is a well established, open-source, digital library tool, but has traditionally had a reputation of only being useful for small, lightweight, or demonstration collections. However recent scalability testing results [2] and large scale digital collections such as the National Library of New Zealand's PapersPast digital library [1] show that Greenstone is capable of serving collections numbering well into the millions of documents.

It is perhaps no surprise that the initial ingest of such large scale collections (using the software without modification) is time-consuming. Real-world VLDLs may take in excess of several days of processing to create. In our scalability testing results we found that Greenstone's importing time is roughly linear with the number of documents. This paper details research in an attempt to improve this performance by applying parallel processing.

The structure of the paper is as follows. We start by outlining the scope of the work reported here, which concentrates on the ingest phase of collection building, and then briefly mention related digital library projects. We then present details of the software changes we made to enable Greenstone to automatically process documents in parallel. Following this, we present and discuss the results of a series of tests that helps to establish where time-consuming steps occur within this process (leading to further adjustments in the software implementation), before concluding with a discussion of the implications of these findings on building large scale collections and future work.

2 Scope

In the work reported here, we focus on the collection indexing phase (*build-time*) of Greenstone—where documents are transformed into a homogeneous format during an import phase and then indexed during a build phase. We leave for future research the opportunity to apply parallel processing to the *run-time* of Greenstone—where the user interacts with the digital library, via a Web browser or other presentation tool, in potentially processor-heavy fashions (for instance when interacting with video or other multimedia).

Of the two build-time tasks mentioned, this paper will specifically explore our work on the process of importing disparate documents into a standardized XML format. We started with importing because:

- it offers clear opportunities for parallelism (given that by-and-large the transform of each document is atomic);
- there is a greater possibility of significant savings in processing time as importing is (anecdotally) slower than indexing (especially for processor intensive documents like media and PDFs);

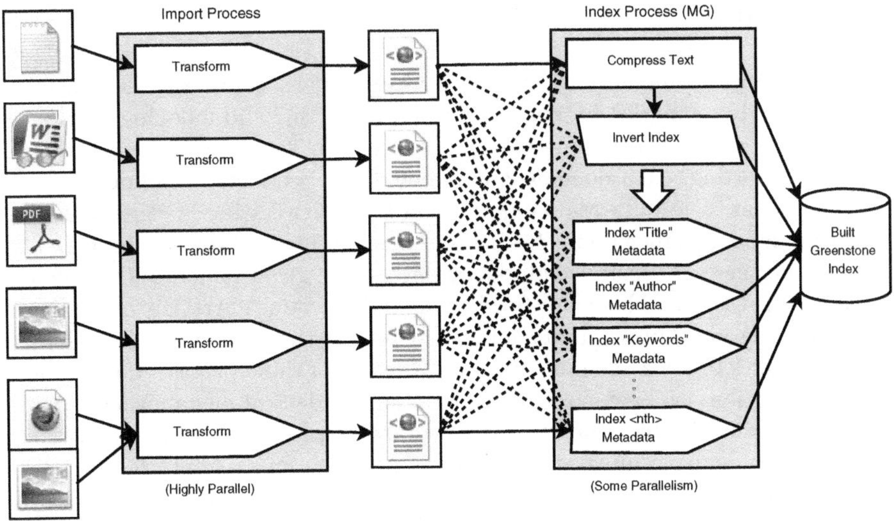

Fig. 1. Potential for parallelization in Greenstone's import and build processes

- it offers a wide window for experimentation (such as altering the number and processing complexity in each group of documents being ingested by each process).

We also believe that importing offers the more novel problem. The accompanying issue of large scale indexing is already being explored (for instance [3]). Furthermore we will need to ascertain whether Greenstone's indexing algorithms and architecture will actually benefit from using multiple processors as suggested by existing literature [6].

3 Related Work

As an example of another large scale digital library project, the HathiTrust Digital Library estimated that an unoptimized re-import of their seven million plus volumes might take ten days.[1] However, by utilizing parallel architecture and optimizing the underlying indexing system's configuration (Solr[2]) in terms of IO interaction they were able to reduce this to around ten days.

It is worth mentioning that, unlike many of its contemporary digital respository projects (such as DSpace and Islandora) that use a relational database for their datastore (and hence only have full-text indexing if that database supports it), Greenstone makes use of a flat database for storing metadata with a separate system for indexing (typically MG, MGPP or Lucene). While this structure proved an early advantage—before relational databases began supporting full text indexes—it now presents unique challenges to large scale collection building.

[1] http://www.hathitrust.org/blogs/large-scale-search/forty-days-and-forty-nights-re-indexing-7-million-books-part-1

[2] http://lucene.apache.org/solr/

4 Method

Figure 1 gives a schematic overview of Greenstone's ingest process—which is divided into two completely separate phases, importing and indexing—and helps to illustrate where the potential lies in the system for supporting parallelism. During importing, each document is handled independently, lending itself well to parallelism. The only real complication to deal with here is metadata that has been defined higher up in the file-system structure and set to inherit to child folders and files—for instance all documents in the given folder, and sub-folders gets assigned the Dublin Core Subject metadata value "Art History." During indexing, the main opportunity for parallelism is the multiple passes that the indexer makes to form the indexing files, as some of these passes—depending on how a collection has been configures—are independent of one another.

More specifically, in order to explore parallel processing opportunities, the decision was made to integrate an existing Message Passing Interface (MPI) library called Open MPI [5] into Greenstone. This mature open source project has been built with high performance in mind and supports a plethora of operating systems, as well as cluster and multi-core configurations. While MPI is arguably a low-level interface, it was adequate for this project and the MPI primitives can easily be translated to other libraries. The Greenstone import process was extended to use Open MPI's architecture to allow for a single controller import process to manage a number of worker import processes, each of which would ingest one batch of files using the serial Greenstone import process. The parallel import process allowed configuration in terms of number of worker threads (parallel jobs) and batch size for files to be processed. Greenstone already had mechanisms for importing the collection's documents in batches (controlled by XML manifest files). It should be noted that a typical, non-manifest, Greenstone import actually has to do more processing; for example, it performs a non-trivial pre-scan of all documents to be imported to locate any applicable metadata files. To avoid these algormithm differences we passed manifest files both to the parallel import processes and the 'control tests' of non-parallel imports processes.

While the majority of the Greenstone importing process (such as the "plugins" used to read the disparate document formats and the "plugouts" used to write the standardized XML files) was found to work fine when applied to multiple processes, there were other aspects that needed customization to support parallel processing. Key amongst these was the interaction between Greenstone and the GDBM database[3] it used to store metadata and processing information. GDBM only allows a single process to write to it. This had not previously been an issue with normal serial processing, but now required the addition of locking/synchronization calls around Greenstone's critical sections that made calls to GDBM in order to prevent errors in the database and abnormal program terminations. Once this work was completed, Greenstone's parallel importing was brought to a point where we could begin performance tests.

[3] `http://www.gnu.org/software/gdbm/`

The initial results from early testing were somewhat disappointing (as detailed in Results) with processing time only being halved despite utilizing multiple cores. At this point we used profiling tools[4] to analyze the import process to determine what factors were limiting the performance of multiple processors. This quickly revealed a number of bottlenecks, many of which were subsequently mitigated by careful selection of built-in Greenstone configuration options and the tests rerun. But even with these fixes in place, performance was still not as good as expected. Real world data suggests that the optimal number of parallel threads for a compute-bound load, in order to maximize processor utilization, occurs at P or $P+1$ (where P is number of computer processors) [4]. Under the assumption that importing might be exhibiting IO bounds rather than processing ones, further tests were designed to determine if IO was the remaining limiting factor in parallel importing.

To test each of the configurations listed below we ran a series of import processes over a collection of 10,000 documents. Each import in the series varied the numbers of parallel worker threads and batch sizes in order to measure the influence of those factors. The documents—plain TXT files—contained carefully generated lorem ipsum[5] text designed to approximate real-world content. The testing was carried out on a modern PC (Intel Core i7) with eight cores (hyper-threaded from four physical cores), 4GB RAM and running CentOS 5.5. The result for each configuration, and for each combination of number of threads and batch size therein, captured the real, user and system time of importing averaged over multiple executions.

The output materials from applicable configurations (explained in the next section) were compared in order to ensure that neither order of processing nor synchronous processing altered or damaged the output (in respect to a standard Greenstone import process). Aside from acceptable differences in timestamps the outputs were the same. The outputs (if any) for the configurations that altered IO were not compared as we already knew their content was compromised by the test.

4.1 Configurations

The following is a list of each of the limiting factors located by profiling and the modifications made to mitigate or remove them:

Standard. The first series of tests were run on an unaltered Greenstone configuration and made use of standard input plugins and plugouts.

Set ID. During import Greenstone uses a hashing process to generate unique identifiers for documents. This is not just processor intensive, but also IO intensive as it would re-read the source file contents (even for large files like videos) to hash upon. To avoid this cost the configuration was changed to assign the filename as the unique identifier instead. This setup was used for the second series of tests.

---

[4] `http://code.google.com/p/perl-devel-nytprof/`

[5] `http://www.lipsum.com/`

Set encoding. In order to convert all documents to a standard encoding (UTF8), Greenstone must first determine the input encoding of each document. Again this was found to be IO dependent as the encoding analysis involves reading in a block of characters of the file and determining the codepage of each one. While not always possible in real world situations, in this experiment we were able to manually assign the encoding of the input documents as we had generated them as ASCII. A configuration with both assigned identifiers and assigned encodings was used for the third series of tests.

No input IO. With the configuration optimized with the above points, we started to investigate whether the act of reading and writing the documents was the limiting factor. We started by replacing the standard input plugin for TXT files with one that returned a hardcoded lorem ipsum document. This document was representative of an average document in the collection and the plugin allowed us to decrease input IO activity during the fourth series of tests. It should be noted that while acceptable for an experiment on importing, replicating the same content multiple times would have had serious consequences if indexing were to be applied.

No output IO. For the fifth series of tests we decided to minimize both input and output IO, and so the output 'plugouts' were replaced with versions that did no file writing (neither the homogeneous XML files nor the GDBM database used for processing information) at all. This decreases the IO activity to nearly nothing, but obviously does not reflect real-world or any practical usage and is instead meant to provide a baseline indicating the fastest possible processing times.

TDB. Finally, during the "no IO" test it was noticed that the interaction with the GDBM database was expensive compared to the output of the XML files. We include below one more test result that shows the 'bleeding-edge' results of an attempt to replace the GDBM database with TDB[6] which supports parallel readers and writers. This sixth series of tests was carried out with both plugins and plugouts restored to default but retained the assigned identifier and encoding configuration.

5 Results

Figure 2 shows a comparison of the average import process times (importing 10,000 documents) of the first five series of tests as the number of worker threads increase. To recap, the series "standard" represents a default Greenstone import configuration, "set ID" and "set encoding" series represent configurations with those metadata values specified (note that the latter has both id and encoding assigned), and "no input IO" and "no IO" represent configurations where IO have been reduced. When the number of worker threads is zero, a serial Greenstone import is used. When processing in parallel a fixed batch size of 100 was used.

[6] Short for Trivial Database, developed by the Samba project. See http://tdb.samba.org for more details.

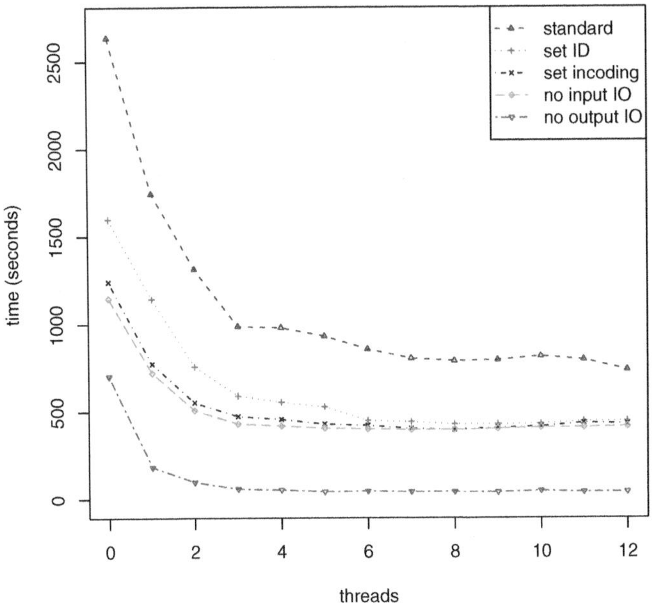

Fig. 2. Comparison of the import process time of five configurations (excluding TDB)

These results show that, in general, increasing the number of threads running in parallel continues to decrease total processing time. Two results that surprised us were that the performance when number of threads equals one actually improved (we predicted performance would drop as a single thread should have all the overhead of parallel processing management with none of the benefit) and that the benefit of adding more threads quickly dropped after threads equal to three (we'd expected performance to keep improving - this is the result we earlier described as disappointing, and hinted that something other than processing capability was hindering progress). While setting simple configuration changes to avoid optional processing and IO causes a slight improvement in performance, the last series clearly highlights that some part of the output IO is dramatically limiting the effectiveness of parallel processing in Greenstone.

Figure 3 illustrates the effects of variation in batch sizes and number of threads on the performance of the import process. The test results shown here made use of the configuration where both unique identifier and encoding are assigned, but are representative, in terms of general shape, for all of the configurations (excluding the one with TDB support). Notice that when the number of worker threads is equal to zero, a serial Greenstone import is used. While a manifest is still provided to this process (as explained above) it always contains the full ten thousand documents and thus the batch size is effectively ignored and the processing time remains essentially constant.

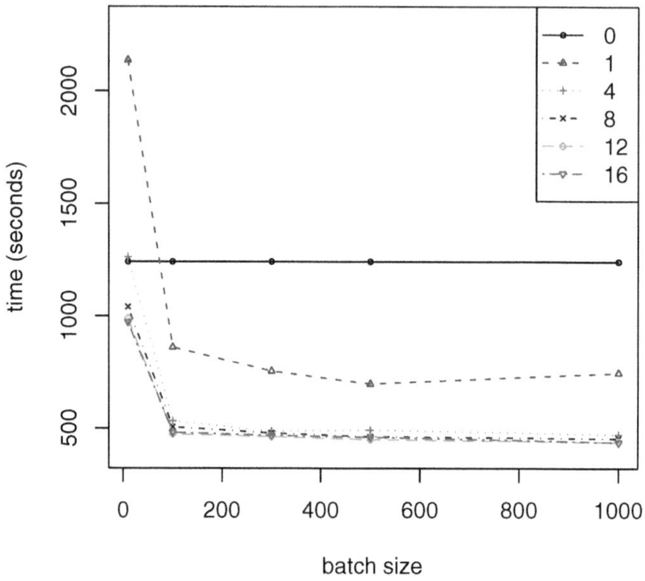

Fig. 3. Influence of number of threads and batch size on import time

The results show that increasing batch size decreases average time up to around a batch size of n/w (where n in the total number of documents and w the number of worker threads) but after that point it causes processing time to (very slightly) increase. With smaller size batches, there is more overhead of worker threads doing IO reading in the XML manifest files, while at larger batch sizes some worker threads are being starved of documents and sitting idle while other threads finish processing.

However it is interesting to notice that in both Figure 2 and Figure 3 the import process achieves a minimum processing time of around six minutes, after which processing time actually increases marginally as more threads are added. While we initially suspected that this was just the overhead of managing more threads, it was these results that led us to take a closer look at what was actually using the time. By reviewing profiling tool output and measuring the system time used (as compared to real time), it was discovered that a significant amount of time was spent waiting for opening and closing connections to the GDBM database.

Figure 4 shows the result of replacing the GDBM database used by default by Greenstone with TDB, which supports parallel readers and writers and thus is better suited for parallel processing. Once again the test used the configuration where both unique identifier and encoding are assigned. From these results it is clear that by using a database that supports parallel access we can avoid the point of diminishing returns created by the basic file locking we implemented around GDBM. The speed-up in changing to this database system is striking—processing times a magnitude smaller than traditional Greenstone. Also interesting is that TDB exhibits the decrease in performance expected (but not seen

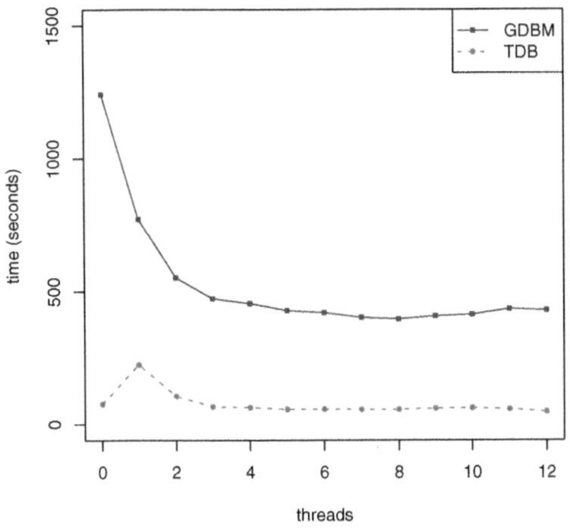

Fig. 4. Timing with comparing GDBM with TDB

in the earlier graph) when threads equals one (and thus the process has all of the overheads of parallel processing but none of the benefits). However there were also some unexpected results; one being that serial Greenstone using TDB performed almost as well as parallel Greenstone using TDB with eight or more threads (with the latter only being a second or two faster) and the other being that TDB's performance seems far more variable than GDBMs (some TDB runs took four times longer than others).

6 Conclusions and Future Work

The results shown above have exposed several questions that we must answer before proceeding with parallel processing support to the global Greenstone community. We applied profiling tools to locate several of the IO or other bottlenecks limiting the performance of parallel processing. While we were able to remove several of these by altering the configuration (for instance, specifying the encoding of the import documents manually rather than leaving Greenstone to automatically determine this), there are still issues with IO which we plan to investigate by reviewing file IO in Greenstone, simulating different drive configurations, and altering filesystem and disk cache settings.

However, most exciting has been the performance gained by moving to TDB as the database. These results are still preliminary and need to be carefully explored. In particular we need to determine if the results for standard, serial Greenstone importing is accurate, since on the surface it seems to marginalize any work on supporting parallel processing (at least for eight cores, although more populous CPUs are just around the corner).

The current results have already demonstrated that there is some benefit in parallelising the import process in Greenstone and that the development of parallelisation support leads to optimisations that benefit the general architecture as well. Given the increasing proliferation of high-core CPUs (with 12-cores available at the time of writing), multi-processor-capable systems and commodity publicly-available high performance computing in the form of pay-per-use virtual utility computing, it is essential that software like Greenstone and its peer systems plan for parallelisation as their architectures evolve.

References

1. Adams, D.: Papers past: browse access for online New Zealand newspapers. Microform & Imaging Review 34, 22–27 (2005)
2. Bainbridge, D., Witten, I.H., Boddie, S., Thompson, J.: Stress-testing general purpose digital library software. In: Agosti, M., Borbinha, J., Kapidakis, S., Papatheodorou, C., Tsakonas, G. (eds.) ECDL 2009. LNCS, vol. 5714, pp. 203–214. Springer, Heidelberg (2009), `http://portal.acm.org/citation.cfm?id=1812799.1812828`
3. Barroso, L.A., Dean, J., Hölzle, U.: Web search for a planet: The Google cluster architecture. IEEE Micro 23, 22–28 (2003), `http://portal.acm.org/citation.cfm?id=776692.776716`
4. Goetz, B.: Java theory and practice: Thread pools and work queues. Tech. Rep. j-jtp0730, IBM, New York, United States (2002), `http://www.ibm.com/developerworks/library/j-jtp0730/index.html`
5. Manghi, P., Pagano, P., Ioannidis, Y.: Second Workshop on Very Large Digital Libraries: in conjunction with the European Conference on Digital Libraries. SIGMOD Rec. 38, 46–48 (2010), `http://doi.acm.org/10.1145/1815948.1815959`
6. Rasmussen, E.M.: Introduction: parallel processing and information retrieval. Inf. Process. Manage. 27, 255–263 (1991), `http://portal.acm.org/citation.cfm?id=117658.117659`
7. Stanfill, C.: Partitioned posting files: a parallel inverted file structure for information retrieval. In: Proceedings of the 13th Annual International ACM SIGIR Conference on Research and Development in Information Retrieval, SIGIR 1990, pp. 413–428. ACM, New York (1990), `http://doi.acm.org/10.1145/96749.98247`
8. Witten, I.H., Bainbridge, D., Nichols, D.M.: How to Build a Digital Library, 2nd edn. Morgan Kaufmann Publishers Inc., San Francisco (2009)

CJK Indexing Prototype for Asian Digital Collections: Developing a Software Tool Where Generations Meet

Alan M. Heath

FamilySearch International, Salt Lake City, Utah, USA
heatham@familysearch.org

1 Who or What Is FamilySearch?

FamilySearch is the largest genealogy organization in the world. FamilySearch consists of a collection of records, resources, and services designed to help people learn more about their family history. FamilySearch gathers, preserves, and shares genealogical records worldwide. FamilySearch is a non-profit organization that offers free access to its resources and service online at FamilySearch.org, one of the most heavily used genealogy sites on the Internet.

2 More Access to More Records for More People

The FamilySearch Records Access program is designed to significantly increase the number of genealogical records available to people around the world. Records contained in the Family History Library and in FamilySearch databases have been gathered from a wide variety of sources worldwide in an ongoing collection effort that has been under way for more than a century. These records are safeguarded in the Granite Mountain Records Vault. FamilySearch is currently undertaken a large digital conversion and capture project, where over 60 million images per month are digitally converted or approximately 370,000 rolls of microfilm per year, the equivalent of about 6 million 300-page volumes. FamilySearch is also enlisting the help of thousands of volunteers to index all those newly digitized records.

3 Indexing by the Numbers

For several years, FamilySearch has been developing an online indexing tool for the Romantic languages. Due to the high cost of indexing, FamilySearch has organized this effort around crowd sourcing volunteers. While crowdsourcing is an overused buzz word for business these days, it is nevertheless a powerful concept for combining the efforts of the community in a worthwhile activity. FamilySearch has found this to be the case in organizing efforts around indexing family history records and essentially bringing to life the names and stories of ancestors previously hidden in the masses of paper records in libraries, archives, public institutions, etc. Online crowds of **FamilySearch Indexing** volunteers, from all walks of life and countries around the world, help index millions of digital images of historic records in seven languages for free access by the worldwide genealogy community. Thousands of new volunteers continue to join the FamilySearch Indexing initiative each month, so the numbers of accessible, free genealogy records will only continue to grow.

C. Xing, F. Crestani, and A. Rauber (Eds.): ICADL 2011, LNCS 7008, pp. 341–346, 2011.

3.1 Number of Records Indexed Daily by FamilySearch Volunteers

Last year:

- Combined Average = 1.3 million/day

This year:

- 1,200,000 records indexed a day
- 570,000 records arbitrated a day
- Combined average = 1.8 million/day.

4 Our Proposed CJK Indexing Solution

FamilySearch has contracted with Lifewood Technology to provide feature and functional development for a Chinese, Japanese, Korean ("CJK") indexing client prototype. This project will be divided into three phases:

Phase 1 – known as the Prototype Phase, aims to:
  a) Demonstrate software at the ICADL Conference in Bejing
  b) Generate support for such project
  c) Gather feedback on CJK indexing tool and for further development of the solution.

Phase 2 – known as the Proof of Concept Phase, will be a working with the Asian library and archival community to seek feedback for improvement on the proof of concept.

Phase 3 – known as the Product Release Phase, aims to implement the feedback received.

5 Our Software Development Partners

5.1 Lifewood Technology Limited

Founded in 2000, Lifewood is a well-established global provider of Business Process Outsourcing (BPO) solutions, committed to offering quality, efficient and cost-effective services to businesses around the world. FamilySearch has successfully contracted with Lifewood over the past many years to key family history records and have established a close working relationship. Lifewood has extensive experience and capabilities in developing their own industrial keying software in many languages, including Chinese and Japanese. Thus, Lifewood was a natural partner for us to assist us in developing the CJK version of our indexing solution.

5.2 National Library of China

We are coordinating with the National Library of China (NLC) to provide technical and cultural expertise during the development phases of the CJK Indexing Solution

Project. In our discussions with the NLC staff we have had guidance about standards and we intend to have continued dialogue and technical guidance in accordance with the "Chinese Collections Full Text Restoration and Full Text Input XML Standards"《中文文献全文版式还原与全文输入XML规范》" .

6 FamilySearch Indexing Solution Roadmap

6.1 Introducing a New Indexing System

One of the needs we see with the new indexing tool is to broaden the crowdsourcing efforts within the larger communities to provide indexing capabilities to the larger community players. So, the information can be shared with a larger set of organizations and that crowdsourcing can continue to grow. The new system was designed to support multiple large projects similar to those currently processed by our system today as well as smaller, local projects created and managed by independent organizations. Some of the key components of the new system include:

- Cloud Infrastructure
  - Horizontal and vertical scalability
  - Support multiple instances and "private label" hosting
- Simple Project Setup
  - Enable Society/Affiliate project setup
  - Enable ad-hoc indexing
- Formal API
  - Better load/integration testing
  - 3rd Party application support/integration.

6.2 System Architecture

Indexing of records is now happening throughout the world by individuals, family groups, societies, archives and various other organizations. The enormous size and scope of this worldwide effort has driven the architecture and design of the next generation of FamilySearch indexing and related technologies. The system architecture below is designed to meet these demands. FamilySearch anticipates large numbers of these systems running throughout the world servicing the needs of all indexers.

FamilySearch has designed the new Indexing system to take advantage of the Internet, while at the same time allowing for off-line work where Internet access is limited or not allowed. Further, the restful web service architecture anticipates the future desire of individuals to creatively find and build new tools to index records, such as mobile devices or games. Backend services to images and other indexing-related information will be provided to help solve the difficult challenges the indexing infrastructure requires.

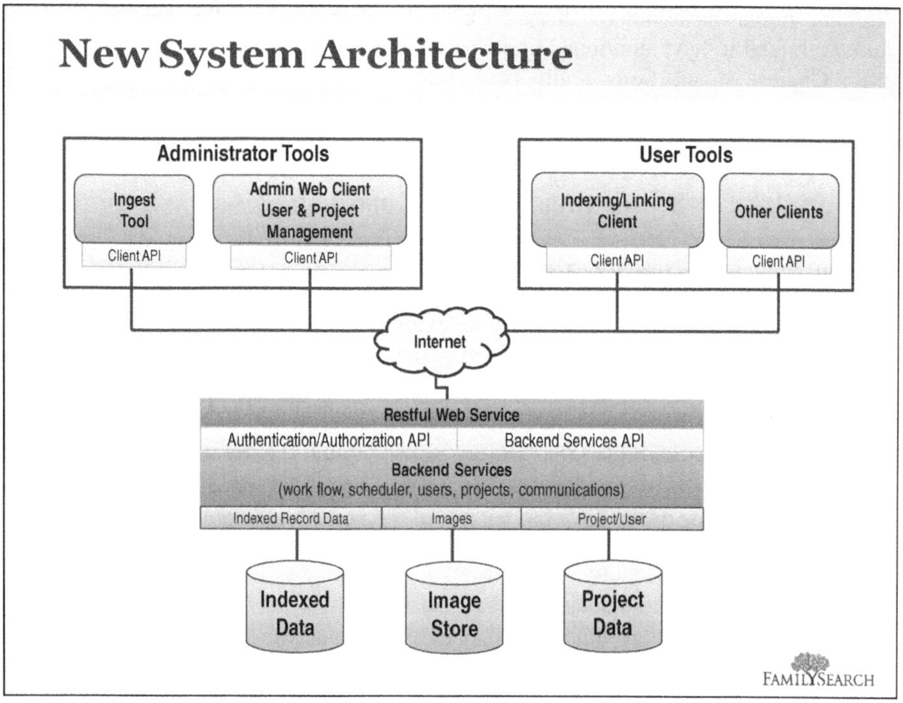

Fig. 1. FamilySearch's New System Architecture

6.3 Developer Access: Integrating with the New System

One of the key components of the new system is a formal API third-party organizations can use to integrate with the new system. This API will be provided as a module of the FamilySearch API.

The Indexing API will provide access to projects, batches, images, records and reports associated with the indexing system. Some possible integration scenarios include:

- Mobile device application
- Browser-based indexing client
- Private-label skinning of admin portal and linking/indexing client
- Integration with social media (publish user and project statistics on other sites)
- Customized reports.

7 The CJK Character Dataset Challenge

According to a study done in 2009 by ZhongYi Electronic Ltd. (中易中标电子信息技术有限公司), which included a full-text indexing project on historical genealogies of Zhejiang Province, ZhongYi found that over 25% of the characters needed were not found in Unicode 4.0 Chinese character dataset.[1]

Our stance and conceptual view in developing this CJK Indexing Solution is that Unicode is deficient when it comes to historical records from Asia, especially when looking at many ancient genealogy records (家谱) for people's given names and geographical place names. The latest version of Unicode includes over 71,000 Chinese characters. This may seem like more than enough; however, when attempting to ascertain the size of extant historical character data set it is still quite small. According to ZhongYi Technology, they have compiled a dataset of over 200,000 historical and modern Chinese characters. Also, some argue that the number of new characters is expected to continuously increase, because Chinese people, especially in Taiwan, are reportedly inventing new characters every day. They do not invent new characters to express new concepts. Rather, new characters are used in personal names. The number of different characters used in personal names is far greater than the number of characters used in recording spoken Chinese.

An argument can be made that if all other languages, except Chinese, are satisfied with Unicode, then Chinese characters will be in trouble. If Unicode distorts, ignores or deletes some aspects of the Chinese writing system, then we should seek to improve it. A writing system should not be changed or handicapped solely for technical reasons.

It has become a routine of the Ministry of Public Security to collect un-encoded characters used for personal names and geography names, from Ministry of Public Security. For example, there are several characters collected since the latest Unicode meeting for CJK characters, known as IRG#35, such as "[illegible]", it is used in a village name in Shandong province: 石[illegible]村. Over the past several years, FamilySearch has worked closely with the Local History Office in Beijing, and we propose working with this office to compile additional historical place name characters not found in the present version of Unicode.

The development and extension of the CJK characters is being done by the Ideographic Rapporteur Group (IRG) (IRG home page), which includes official representatives from China, Hong Kong , Macao, Singapore, Japan, South Korea, North Korea, Taiwan and Vietnam, plus a representative from the Unicode consortium. The IRG is very carefully cataloging, reviewing, and assessing CJK characters for inclusion into the standard. The only real limitation on the number of CJK characters in the standard is the ability of this group to process them.

In our prototype CJK Indexing Solution, text recognition will be based on the large character set (71,000 characters) of the Unicode 5.0 standard [文字认同基于符合 Unicode 5.0 标准的大字符集（7.1万字）].

[1] Powerpoint Presentation given by Fei Lan, VP of ZhongYi Technology, Ltd. in 2010 to FamilySearch. See also:
`http://www.china-e.com.cn/main/service/jiapu.htm`

In terms of handling unrecognizable or un-encoded characters in the prototype CJK Indexing Solution, we propose to tag the character with the following identifiers:

1. 无法识别的模糊字 – unreadable character
2. 异体字，未識字or 无法认同的集外字 – variant or unrecognizable character

The CJK Indexing Solution can compile a list of variant or unrecognizable characters for review. We propose that a committee or interested attendees from ICADL be organized to review this list of unrecognizable characters for vetting and formal submittal to IRG for approval.

8 The Conference Presentation

We will present our online indexing tool for Chinese/Japanese/Korean records and freely share this application with librarians, archivists, and records custodians. By October of this year, the software code will ready for us to present a demo version of this software for the Asian library community.

We will demo the software's design principles, functionality, features, and share our future vision for the development of this solution. Lastly, and one of the most important aspects of this presentation, we will seek feedback from individuals and organizations that are willing to work with us in developing this online indexing tool.

Effective Approaches to the Evaluation and Selection of a Discovery Tool

Huibin (Heather) Cai[1], Tianfang Dou[2], and Airong Jiang[2]

[1] McGill University Library, Montreal, Canada
heather.cai@mcgill.ca
[2] Tsinghua University Library, Beijing, China
{doutf,jiangar}@lib.tsinghua.edu.cn

Abstract. Evaluating and selecting a discovery tool can be a tough process, given that the discovery tools available in the market have many features in common, and from an end user's point of view, they do pretty much the same things. The more or less aggressive advertisements make the evaluation and selection process even more challenging. How to choose the most appropriate product to best serve your users' needs and protect your library's investment? Which key factors should be taken into consideration when making this decision? This article is intended to suggest some effective approaches to the evaluation and selection of a discovery tool. The authors hope that by taking these approaches, libraries that are looking for a discovery tool or switching to another one will be able to make informed and reasonable decisions.

Keywords: resource discovery, architecture framework, discovery system, metadata quality, user interface.

1 Introduction

Discovery tool[1] became popular in recent years as it promises simple and fast access to all library content across print, electronic and digital platforms. There are currently four major players in the market: Primo from Ex Libris, Summon from Serials Solutions, EBSCO Discovery Service from EBSCO, and WorldCat Local from OCLC [1-10]. The "Big Four" have many features in common, and from an end user's point of view, they do pretty much the same things. The more or less aggressive advertisements make the evaluation and selection process more challenging.

Implementing a discovery tool is a big decision, not only because of a large investment involved, but also due to its significant impact on library systems and services [11]. It is interesting to observe that, while some libraries can never be able to make a decision, others have already switched to their second choice. This scenario indicates that there are some issues with the approaches to the evaluation and selection process.

In this article, we will try to answer the following questions: How to choose the most appropriate product that can best serve your users' needs and protect your library's investment? Which key factors should be taken into consideration when

[1] The term "discovery tool" used in this article is interchangeable with "discovery system", "discovery layer" or "web-scale discovery service", depending on the context.

C. Xing, F. Crestani, and A. Rauber (Eds.): ICADL 2011, LNCS 7008, pp. 347–356, 2011.

making this decision? We will begin with our perceptions of discovery tools and the evaluation priorities, followed by a proposed workflow, and finally conclude with some further thoughts.

2 Perceptions and Evaluation Priorities

Libraries have tremendous amount of valuable scholarly contents, but they are scattered in different silo systems, making them difficult to find and use. Federated search has not been a perfect solution to address the issue [12]. The transformation of federated search to discovery service is a result of the large-scale digitization trend: because obtaining metadata from content providers is no longer difficult, vendors are now able to build a large pre-indexed metadata repository to support fast concurrent searches through seamless integration of local and remote systems.

2.1 Architecture Framework

The architecture framework of a discovery tool is a fundamental factor that deserves the most careful thoughts and thorough evaluation, as it cannot be easily modified or changed. It defines a discovery tool's capabilities and potentials in terms of functionality, performance, and extensibility, etc. A well-designed architecture seamlessly integrates print and electronic resources and yields desired features.

There are currently two types of architecture models: the cloud-based model (Fig. 1) and the hybrid model (a mixed model of cloud-based and locally hosted data center) (Fig. 2). In a pure cloud-computing environment, services are provided through the SaaS (software as a service) platform. In this model, data are stored in the cloud, servers are maintained by the cloud provider, and services are provided through subscription. This model may reduce your total cost of ownership, but may also bring some critical issues, such as: risk of losing control on local data and sensitive information [13], service quality relying on external network, and limited control over user interface, etc. Summon is a typical cloud-based discovery product [9].

The hybrid architecture has two aspects: the first aspect is hybrid data, i.e. data is indexed and stored both locally and remotely: local bibliographic data and holdings information are stored in local servers for pre-indexing; a repository containing massive metadata is deployed in the cloud as central indexes. Local servers interact with local indexes and central indexes to produce and present a unified resource listing.

The second aspect is hybrid service. We know metadata-based pre-index repository enables fast search and increases search efficiency significantly, but since some publishers are reluctant to give away their metadata to the vendors, so solely relying on the repository might exclude some local resources that are not included in the repository. The hybrid service model can take the advantage of federated indexes provided by MetaLib, thus augments the metadata repository. Primo is a product using the hybrid architecture [8].

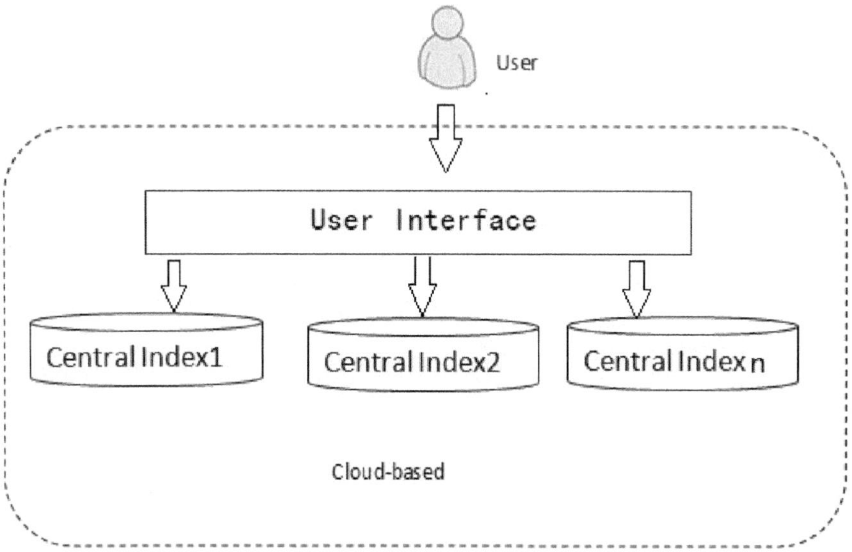

Fig. 1. Cloud-based architecture

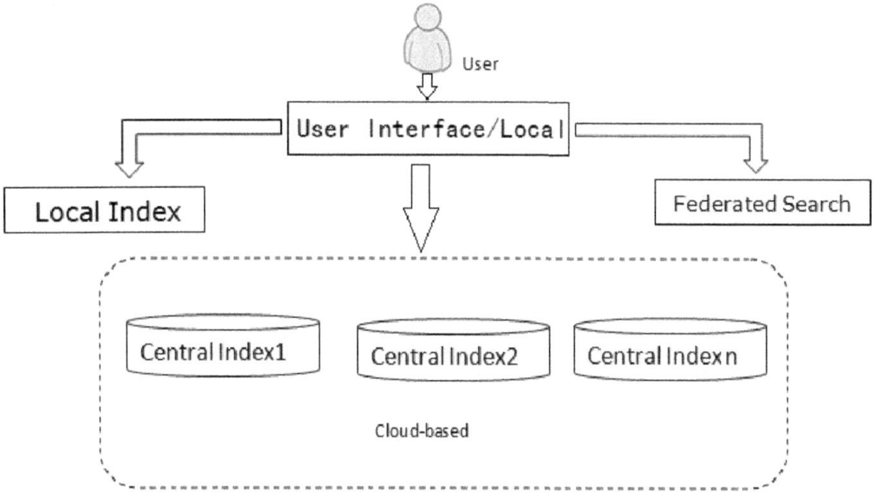

Fig. 2. Hybrid architecture

2.2 Depth of Integration

To provide a single search interface, a discovery tool needs to integrate with local ILS [15]. The depth of the integration defines a discovery tool's service level, or usefulness, and is thus extremely important. Different discovery products have different depths of integration.

There are two depths of integration: basic integration is data integration, that is, all metadata are first uploaded to a data center, usually hosted by the vendor, and then re-indexed for search and retrieval. This depth of integration allows an end user to search a library's entire collection and to find out the availability status for a specific item, but the user cannot perform actions on the item, such as renewing a book or placing an order, and so on. The depth at this level may be sufficient for resource discovery, but not for library services.

Advanced integration is data and service integration, represented by Primo's "OPAC functionality in the Primo UI" [14]. When users find a print item in the Primo UI, they can act on that item the same way as in a traditional OPAC. With the advanced integration, users do not have to switch between a discovery tool and an OPAC to complete certain tasks, and staff do not have to maintain two separate user interfaces.

2.3 Resource Coverage

The coverage of a pre-indexed repository defines a discovery tool's search scope. A larger size of a repository offers better chance of finding the information; however, each repository has its own coverage with the vendor's strengths in its specialized subject areas (Fig. 3). It is important to know the composition of a repository and how well it matches of the composition of your library's electronic resources collection. Simply comparing the size of the repositories may lead to a mistake [16].

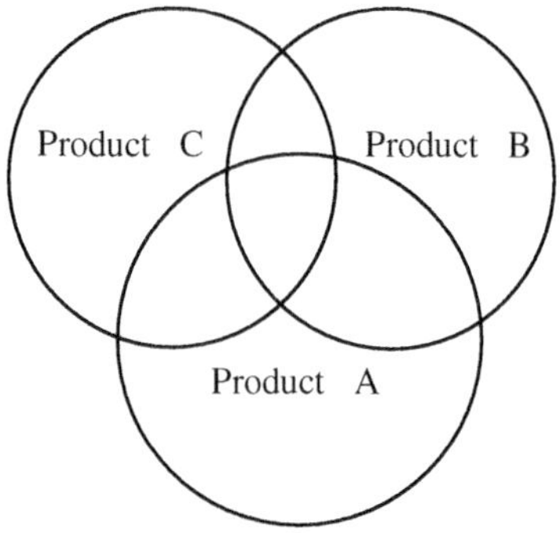

Fig. 3. Each repository has its own coverage

On the other hand, all repositories in the current discovery tools are growing quickly through the vendors' metadata supplying channels. These channels are usually stable and reliable, because the content providers are willing to share their metadata and full-text indexes, in order to generate more traffic to their web site and make their resources more discoverable. With this in mind, we can foresee that eventually all repositories will become equally large.

Vendors do offer matching tests to determine the compatibility between their products and a library's e-resources collection, but the process is not transparent and the bases used for calculation are different from one vendor to another [16]. Ideally, the test should be done in a local system environment, but there is no effective or easy method to accomplish that. Tsinghua University Library recently conducted a compatibility test with two chosen discovery tools [17]. They chose a famous Chinese scholar, Andrew C. Yao (a Turing award recipient) as subject, pulled a complete list of his articles (118 in total) from Tsinghua University's institutional repository, and then searched the two discovery tools for purpose of comparison. Table 1 shows the compatibility test results.

They also compared the results with Google Scholar and came up with some interesting findings:

1) Product B has more matching entries than Product A;
2) Google Scholar has more matching entries than Product A and B, but has much more noisy entries as well.

Table 1. Compatibility test results (Base: 118)

Product	Returned entries	Dedup & filtering	Matching entires	Compatibility (%)
Product A	29	5	24	20
Product B	56	6	50	42
Google Scholar	256	159	69	58

This sample test provided some valuable insight about usefulness of discovery tools in general in comparison with Google Scholar.

"The success of discovery product always lies in the coverage of the content which the library subscribes" [18]. Multiligual support is part of the coverage issue. For example, none of the currently available discovery tools can provide full coverage of metadata and fulltext indexes for Chinese content [16].

2.4 Metadata Quality

Metadata quality refers to the integrity and normalization of metadata. Common problems with metadate integrity include incomplete or insufficient metadata, i.e. "thin" metadata. When coming from aggregators rather than directly from publishers, metadata are more likely to be "thin" metadata, which typically only contain basic elements such as title and author, but not subject. Thin metadata is far less useful than thick metadata to the users. Turning "thin" metadate into "thick" has been a significant issue.

Fig. 4 and Fig. 5 show how metadata normalization problems affect search results.

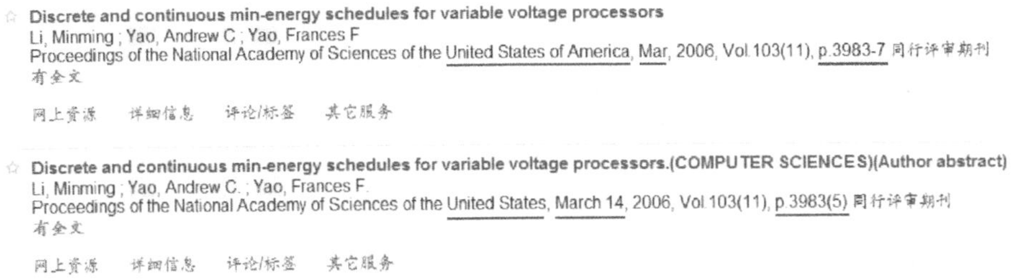

Fig. 4. Duplicate entries found in Primo

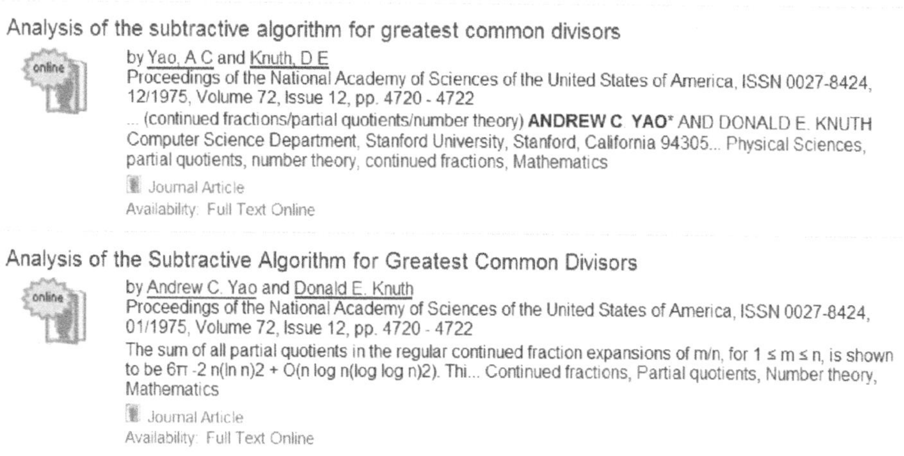
Analysis of the subtractive algorithm for greatest common divisors

by Yao, A C and Knuth, D E
Proceedings of the National Academy of Sciences of the United States of America, ISSN 0027-8424, 12/1975, Volume 72, Issue 12, pp. 4720 - 4722
... (continued fractions/partial quotients/number theory) **ANDREW C. YAO*** AND DONALD E. KNUTH Computer Science Department, Stanford University, Stanford, California 94305... Physical Sciences, partial quotients, number theory, continued fractions, Mathematics
Journal Article
Availability: Full Text Online

Analysis of the Subtractive Algorithm for Greatest Common Divisors

by Andrew C. Yao and Donald E. Knuth
Proceedings of the National Academy of Sciences of the United States of America, ISSN 0027-8424, 01/1975, Volume 72, Issue 12, pp. 4720 - 4722
The sum of all partial quotients in the regular continued fraction expansions of m/n, for 1 ≤ m ≤ n, is shown to be 6π -2 n(ln n)2 + O(n log n(log log n)2). Thi... Continued fractions, Partial quotients, Number theory, Mathematics
Journal Article
Availability: Full Text Online

Fig. 5. Duplicate entries found in Summon

As seen above, it is the different presentation formats of the metadata (year, volume, and page numbers, etc.) causing the duplicate entries of the records. Current discovery tools are all suffering from non-standard metadata. Libraries should advocate metadata normalization for discovery tools. In practice, libraries should follow the standard workflow when loading vendor data to the systems, to minimize normalization problems.

Metadata quality is a community effort. Libraries should continue to pay attention to metadata quality and constantly push the vendors to improve metadata quality and enrich metadata.

2.5 User Interface

With massive amount of metadata, search results sets tend to become larger. The ability to quickly narrow down search results is essential. Common features include FRBRization, use of facets, and aggregation, etc.

A good system should also help users to expand and adjust search queries, to make a suggestion to a failed search (such as the popular "Did you mean" feature) or to form a new search strategy.

Based on user behaviour studies and data mining technology, recommendation service represents a new technology trend and an important application in library services. Ex Libris bX scholarly article recommender service [19] introduces a brand new concept and perspective to the library world: search is no longer the only way of finding information; pushing the right information to the users at the point of need. bX is now embedded in the Primo UI, enriching the user interface and enhancing the user experience.

Web 2.0 features such as book cover displays, book reviews, and visualization applications are all valuable additions to the user interface.

Although single search box is a basic feature of a discovery tool, it may not satisfy all user groups. For example, advanced searchers may find combined multiple-field search and Boolean search more useful. Therefore the two features should still be kept and enhanced. There are issues with user interface here and there, but those issues are not fundamental and can be resolved eventually.

2.6 Additional Evaluation Criteria

Additional criteria, such as back office management, interoperability with existing systems, openness and API, statistics module, maintanence workload, and vendor support are equally important. The importance may not be realized until reaching the implementation stage or even after going live [11]. Due to length restriction of this article, we can't discuss them in detail. Readers should consult appropriate resources for more information.

3 Proposed Workflow

"If you fail to plan you plan to fail." The evaluation and selection process may take a long time, ranging from several months to a couple of years. Having a workflow in place would help ensure a smooth and effective process. The following workflow is proposed for your reference.

3.1 Preliminary Work

The needs for a discovery tool may be assessed at the library or consortium level with a provision of your organization's long-term goals. Use of your LibQual+ survey results and usability testing reports may help you understand the urgency. Needs assessment involves communicating with staff and users and soliciting input from them. Keep in mind that it is the library directors who make the final decision, so their understanding and support is vital.

A review of your library's budget and human resources may give you an insight into the possibility and the approach to take: to buy it or build it yourself? In either case, the total cost of ownership may end up the same, because of the cost of hiring on-site developers. The discussion below will be limited to the "buying" approach.

3.2 Forming a Working Group

The first thing to do is to form a working group. The purpose of the working group is to investigate products, communicate with staff members, seek input from the user community, and make recommendations to the library management. The group may be appointed and may be composed of several members from the relevant units of your library - experience tells us a group larger than 5~6 may decrease communication and work efficiency. This group will be charged to work on the evaluation and selection.

3.3 Building Your Own Evaluation Criteria

Consulting evaluation criteria developed by other sites might give you some idea about how to get started, ultimately it is your users, your needs, and your choice, so you really should develop your own evaluation criteria. In practice, you will need to set up meetings with staff and users to collect a list of desired features, weight them according to importance.

3.4 Narrowing Down the Scope

After a couple of quick mappings between your evaluation criteria and the available products, you should be able to eliminate those that are obviously inappropriate for your library. The remaining ones then become the candidates, usually no more than 3, and will be evaluated seriously.

3.5 Reviewing Existing Customers

Reviewing existing customers is a quick way of gaining first-hand experiences and product-neutral opinions. For each candidate, select a few sites that share common characters with your library: the scale of the library, the composition of the library systems (especially the ILS), etc. This approach makes your comparisons more meaningful. For best results, you can prepare a set of questions for the reviews and document the answers for future reference.

3.6 Setting Up Vendor Demos

Vendors love to be contacted for demos. You only need to send a request and then schedule a time for the demo. If you can recruit some students and faculty members to participate, that would be ideal as they are the end users of the service. Be sure to ask for feedback, which is the purpose of the demo.

3.7 Testing with Trials

When you feel serious about a product, give it a try. You should get a good sense of what it can do for you after you have installed the trial and tested with your local systems and data. The feedback from the end users is almost always different from the librarians, so you might want to conduct two separate surveys, one with the end users, and the other with the staff. If you have the time and energy, you could even make the trials available to the public to collect feedback from a larger user group.

3.8 Making Recommendations

The last step is to report your findings and make recommendations to the library management. Your goal is to help them understand the options, i.e. the potential benefits and risks of each option. Try to use as little technical jargon as possible since they might not be familiar to the management. Keep this in mind: your directors are very busy. Think about possible ways to save their time reading, such as including a summary of the report, adding tables and charts, etc. so your report can be easily understood and your ideas can be quickly comprehended, consequently speeding up the decision-making process.

4 Conclusion

Library resource discovery is an on-going and collaborative effort. The systems vendors, the content providers, the libraries, and the users need to work together to improve the products.

The discovery tool of your choice should be technologically advanced, effective and with potential room for improvement and growth. It should not only address today's problems but also tomorrow's problems.

There is no reason for a library to exist except serving its community. We hope that, by taking the suggested approaches, libraries looking for a discovery tool or switching to another one will be able to make informed and reasonable decision, consequently providing a better service to the community they serve.

References

1. Breeding, M.: Automation marketplace 2010: New Models, Core Systems (2011), http://www.libraryjournal.com/article/CA6723662.html (June 6, 2011)
2. Breeding, M.: Automation Marketplace 2011: The New Frontier (2011), http://www.libraryjournal.com/lj/home/889533-264/automation_marketplace_2011_the_new.html.csp (June 7, 2011)
3. Breeding, M.: Web Scale Discovery Services (2010), http://www.alatechsource.org/taxonomy/term/106/web-scale-discovery-services (June 11, 2011)
4. Breeding, M.: Building Comprehensive Resource Discovery Platforms (2011), http://www.alatechsource.org/blog/2011/03/building-comprehensive-resource-discovery-platforms.html (June 5, 2011)
5. Vaughan, J.: Web scale discovery services. Library Technology Reports 47(1), 5–61 (2011)
6. OCLC. Perceptions of Libraries, 2010: Context and Community, http://www.oclc.org/us/en/reports/2010perceptions.htm (June 7, 2011)
7. EBSCO Publishing. EBSCO Discovery Service, http://www.ebscohost.com/discovery/eds-content (June 1, 2011)
8. Ex Libris. A Solution for Current and Future Library Architecture, http://www.exlibrisgroup.com/category/PrimoArchitecture (June 5, 2011)
9. SerialsSolutions. Summon™, http://www.serialsolutions.com/summon (June 3, 2011)
10. OCLC. WorldCat Local, http://www.oclc.org/worldcatlocal/ (June 4, 2011)
11. Cai, H.: Impact of resource discovery tools on library systems and services. Presented at the 7th Advanced Digital Library Seminar, Harbin, China, July 21-23 (2010) (in Chinese), http://202.118.250.137:88/document (June 11, 2011)
12. Rochkind, J.: (Meta)search Like Google. Library Journal (February 2007)
13. SaaS Directory. Advantages & Disadvantages of Cloud Computing, http://www.saasdir.com/news/showNews.aspx?ID=33174 (June 4, 2011)

14. Kaldenberg, B.: Primo Version 3 Collaborative Testing Report to IGeLU of testing performed, January 24-28 (2010), http://gep.ub.fu-berlin.de/wordpress/wp-content/uploads/2010/09/coll_testing_kaldenberg.pdf (June 10, 2011)
15. DLF. DLF ILS Discovery Interface Task Group (ILS-DI) Technical Recommendation, http://old.diglib.org/architectures/ilsdi/DLF_ILS_Discovery_1.0.pdf (June 11, 2011)
16. Jiang, A.: The development of resource integration in the context of the cloud service framework. Presented at the 2010 Information Technology Application Conference, held by the Academic Libraries Committee of the Ministry of Education of the People's Republic of China, Shanghai, China, December 16-17 (2010) (in Chinese), http://www.library.fudan.edu.cn/2010meeting/meeting/reports/clouds008.pdf (June 3, 2011)
17. Dou, T.: The foundation, exploration and prospects of data services. Presented at the China Academic Libraries Forum and the Advanced Digital Library Seminar, Chengdu, China, June 22-24 (2011) (in Chinese), http://www.sal.edu.cn/2011/Richeng.asp (June 10, 2011)
18. Pradhan, D.R., Trivedi, K., Arora, J.: Searching Online Resources in New Discovery Environment: A State-of-the-Art Review, http://shodhganga.inflibnet.ac.in/dxml/bitstream/handle/1944/1623/14.pdf?sequence=1 (June 10, 2011)
19. Ex Libris. bX Recommender Service Overview, http://www.exlibrisgroup.com/category/bXOverview (June 2, 2011)

Design of Automatic Mapping System between DDC and CLC

Yihua Zhang, Jia Peng, Di Huang, and Fang Li

Shanghai Jiao Tong University Library,
Shanghai 200240, China
{yhzhang,jpeng,dhuang,fli}@lib.sjtu.edu.cn

Abstract. Dewey Decimal Classification (DDC) and Chinese Library Classification (CLC) are two widely-used library classification schemes internationally and in China. The construction of interoperability between DDC and CLC is urgent and significant to information resources organization and retrieval, especially the classification indexing of electric resources in foreign languages. This paper aimed to design an automatic mapping system based on a statistical mapping table and a manual one. Both tables were established one-direction, from DDC to CLC. The statistical mapping table was created based on the frequency of the mapping relations with a sample of USMARC records, which contained DDC and CLC numbers. The manual mapping table provided a CLC number for each DDC section. A mapping strategy was generated according to the mapping results comparison of a second sample. A third sample was utilized to analyze the results of the automatic mapping system with two established mapping tables and the specific mapping strategy. In conclusion, possible improvements of the system in the future were discussed.

Keywords: Interoperability, classification schemes, automatic mapping system, DDC, CLC.

1 Introduction

Distributed digital libraries and online information resources are recently growing rapidly. Information resources from different sources are organized with different knowledge organization systems (KOS). In order to provide better user experience of searching, browsing, and one-stop service in the digital library, it is urgent to establish interoperability models of different KOS with high reliability.

Research projects of interoperability among different KOS have started since 1980s. One early project of Online Computer Library Center (OCLC) was to match the Educational Resources Information Center (ERIC) Thesaurus to the Library of Congress Subject Headings (LCSH) [1]. Another project was to link LCSH to the Dewey Decimal Classification (DDC) scheme. The mapped LCSH are provided and updated in the electronic version of the DDC and an aid for catalogers [2]. OCLC actually provides wider Terminology Services, which focus on the interoperability of different types of controlled vocabularies [3]. Renardus Service (EU Project)

C. Xing, F. Crestani, and A. Rauber (Eds.): ICADL 2011, LNCS 7008, pp. 357–366, 2011.

utilized DDC as a switching language among different classification schemes [4]. The correlations between Library of Congress Classification (LCC) and DDC are provided on Classification Web, which is of great use to catalogers [5]. The Unified Medical Language System (UMLS) and The High-Level Thesaurus Project (HILT) both involve multiple KOS. The UMLS brings together many health and biomedical vocabularies and standards to enable interoperability between computer systems [6]. The HILT aims to research, investigate and develop pilot solutions for problems concerning cross-searching multi-subject scheme information environments [7].

In China, users are now confronted with plenty of information resources organized with foreign KOS. However, there are only a few pilot projects or studies of interoperability between Chinese and foreign KOS. For instance, Dai and Hou have tried to construct an automatic mapping system between Chinese Library Classification (CLC) and DDC by computing the word similarity of each class concept [8]. The pilot projects have generally been limited to a certain range of disciplines, and the results have not been widely applied. Thus a comprehensive interoperability system is urgently required to provide better user experience. This paper is an experimental study to link DDC to CLC in complete disciplines.

2 Design Ideas of Automatic Mapping System

DDC, LCC and LCSH are mostly applied KOS and also the focus of the recent interoperability research projects [9]. Considering the feasibility of our experimental study, DDC was first selected to match with CLC. DDC and CLC are two comprehensive classification schemes with hierarchical structures based on the concept of logic and knowledge. In general, the concepts of both classification headings are similar.

Mapping is a popular methodology in the research projects and usually divided into direct mapping and co-occurrence mapping. Direct mapping is to establish equivalence between terms in different controlled vocabularies or between verbal terms and classification numbers. Co-occurrence mapping is to establish mappings from the co-occurrence of terms from different schemes in the same metadata or catalog record. As the method of co-occurrence mapping can avoid the complex comparison of the vocabulary terms, a statistical mapping table was established by means of it in our study. However, due to a limited record sample size, the complete classification scheme can not be covered by the statistical mapping table. A manual mapping table with all DDC sections was created as a complement for the statistical one.

The reliability of the automatic mapping system lies on the two mapping tables and how to apply them. The precision of the statistical mapping table is most essential and has a great impact on the mapping results. As the precision depends mainly on the record sample size, the system reliability can be continuously improved with the increase of sample size.

In this study, only one direction mapping, from DDC to CLC, was built up. However, the opposite direction can be similarly realized in the future. Furthermore, LCC and LCSH can be integrated to enhance the functionality of the conversion system.

3 Functions and Implementation of Automatic Mapping System

ISO USMARC is mostly-used data format in recent cataloging of foreign resources. Therefore the automatic mapping system was designed to insert CLC numbers into USMARC records with DDC numbers for practical applications. The system input and output files were both USMARC data packets. In order to produce a proper CLC number and add it to the original record, the system required three main functions: disassembly of USMARC records, automatic mapping from DDC to CLC, and reconstruction of USMARC records. A system flow chart was illustrated in Figure 1.

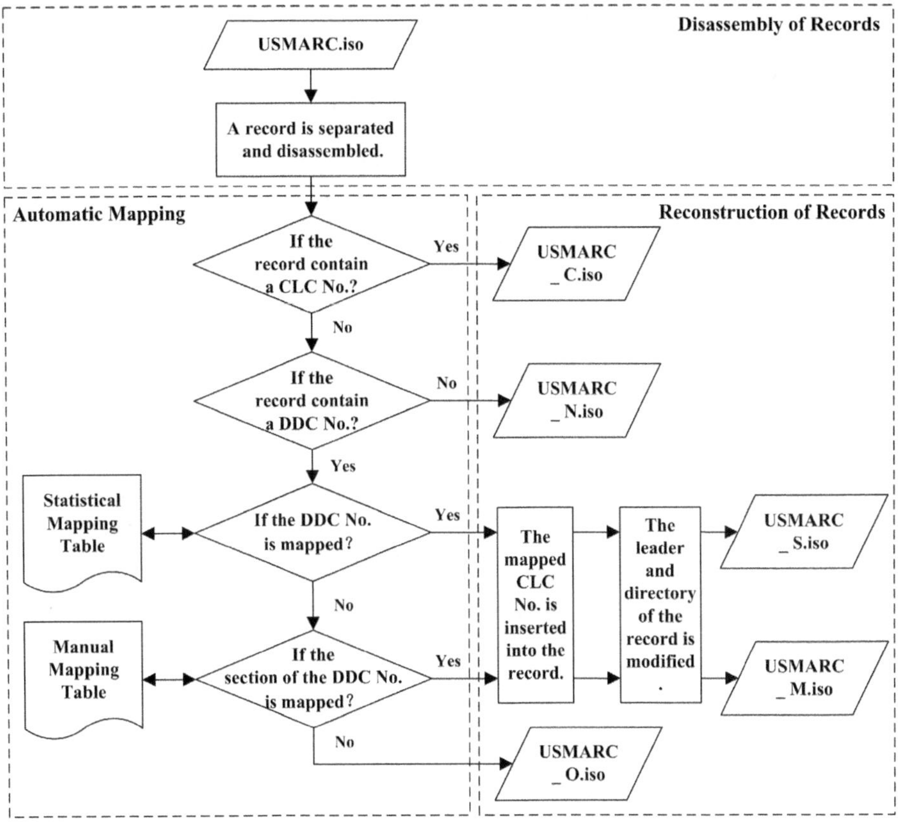

Fig. 1. System flow chart

3.1 Disassembly of USMARC Records

A USMARC data packet, e.g. USMARC.ISO, was composed of a number of USMARC records. The disassembly function module was applied to separate the records and then the fields and subfields so that the DDC number of each record was extracted.

3.2 Automatic Mapping

The automatic mapping function was the key point of the system. It decided the rule of producing a CLC number for a given DDC number. There were two mapping tables, statistical and manual, in the automatic mapping module. A DDC number was mapped through these two tables in turn according to a specified mapping strategy and a CLC number was then obtained.

Statistical Mapping Table. The statistical mapping table was a set of DDC-CLC mapping relations with their frequencies. The same mapping relations were counted and clustered for record data. A certain DDC number could be classified in several different CLC numbers, i.e. a one-to-many DDC-CLC relation. All the different relations for one DDC number were collected in the table. However, only the CLC number with the highest frequency was used and selected in the mapping process. A pilot statistical mapping table was established with a sample in Section 4.1.

Manual Mapping Table. The manual mapping table was established by hand to provide one recommended CLC number for each assigned DDC section. Similarly, one DDC section could be related to several CLC numbers. In order to cover most cases, a superordinate class was generally selected by experience in stead of a longer CLC number. For example, the DDC number of 811 represents American poetry in English. The CLC number of I712.2 was provided correspondently.

Mapping Strategy. A mapping strategy was a procedure that specified how to apply the tables to map from a DDC number to a CLC one. It included mapping sequence, mapping conditions, modification of mapped CLC numbers, etc. Different mapping strategies would lead to different mapping results.

The precision of mapping results from the two mapping tables was separately tested and compared in Section 4.2. It was found that the statistical mapping table was more reliable. Therefore the mapping sequence illustrated in Figure 1 was "first statistical second manual". Before to be mapped a record must be checked whether it contained a CLC number and/or a DDC number. A record with only a DDC number was the target record of the automatic mapping system. The target records were first mapped through the statistical mapping table with the mapping condition of the highest frequency. In the case of more than one CLC number with the highest frequency, the first emerged CLC number in the table was selected. If the DDC number of the record did not exist in the statistical table, the section of the DDC number was then mapped through the manual mapping table.

The applied mapping strategy was really simple so that it was easy to obtain the preliminary results. It should be adapted to experiment more and compare different results so that a more reasonable mapping strategy would be found out. For example, in the case of more than one DDC-CLC relation with the highest frequency for some DDC number, these relations should be better utilized and processed before a mapped CLC number be generated.

3.3 Reconstruction of USMARC Records

If a CLC number was mapped for a certain record, the reconstruction function module inserted the CLC number to the corresponding record and modified its leader and directory according to ISO USMARC. All the records were repacked in five files. As shown in Figure 1, the records with CLC numbers mapped from the statistical mapping table were saved in USMARC_S.iso, while the records with CLC numbers mapped from the manual mapping table in USMARC_M.iso. USMARC_C.iso consisted of the records with CLC numbers without change, and USMARC_N.iso of those without DDC numbers. The other records except for the above four occasions were collected in USMARC_O.iso.

4 Experiments and Results

With the general design of the automatic mapping system, experiments were required to examine the feasibility of the system. Three independent sample data were prepared for different purposes. All the sample data were USMARC records with both DDC and CLC numbers. Sample X was used to establish the statistical mapping table. Sample Y was applied to examine and compare the precision of the results from the two mapping tables in order to specify the mapping sequence. A third sample Z was required to obtain and analyze the mapping results of the automatic mapping system with the specific mapping strategy.

4.1 Sample X – Establishing the Statistical Mapping Table

DDC and CLC are two comprehensive classifications. The size of a sample data to cover most possible DDC-CLC relations must be extremely large. Sample X contained totally 34648 USMARC records. It was used to establish the pilot statistical mapping table due to the small sample size.

The sample records were counted and clustered according to the DDC-CLC relations. The 34648 records were distributed in 27009 different DDC-CLC relations. All the relations were sorted by frequency in descending order in the statistical mapping table. An ID number was given to each DDC-CLC relation.

Table 1 showed two typical examples. The DDC number of 332.6 was related to the CLC numbers of F830.59, F069.9, F279.712, etc. The 332.6-F830.59 relation had the highest frequency of twenty-one. Therefore, F830.59 would be the mapped CLC

Table 1. The pilot statistical mapping table

ID	DDC	CLC	Frequency
...	...	...	...
27	332.6	F830.59	21
...	...	...	...
9204	332.6	F069.9	1
9205	332.6	F279.712	1
...	...	...	...
17693	616.6	Q319	1
17694	616.6	R167	1
17695	616.6	R69	1
...	...	...	...

number for 332.6. Similarly the DDC number of 616.6 was related to several CLC numbers. However, the frequency was one for all these relations. The first emerged CLC number of Q319 would be the mapped number. In order to facilitate the mapping process, the selection of a CLC number was a bit arbitrary and the data of other DDC-CLC relations were not utilized. It could be modified in the future study and improved experiment results could be compared with the preliminary results.

4.2 Sample Y – Comparing the Results from Two Mapping Tables

Before a mapping sequence was specified, the precision of the mapping results from two tables should be checked and compared. To be brief, following abbreviated forms were defined:

- A DDCo number: an original DDC number of the sample records.
- A CLCo number: an original CLC number of the sample records.
- A CLCs number: a CLC number mapped from the statistical table.
- A CLCm number: a CLC number mapped from the manual table.
- A CLCn number: a new CLC number mapped either from the statistical table or the manual one, i.e. a CLCs or CLCm number.
- P1: the percentage of sample records, of which CLCn numbers and CLCo numbers contain the same CLC top-level classes.
- P2: the percentage of sample records, of which CLCn numbers and CLCo numbers contain the same CLC second-level classes.
- P3: the percentage of sample records, of which CLCn numbers and CLCo numbers contain the same CLC third-level classes.
- P4: the percentage of sample records, of which CLCn numbers and CLCo numbers contain the same CLC fourth-level classes.
- Pi: the percentage of sample records, of which CLCn numbers and CLCo numbers are identical.

Sample Y of 2000 records with both DDC and CLC numbers was mapped respectively through the statistical table and the manual one. And 2000 CLCs numbers and CLCm numbers were obtained. The CLCs and CLCm numbers were then compared with the CLCo numbers seperately. Figure 2 illustrated the percentages of the sample records of different similarities between the mapped CLC numbers and the original CLC numbers. P1 was 87.9% for CLCs, i.e. 87.9% of the CLCs numbers contained the same top-level classes as the CLCo numbers. P1 of CLCm was 86.4%, which was slightly lower than that of CLCs. With the increase of the order of classes, the percentages for CLCm dropped deeply. Low proportion of the CLCm numbers was completely the same as the CLCo numbers. The percentages provided a broad overview of the mapping results. As expected CLCs was closer to CLCo than CLCm. The most obvious reason was that the length of CLCm was much shorter than that of CLCs. The former was averagely 2.84 digits, while the latter was 4.84 digits.

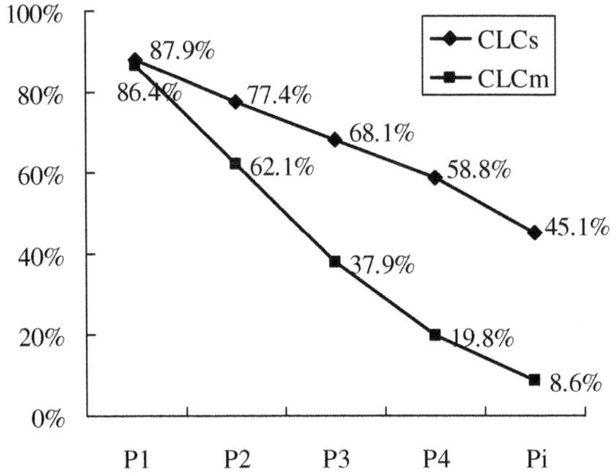

Fig. 2. Percentages of the sample records of different similarities between CLCn and CLCo

Although the Pi values of both CLCs and CLCm were low, the mapped CLC numbers could provide more useful information. If a CLCn number was a superordinate class of the CLCo number, it was thought as respectable. Meanwhile a longer CLCn number was certainly more desirable. In order to measure the precision of the CLCn numbers, two indicators were defined as follows.

- A CLCd number: a desirable CLCn number, which is the same as the CLCo number or a superordinate class of the CLCo number.
- i_proportion: the percentage of sample records, of which the CLCn number is desirable, i.e. the proportion of CLCd in CLCn.
- i_length: the average number of digits of the CLCd numbers for the sample.

The indicator i_proportion was a measure that how much the CLCd numbers accounted for of the all the mapped numbers, while the indicator i_length measure how specific the CLCd numbers were. Generally the shorter i_length is the higher i_proportion. However, shorter CLCd numbers could provide less information about topics of resources. Therefore it was a trade-off between i_proportion and i_length. A minimum limit ought to be set for both indicators in order to assure the mapping quality and to decide whether the mapping results were acceptable. Figure 3 showed that the indicator values of the CLCs and CLCm numbers of sample Y. The i_proportion value of CLCm was slightly higher than that of CLCs, but the i_length value of CLCs was nearly the twice as high as that of CLCm.

With the values of the above percentages and indicators, the CLCs numbers were considered more precise than the CLCm numbers. Therefore the mapping strategy "first statistical second manual" showed in Figure 1 was generated. As the statistical mapping table could not cover all the possible DDC numbers, the manual mapping table was a necessary complement to ensure that a record with a DDC number could be mapped a CLC number.

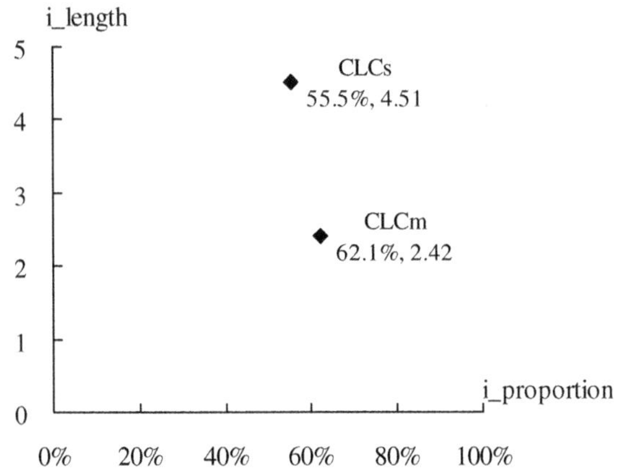

Fig. 3. Two indicators of the results from two mapping tables

4.3 Sample Z – Analyzing the Results of the Mapping System

With two established mapping tables and the specific mapping strategy, a third sample was required to check the effect of the automatic mapping system. This sample also contained both DDC and CLC numbers so that the comparison and analysis of mapping results could be possible.

Sample Z consisted of 4751 records. The sample records were mapped first through the statistical mapping table and then through the manual one. The input data packet was named SampleZ.iso. Three output files, SampleZ_S.iso, SampleZ_M.iso and SampleZ_O.iso, were obtained. SampleZ_C.iso was not given as each record was mapped to obtain a CLC number in spite it already contained one. There was no SampleZ_N.iso as the selected sample records all contained DDC numbers.

SampleZ_S.iso, SampleZ_M.iso and SampleZ_O.iso consisted of 3200, 1537 and 14 records respectively. The mapping results of Sample Z were generally similar to those of Sample Y, but a bit worse. CLCs, CLCm and CLCn were all compared with CLCo. The percentages and two indicators were calculated in Table 2. The i_proportion value was 45.3% for the CLCn numbers, meaning 45.3% mapped numbers were desirable. The i_length value of these desirable numbers was 3.44 digits. SampleZ_O.iso contained 14 records with special cases. For example, the DDC number was [E] or [Fic], which was used to be applied. The special cases must be separately processed.

The indicator i_proportion of the mapping results was less then a half, which was thought as low. Prom a perspective of P1, nearly 20 percent CLCn numbers were

Table 2. Mapping results of sample Z

	No. of Records	P1	P2	P3	P4	i_proportion	i_length
CLCs	3200	83.1%	71.4%	59.9%	46.5%	44.7%	4.02
CLCm	1537	77.6%	48.0%	24.0%	9.4%	46.5%	2.28
CLCn	4737	81.3%	63.8%	48.3%	34.5%	45.3%	3.44

different from the CLCo numbers in main CLC classes. The distribution of the sample records in ten main DDC classes was examined. It was indicated in Table 3 that there were large number of records in Class 300 (Social Sciences), Class 500 (Science) and Class 600 (Technology). The percentages of the records with different main CLC classes were high in Class 300 and 600. Due to the large size and high proportion, these two DDC classes had a great impact on the mapping results. They should be a focus in the future improvement of the automatic mapping system.

Table 3. Records with different main CLC classes and their distribution in DDC classes

Main Class	Disciplines	No. of Records	No. of Records with Different main CLC Classes	Percentage
000	Computer science, information & general works	701	68	9.7%
100	Philosophy & psychology	98	17	17.3%
200	Religion	14	1	7.1%
300	Social sciences	769	199	25.9%
400	Language	70	6	8.6%
500	Science	1134	139	12.3%
600	Technology	1752	410	23.4%
700	Arts & recreation	43	7	16.3%
800	Literature	84	8	9.5%
900	History & geography	72	31	43.1%
Total		4737	886	18.7%

5 Discussions and Conclusions

In order to reorganize distributed digital resources to improve user experience of information retrieval system and facilitate user search and access to discrete heterogeneous information resources, the interoperability of KOS becomes a key topic of knowledge organization.

This study selected two widely used classification schemes internationally and in China and established their relations mainly based on co-occurrence mapping. As far as the mapping effect is concerned, the automatic mapping system is far from application. The reason is probably the small size of sample data in establishing the statistical mapping table and only DDC third-level classes covered in the manual mapping table.

Improvements are naturally necessary to realize the application of the system. First of all, the sample size of record data is the significant condition for the high quality of the mapping system and therefore should be increased for refining the statistical mapping table. Secondly, rather than one DDC-CLC relation with the highest frequency, more relations in the statistical mapping table should be utilized for generating a mapped CLC number. Thirdly, the manual mapping table should be modified, for example, to the DDC fourth-level classes, especially in two important main classes. Moreover, suitable mapping strategies ought to be compared and selected, which can also improve the two indicators of the mapping results, but slightly. Further work includes realizing the mappings from CLC to DDC and introducing LCC and LCSH into the conversion system.

References

1. Vizine-Goetz, D., Hickey, C., Houghton, A., Thompson, R.: Vocabulary Mapping for Terminology Services. Journal of Digital Information 4(4) (2004), http://journals.tdl.org/jodi/article/viewArticle/114/113
2. OCLC LCSH/DDC Mappings, http://www.oclc.org/dewey/updates/numbers/
3. Terminology Services, http://www.oclc.org/research/activities/termservices/
4. Koch, T., Neuroth, H., Day, M.: Renardus: Cross-browsing European subject gateways via a common classification system (DDC). In: McIlwaine, I.C. (ed.) Subject Retrieval in a Networked World: Proceedings of the IFLA Satellite Meeting, Dublin, OH, pp. 25–33 (2001)
5. Classification Web, http://www.loc.gov/cds/classweb/
6. U.S. National Library of Medicine: Unified Medical Language System (UMLS), http://www.nlm.gov/research/umls/
7. High-Level Thesaurus Project (HILT), http://hilt.cdlr.strath.ac.uk/
8. Dai, J.B., Hou, H.Q.: Construction and Use of Automatic Mapping System between CLC and DDC. Journal of the China Society for Scientific and Technical Information 25(5), 594–599 (2006)
9. Si, L.: Interoperability and Its Implementation among Knowledge Organization Systems. New Technology of Library and Information Service (3), 29–34 (2007)

Digital Library Research (1990-2010): A Knowledge Map of Core Topics and Subtopics

Son Hoang Nguyen and Gobinda Chowdhury

Information & Knowledge Management,
University of Technology Sydney, NSW, Australia
Hoang.S.Nguyen@student.uts.edu.au,
Gobinda.Chowdhury@uts.edu.au

Abstract. In the paper, we briefly present a methodology and a digital library (DL) research knowledge map of core topics and subtopics (1990-2010) which can be used to create a visual knowledge map of DL research, and also to develop a new DL curriculum. The knowledge map can be an essential knowledge platform for DL researchers, educators, students and practitioners.

Keywords: Digital Library Research, Knowledge Map, Digital Library Ontology.

1 Introduction

Various research topics in DLs have been identified by researchers in the late 1990s and early 2000s that have been widely used in DL research and education around the globe. While reviewing research and development in DLs in the nineties, Chowdhury and Chowdhury grouped DL research into 16 areas [1]. More recently, two research groups have attempted to find out the core topics in DLs: the first research was conducted by Pomerantz et al (2006) on a sample of 1064 DL publications (1995-2006) that produced 19 modules (core topics) and 69 related topics [3]; and the second was conducted by Liew (2008) with 557 publications (1997-2007), producing 5 themes (core topics) and 62 related or subtopics [2]. The basic tenet of the research reported in this paper is that there is the need for a more comprehensive research with a larger sample covering the field of DL research from the beginning. The major research objective is to identify the core topics and subtopics of DL research (1990-2010) to build a DL knowledge map.

2 Methodology

Based on literature review and analysis of the call for papers from 37 conferences volumes of 3 major international DL conferences: JCDL(2001-2010), ECDL(1997-2010) and ICADL (1998-2010), a list of 15 core topics and 210 subtopics was created. Searches were conducted on the SCOPUS database using the list of topics to validate and expand the list. The resulting list of topics was structured into a DL knowledge map comprising a list of 21 DL core topics and 1015 subtopics. Some techniques of the thesaurus building and classification were used for coding and categorizing the terms. The research process followed 4 major steps:

C. Xing, F. Crestani, and A. Rauber (Eds.): ICADL 2011, LNCS 7008, pp. 367–371, 2011.

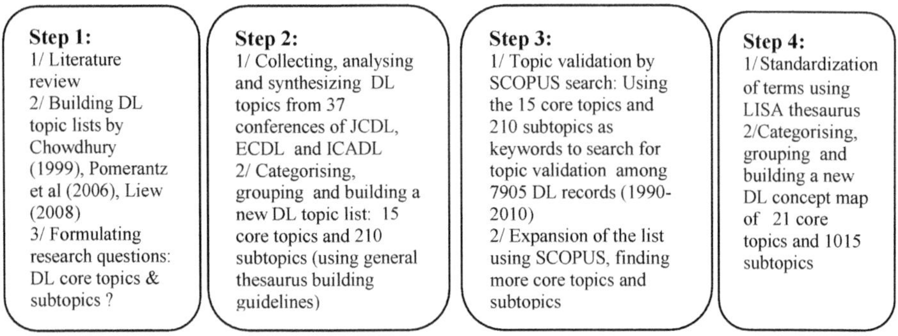

3 Findings

Due to page limitation, only a brief knowledge map of DL research (1990-2010) with 21 core and some of 1015 subtopics is presented (Table 1). For each core topic, various clusters of subtopics and the corresponding number of publications is shown, e.g., Digital Collections(General)(363): there are 363 publications on this subtopic.

Table 1. A Brief Digital Library Knowledge Map (1990-2010)

Core Topic 1: Digital Collections (48 subtopics): Cluster 1. Collections (363): Resources(603), Digital documents(41),Digital information(57), etc Cluster 2. Acquisition (432): Collection development(35),Collection development policy(3), etc Cluster 3.Database(General)(1210): Web Databases(13),Video Database(14),Image Database(29), etc Cluster 4.Collection management(50): Collection Evaluation(2),Information Evaluation(2), etc Cluster 5.Multimedia(General)(496):Multimedia collections(6),Multimedia contents(6), etc
Core Topic 2: Digital Preservation (46 subtopics): Cluster 1.Preservation(General)(174): Preservation management(2),Preservation Policy(2),etc Cluster 2.Storage(General)(634):Storage systems(13),Storage devices(2), Storage media(4), etc Cluster 3.Archives(General)(281): Archives management(30),Web Archiving(6),Online Archive(5), etc Cluster 4.Repositories(General)(217): Institutional Repositories(40),Online repositories(3), etc
Core Topic 3: Information Organization (141 subtopics): Cluster 1.Metadata(753):Interoperability(metadata)(81),Metadata management (12), etc Cluster 2.Structured Documents(14):Markup languages(81),SGML(14), XML(330), HTML(119), etc Cluster 3.Bibliographic(General)(161):Bibliographic data(6),Bibliographic records(11),etc Cluster 4. Discovery(General)(84):Resource Discovery(12),Information Discovery(11), etc Cluster 5.Information Organization(General)(8):Classification(256),Classification systems(15), etc Cluster 6.Conceptual(General)(47):Conceptual design(9),Conceptual graph(2), etc Cluster 7.Hierarchy(General)(24):Concept hierarchies(3),Topic hierarchy(2),Hierarchical systems(69), etc Cluster 8. Annotation(General)(125):Document annotation(4),Content annotation(2),Digital annotation(2), etc Cluster 9. Compression(General)(87):Data compression(31),Image compression(53),Compression ratio(5), etc Cluster 10. Video processing(3):Video editing(4),Video recording(24),Video Streaming(15), etc Cluster 11.Information analysis(263): Documents Analysis (15),Text Analysis (5),Data Analysis(31), etc Cluster 12.Recognition(General)(302): Character recognition(101),Handwriting recognition(7), etc Cluster 13.Information processing (25):Text Processing(145),Image processing(223), etc

Table 1. *(Continued)*

Core Topic 4: Information Retrieval (78 subtopics):
Cluster 1.Information Retrieval(General)(1376):Bibliographic retrieval systems(113), Interoperability(IR)(35),etc.
Cluster 2. Multilingual(IR)(19),Multicultural(IR)(1), etc.
Cluster 3.Search(General)(768):Search engines(496),Searching(386),Full text search(5),Search strategies(10), etc
Cluster 4 Query(General)(474):Query Language(298),Query Expansion(15),Query Processing(55), etc
Cluster 5. Browsing(General)(95):Video browsing(7),Document browsing(4),Web Browsing (3)
Cluster 6.Recommendation(General)(51):Recommendation system(17),Recommender systems(57)
Cluster 7.Filtering(General)(89): Information filtering(17),Collaborative filtering(42)
Core Topic 5: Access (14 subtopics):
Cluster 1.Access(General)(319):Access control(58),Open access(45),Multi-lingual access (3), etc
Core Topic 6: Human - Computer Interaction (61 subtopics):
Cluster 1.Interactions(General)(279):Human-Computer Interaction(General)(168),User Interaction(10), etc.
Cluster 2.Human engineering(70):Technology Acceptance Model(11),Human Factors(36), etc
Cluster 3. Visualization(262):Knowledge Representation(51),Information Representation(3), etc
Cluster 4. User Interfaces(790):User-Computer Interface(30), User interface evaluation(3), etc.
Core Topic 7: User Studies (59 subtopics):
Cluster 1. Users(1208):User Communities(15),Students(267),Teachers(14),Children(30),Adults(14), etc.
Cluster 2. Usability(76):Usage(55),Usage patterns(6),Usability Evaluation(7),Usability engineering(30), etc
Cluster 3. Information Needs(26):User needs(10),User interests(11),User requirements(12), etc
Cluster 4. User studies(General)(97):User evaluation(19),User profile(23),User behavior(19), etc
Core Topic 8: Architecture - Infrastructure (144 subtopics):
Cluster 1. Computing(General)(509):Client server(84),Distributed Computing(18), Cluster Computer(2), etc.
Cluster 2. Algorithms(General)(895):Learning Algorithm(53),Mathematical Model(457), etc
Cluster 3. Infrastructure(General)(95):Global Information Infrastructure(2),Cyberinfrastructure(10), etc
Cluster 4. Software(General)(1203):Digital library software(28),Software infrastructure(5), etc
Cluster 5. Architecture(General)(472):Digital Library Architecture(11), Interoperability(Architecture)(184), etc.
Cluster 6. Internet(699):Network(875),Web(1441), Web 2.0(33),Semantic Web(137), etc.
Cluster 7. Data sets(80): Data Model(29),Learning Object(41),Data Warehousing(3), etc
Cluster 8. Digital Objects(83):Object oriented (213),Object oriented programming(196), etc
Cluster 9: Information systems(393),Database systems(1047),Information systems design(4), etc
Cluster 10.Heterogeneous(General)(58):Heterogeneous Collections(4),Heterogeneous systems(8),etc
Cluster 11.Integration(General)(148):Digital library integration(12),Digital library federation(5),etc
Cluster 12.Distributed digital libraries(24):Distributed Collections (2),Distributed portal(3), etc
Cluster 13. Fuzzy linguistic(9), Fuzzy logic(14), Fuzzy systems(9)
Cluster 14.Agents(General)(165):Multi agent systems(50),Agent based(13),Intelligent agent(44)
Core Topic 9: Knowledge Management (58 subtopics):
Cluster 1. Knowledge management(General)(185):Data management(38),Content management(45), etc
Cluster 2. Knowledge process(2):Knowledge building(1),Knowledge engineering(73), etc
Cluster 3. Collaboration(102):Collaborative work(5),Collaborative knowledge(4), etc
Core Topic 10: Digital Library Services (30 subtopics):
Cluster 1. Services(General)(1134):Information dissemination(278),Information services(572), etc
Core Topic 11: Mobile technology (22 subtopics):
Cluster 1. Mobile library(3):Mobile services(5),Mobile content(1),Mobile information(3), etc
Cluster 2. Mobile(General)(147):Mobility(15),Mobile devices(31),Mobile application(3),etc
Core Topic 12: Social Web(Web 2.0)(21 subtopics):
Cluster 1. Library 2.0(110):Librarian 2.0(15),Library user 2.0(1),Information literacy 2.0(2)
Cluster 2. Web 2.0(37), Social Web(2),Social search(1),Social Networking(9), etc
Cluster 3. User generated content(3),Crowdsourcing(2),Wisdom of crowds(1), Folksonomy(7), etc

Table 1. (*Continued*)

Core Topic 13: Semantic Web (Web 3.0) (30 subtopics): Cluster 1.Semantic digital library(13),Library 3.0(3) Cluster 2. Semantic Web(137),Web3.0(2):Semantic technology(16),Semantic search(4), etc Cluster 3.Ontologies(General)(258):Ontology services(1),Ontology development(2),Ontology semantics(21),etc
Core Topic 14: Virtual Technologies (20 subtopics): Cluster 1.Library 3D(7):Virtual library(74),3D digital preservation(2), Web 3D(3),Virtual Instrument(10), etc Cluster 2.Virtual reality(282):Virtual worlds(12),Virtual environments (33),Virtual community(7), etc
Core Topic 15: Digital library management (53 subtopics): Cluster 1. Policy(General)(96):Information policy(6),Digital Library policy(1) Cluster 2.Planning(General)(145):Digital library planning(2),Strategic planning(45),Project Planning(9) Cluster 3. Finance(10):Cost benefit analysis(12),Benefits(20),Pricing(5),Budget(14),Investment(23), etc Cluster 4.Human resources(6):Staff(20),Digital librarians(5),Information professionals(14), curators(2) Cluster 5. Organization and management (23),Digital library performance(5),Digital library project(40), etc. Cluster 6.Evaluation(General)(310):Performance Evaluation(16),Performance Measure(3),etc. Cluster 7. Quality control(53):Quality Assurance(46),Quality Model(3),Quality Indicator(4), etc Cluster 8. Risk management(11),Risk assessment(18)
Core Topic 16: Digital Library Applications (64 subtopics): Cluster 1. Research(623):Societies and institutions(298),Scholarly communication(27), etc Cluster 2. Education(General)(645):Distance education(90),Teaching(197),Higher education(35),Classroom(16),etc Cluster 3. Learning(General)(621):Learning systems(304),Learning Management System(6),etc Cluster 4. E-government(9),E-governance(4), E-science(24), E-discovery(1),Electronic administration(1), etc Cluster 5. Natural Science(23):Digital Earth(4),Geospatial(18),Information Industry(2),etc Cluster 6. Social Sciences(21):Art(52),Culture(31),Humanities(19),Museums(53), News(27),etc
Core Topic 17: Intellectual Property, Privacy, Security (28 subtopics): Cluster 1. Intellectual Property(General)(55):Intellectual Property Protection(1),Copyright(107), etc Cluster 2. Security(General)(223):Security systems(17),Data security(10),Security model(1),Security policy(1),etc Cluster 3.Privacy(General)(38),Privacy policies(1),Privacy protection(6)
Core Topic 18: Cultural, Social, Legal , Economic Aspects (25 subtopics): Cluster 1.Cultural Aspects(103):Cultural Heritages(70),Multicultural Digital Library(1),Heritage(96),etc Cluster 2. Social Aspects(221):Societies and institutions(285),Citizen science(3), Pedagogical(Aspects)(8), etc Cluster 3.Legal Aspects(17):Censorship(2),Trust(8),Copyright law(16),Law(85) Cluster 4. Economic Aspects(46): Electronic commerce(122)
Core Topic 19: Digital Library Research & Development (48 subtopics): Cluster 1. Interdisciplinary(General)(12):Interdisciplinary collaborations(1),Interdisciplinary research(4), etc Cluster 2.Research and Development (91):Digital library research(17),Digital library concepts(2), etc Cluster 3. International cooperation(20):International digital library(2),Global collaboration(3), etc
Core Topic 20: Information Literacy (20 subtopics): Cluster 1. Information Literacy(General)(40):User education(4),Critical thinking(6),Critical evaluation(3), etc
Core topic 21: Digital Library education (5 subtopics): Cluster 1.Digital Library education(General)(148):Digital Library Curriculum(1),Digital Library program(20), etc

4 Conclusions and Further Work

The resulting DL knowledge map can be used for a variety of purposes, e.g. to show the major areas of research in DL, and to demonstrate the interdisciplinary nature of DL. The DL knowledge map can also be used to develop a visual knowledge map of DL and it can be mapped onto a live database like SCOPUS, Web Of Knowledge, etc. to facilitate information search and retrieval on DLs as well as more critical analysis of the field. It can also be used to develop a DL curriculum by identifying the traditional information science topics as well as new interdisciplinary knowledge and skills that are enriching DL as a field of study and thus should form part of a new DL curriculum. Further research based on the DL knowledge map is in progress showing the evolution of the DL as field of study. A visual knowledge map of DL is also being created using a visual ontology development tool like Protégé to map the relationships among DL topics, authors, institutions, research domains, etc.

References

1. Chowdhury, G.G., Chowdhury, S.: Digital library research: major issues and trends. Journal of Documentation 55(4), 409–448 (1999)
2. Liew, C.L.: Digital library research 1997-2007. Journal of Documentation 65(2), 245–266 (2008)
3. Pomerantz, J., Wildemuth, B.M., et al.: Curriculum Development for Digital Libraries. In: JCDL 2006, p. 10 (2006)

A Case Study for Multilingual Support: Applying the AAT-Thesaurus to TELDAP's Multilingual Project

Hsueh-Hua Chen[1], Shu-Jiun Chen[2], Shin-Yen Lee[1], and Jessamine Cheng[1]

[1] National Taiwan University Library, Taipei, Taiwan
{sherry,leeshy,jessaminecheng}@ntu.edu.tw
[2] Academia Sinica, Taipei, Taiwan
{sophy,chingju}@gate.sinica.edu.tw

Abstract. The translation of cultural-laden words has been a formidable project. This paper discusses how AAT-Thesaurus, with its classification of semantic equivalence and carefully structured terms, helps the Taiwan e-Learning and Digital Archives Program (TELDAP) conquer the difficulty of untranslatability.

Keywords: multilingual support, thesaurus, digital archives.

1 Introduction

The Taiwan e-Learning and Digital Archives Program (TELDAP) aims to sustainably maintain Taiwan's valuable cultural assets and has been working on the multilingual translation of approximately 3 million Chinese digitized items with the aid of the Art & Architecture Thesaurus Taiwan (AAT-Taiwan), a project that builds a structured English-Chinese thesaurus to facilitate access to information on art, architecture, and material culture. This study looks into the basic methodological framework upon which the AAT-Taiwan is created, and how it has helped TELDAP's multilingual development.

2 The Methodological Framework of AAT-Taiwan

The framework for AAT-Taiwan includes a process of four modules: *translation, mapping, localization*, and *creation*. In *translation*, each term went through editing, proofreading, and sometimes fervent debate to reach its final corresponding term. Next, the fruits of translation entered the *mapping* process in which relationships between Chinese and English terms are identified and the pair of terms classified into the following: "exact equivalence," "inexact equivalence," "partial equivalence," "single to multiple equivalence," and "no equivalence." To further comply with the principle of literary warrant, *localization* came in as a necessary step. This was the stage when synonyms were collected, then a descriptor was chosen from them, and visualization aids were added. Finally, for terms of "non-equivalence" or "little-equivalence" to be qualified to convey meanings, hierarchical relationships and conceptual structures [see Figure 1] were generated so that the translated terms are not merely arbitrary combinations of words but systematic and reasonable creations that bear scrutiny and exploration.

C. Xing, F. Crestani, and A. Rauber (Eds.): ICADL 2011, LNCS 7008, pp. 372–373, 2011.

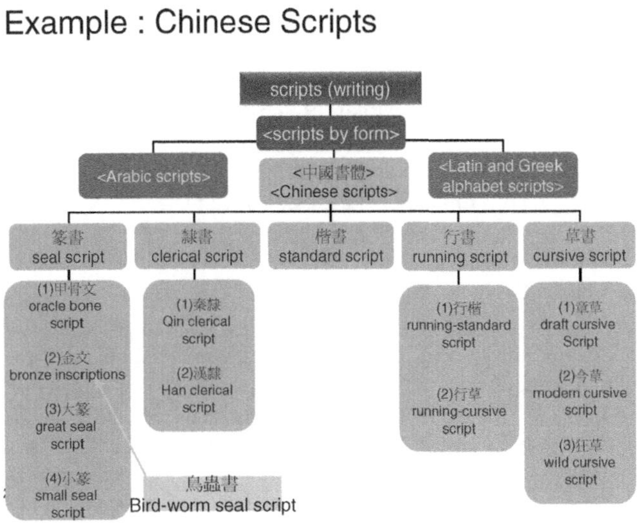

Fig. 1. An example of Hierarchical Relationship of Terms

3 Features of AAT-Taiwan

The application of AAT-Taiwan has been a wonderful experience for TELDAP translators. AAT is useful particularly due to its classification of semantic equivalence that assists translators in figuring out what is lost in translation and how the readability of translations can be improved. Furthermore, AAT's structural and pictorial references have been proved beneficial in helping TELDAP translators cross the limitation of untranslatability, in particular when the source language is itself culture-laden and difficult to absorb even to modern speakers of Chinese. The AAT's sources are reliable because it contains at least three authoritative sources, making further inquiry into the information time-saving. In Taiwan the issue over which system of transliteration should be made official has not been settled, and AAT suits a wide range of needs as it has at least Hanyu Pinyin, Pinyin without Tones, and Wade-Giles transliteration available. It is hoped that TELDAP's experiences may contribute to future implementations of multilingual support in creating better digital resources for a wide range of international users.

Acknowledgements. This study is supported by the National Science Council's Taiwan e-Learning and Digital Archives Program (Project No. NSC-98-2631-H-002-015).

The World Digital Library

Allison B. Zhang

World Digital Library,
Library of Congress, United States
azha@loc.gov

Abstract. The World Digital Library (WDL) is a website, *www.wdl.org*, which provides free access to primary cultural heritage materials that tell the stories and cultural achievements of all countries. The WDL is a joint effort of the Library of Congress, United Nations Educational, Scientific and Cultural Organization (UNESCO), and over 120 participating institutions around the world. This poster session will introduce the WDL by presenting brief history of the WDL, its objectives, contributors, content, capacity building, and features. The poster will also show how the WDL web site is used by the people around the world.

Keywords: World Digital Library, cultural heritage materials, Library of Congress, UNESCO, intercultural understanding.

1 History of WDL

In June 2005, United States Librarian of Congress James H. Billington proposed the establishment of the World Digital Library in a speech to the United States National Commission for UNESCO. The basic idea was to create an Internet-based, easily-accessible collection of the world's cultural riches that would tell the stories and highlight the achievements of all countries and cultures, thereby promoting cross-cultural awareness and understanding. UNESCO welcomed the idea as a contribution toward fulfilling UNESCO's strategic objectives.

In December 2006, UNESCO and the Library of Congress co-sponsored an Experts Meeting with key stakeholders from all regions of the world.

In April 2009, the WDL is launched to the international public, with content about every UNESCO member country.

2 Objectives

The principal objectives of the WDL are to:

- Promote international and intercultural understanding;
- Expand the volume and variety of cultural content on the Internet;
- Provide resources for educators, scholars, and general audiences;
- Build capacity in partner institutions to narrow the digital divide within and between countries.

C. Xing, F. Crestani, and A. Rauber (Eds.): ICADL 2011, LNCS 7008, pp. 374–376, 2011.

3 Contributors

The WDL was developed by a team at the Library of Congress, with contributions by partner institutions in many countries. By the end of July, 2011, the WDL has 132 partners from 71 countries, among them, 13 countries are from the Asia-Pacific areas, representing libraries, archives, museums, and ministries of culture.

The WDL is supported by the UNESCO and IFLA, and received financial support from a number of companies and private foundations, including Google, The Qatar Foundation, The Carnegie Corporation of New York, Microsoft, Inc., and so on.

4 Content and Capacity Building

The WDL makes it possible to discover, study, and enjoy cultural treasures representing about 70 countries, over 40 languages, on one site, in a variety of ways. These cultural treasures include, but are not limited to, manuscripts, maps, rare books, musical scores, recordings, films, prints, photographs, and architectural drawings.

In addition to building a website, the WDL is also an effort to share knowledge and build capacity around the world by:

- assisting in the establishment of digital conversion centers;
- encouraging digitization where little or no digital content is being produced;
- facilitating cooperation between experienced institutions and those just getting started.

By providing training, equipment, and technical support to partner institutions in countries such as Egypt, Iraq, and Uganda, the WDL is bridging the global digital divide and ensuring that rare and unique cultural treasures previously accessible only to a tiny minority of researchers are now available for worldwide use.

5 Features

The WDL provides many user-friendly features, including:

- Interface in seven languages: The metadata, navigation, and curator videos are translated into seven languages: Arabic, Chinese, English, French, Portuguese, Russian, and Spanish. This feature brings WDL closer to the goal of being truly universal.
- Consistent, high-quality metadata.
- Curatorial and scholarly interpretation: Among the most impressive features of the WDL are descriptions of each item, answering the questions: "What is this item and why is it significant?"
- Multiple methods to browse, search, discover, and compare content.
- Enhanced display and zoom features.
- Exposure to search engines: individual items are indexed in search engines.
- High quality images and PDF documents for easy downloading.
- Web 2.0 features
- Accessibility

6 How the WDL Is Used?

From April 21, 2009 to April 20, 2011, in two years, the WDL had over 100 million page views; over 15 million visitors, among them, 89% of traffic is from outside of the Unites States and 76% of users access the site in languages other than English. Over 3.5 million web sites linked to WDL. The WDL staff received and responded over 5,700 comments and questions. Below are a few comments from the users.

"Thank you! This is wonderful, beautiful and extraordinary. You have given me and my children a marvelous gift."

"I cannot thank you enough for having created the Digital Library. This is truly the sort of thing the Internet is most useful for…I have just had the pleasure of being able to look through an entire book, Stories from Hans Anderson…I am familiar with some of the illustrations…but seeing them in their original context was an exciting experience. It was almost as good as being able to touch the actual book. Thank you again. I look forward to having more such discoveries in future."

Embryo App for iPhone/iPad/iPod Touch

Ying Sun[1], Florence Haseltine[2], John Cork[3], Elizabeth Lockett[4], Florence Chang[1], and Lucie Chen[1]

[1] National Library of Medicine at National Institute of Health
6707 Democracy Blvd. Suite 510, Bethesda, MD 20892 USA
yingsun@mail.nih.gov
[2] National Institute of Child Health and Human Development, USA
[3] Louisiana State University Health Sciences Center, USA
[4] National Museum of Health & Medicine, USA

Abstract. Scientists and educators worldwide use the Carnegie Collection of Embryology to define normal human embryo development for decades. The Embryo App for iPhone/iPad and iPod Touch, developed by National Library of Medicine (NLM), utilizes mobile telecommunication and multimedia technologies to add interactive capabilities to the embryo data, enhancing our understanding of embryo development. Embryo is a collaborative project between NLM, the *Eunice Kennedy Shriver* National Institute of Child Health & Human Development (NICHD), Louisiana State University Health Sciences Center (LSUHSC) and National Museum of Health & Medicine's Human Developmental Anatomy Center (NMHM & HDAC). This App is part of the National Library of Medicine's program to fulfill the NLM's role as a provider of medical, science and health care information with mobile technologies. The Embryo App provides a new venue for us to access and interact with the Carnegie Collection of Embryology for the first time.

Keywords: Mobile app, iPhone, database, Embryo, 2D images, 3D animations.

1 Introduction

The Carnegie Collection of Embryology's primary focus is on normal development in the first eight weeks. The 23 Stages of human embryonic development used to describe early embryonic development. A database, called the Virtual Human Embryo, has been created by LSUHSC to provide digital serial sections of human embryos from the collection. Mobile devices are being utilized by the scientific community to expand outreach of valuable information. NLM joined the team, led the development and launching the Embryo App. The app incorporates the information about embryos into a publicly-available, interactive format. It gives users instant access and a tactile functionality to existing data.

2 Method

The mobile app was designed and developed using Xcode, iOS SDK 4.2. Many of the powerful features of Objective-C, Cocoa Touch, the various iOS SDK libraries and

C. Xing, F. Crestani, and A. Rauber (Eds.): ICADL 2011, LNCS 7008, pp. 377–378, 2011.

frameworks were used, such as model-view-controller, delegate pattern, core animation, and performance considerations. By integrating Facebook® iOS SDK library into Embryo App, it is possible for user to share the embryonic information with social media network. Property list files are utilized to store the embryo data.

3 Features

The Embryo App includes the following features:

- Movies with accompanying audio on fertilization and in vitro fertilization were available on the Embryo App (See Fig.1).
- 2D photographs illustrate the 23 stages of human embryonic development. User can implement pinch zoom in to enlarge the images.The App allows the sharing the information using email, Facebook® (See Fig. 2).
- 3D reconstructions, animations for embryo stages of the first eight weeks (Fig. 4 demonstrates the 3D reconstructions for embryo at stage 14).
- Pregnancy and ovulation calculator are provided (See Fig.4).
- Extended resources from the NLM, NICHD, LSUHSC and NMHM can be viewed via mobile web inside the app.

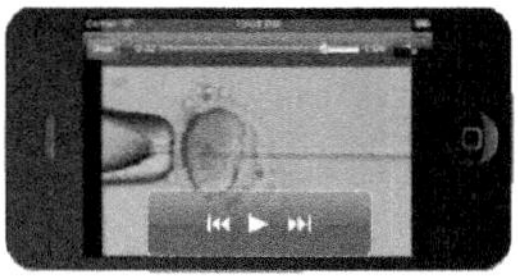

Fig. 1. Screenshot of ICSI

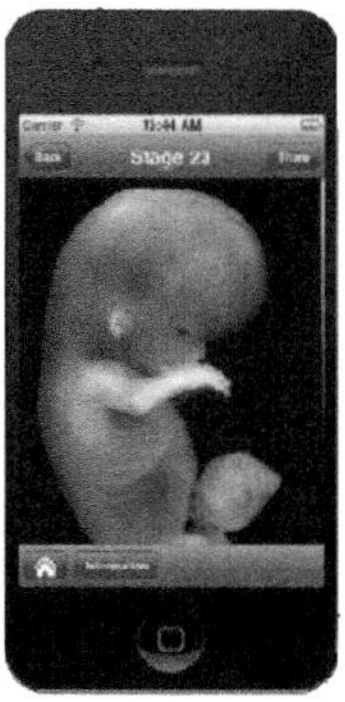

Fig. 2. 2D images

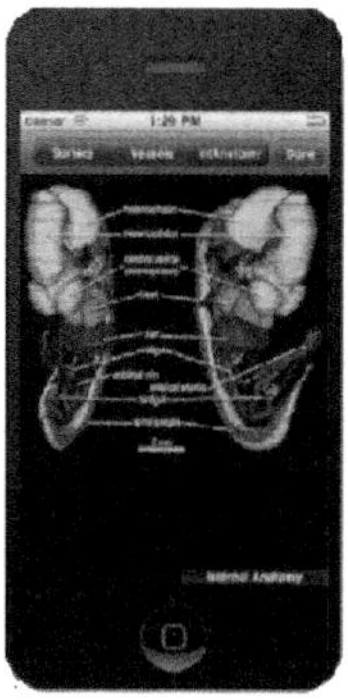

Fig. 3. 3D reconstructions

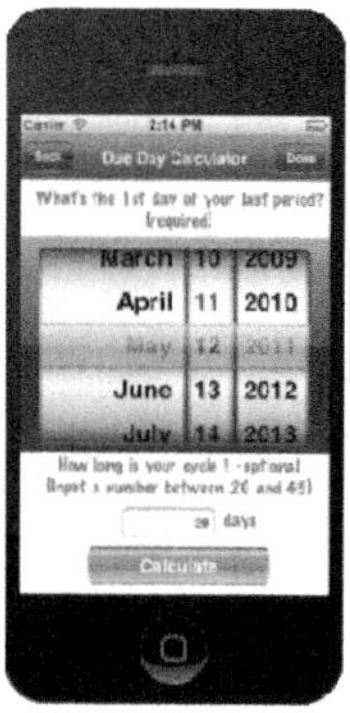

Fig. 4. Calculator

4 Conclusion

Mobile Health or mHealth broadly encompasses the use of mobile technologies for health research and healthcare delivery. The new Embryo App will provide a valuable reference for scientists, enable students to learn human embryology from images of actual embryos, and is a resource for the general public wishing for more information on a progressing pregnancy. The Embryo App was released at Apple - iTunes® on March 8th, 2011. Evaluation and feedback from the use will give us opportunities to expand the outreach of NLM. This project adds to the growing collection of mobile apps expanding the scientific resources provided by NLM. The Embryo mobile app can be downloaded at http://itunes.apple.com/us/app/embryo/id422337604?mt=8.

A Survey on E-Book Utilization in University Libraries

Ying Yang*, Jiayan Yang, and Xuemei Luo

Library of University of Electronic Science and Technology of China,
Chengdu, 611731, Sichuan, P.R. China
yyhelen@uestc.edu.cn

Abstract. The development of network and electronic technology has brought about the blooming of the digital resource, and the library is now in the age of coexistence of paper and digital resources. In this paper, we present the detailed results of a survey on e-books utilization among the students and faculties in University of Electronic Science and Technology of China (UESTC). We hope that this survey can provide a valuable reference for e-book prompting in university libraries.

Keywords: e-book usage, survey, university library.

It is noted according to the researches in foreign countries that the use of e-books becomes more and more frequently as the growing of patrons understanding on e-books[1,2]. The further issue needs to pay attention to is how to, under the national and social conditions of China, make more patrons know about e-books and effectively use them. Therefore, we have conducted a survey among the students and faculties in UESTC, aiming at gaining a general acquirement of the patrons' habit and predilection for e-books, and boosting e-books spreading.

The survey was conducted using Wenjuanxing Online Survey System (http://www.sojump. com/), 618 samples in total were collected duration the survey that lasts three weeks from 1st to 21st, march, 2011. In the following paragraph of this section, the results of the survey are presented and analyzed.

Fig. 1(a) illustrates the identity composition of the respondents. Fig. 1(a) shows their using frequency of e-books. It is observed that less than 50% of the respondents are frequent users of e-books, 41% of them use e-books occasionally, and 12% of the respondents never use e-books. The figures indicate the fact that e-books are not quite extensively used among the patrons in UESTC, spaces and necessity exists for further e-books spreading. Fig. 1(c) gives the utilization condition of three famous domestic e-books database, SuperStar, Shusheng, and Fangzheng APABI. The results show that Superstar is the most favorable e-books provider in UESTC. Fig. 1(d) shows the patrons' desired devices for e-books reading. It is observed that 43.7% of the respondents prefer tablet PC for e-books reading, the possible reason is believed to be the successful commercial advertising of the famous Apple Ipad. 38% respondents like conventional e-books reader, and it implies that e-books reader be come more and more popular since it was invented in 2000. Moreover, due to the excellent portabil ity, 35.6% respondents choose smartphone for e-books reading. Fig. 1(e) shows the

* Corresponding author.

C. Xing, F. Crestani, and A. Rauber (Eds.): ICADL 2011, LNCS 7008, pp. 379–380, 2011.

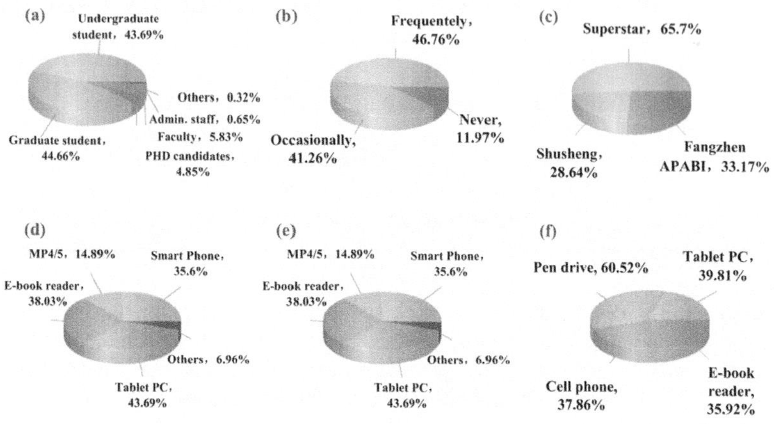

Fig. 1. Results of the survey

loan carrier for e-books suggested by the respondents. According to the data, we find that patrons prefer to borrow e-books with their most frequently used devices. This conclusion can explain that more than 60% of the respondents chose cheap pen drives for e-books borrowing. E-books readers are not quite popular since their high price which is usually above▯ 1000. Moreover, the advantages of e-books reader such as outstanding reading experience and friend-interface are not widely accepted by the customers. However, we believe that e-book reader will be accepted by more users as the price reduction and further advertising. Fig.1(f) shows respondents' expected functions of e-books reading devices. Among the others, easy operation is chose by 67.3% of the respondents as a necessary character of e-book reading devices. It is noted that, although among highly educated patrons in universities, convenience and easy operation are the primary consideration for users. Meanwhile, 64.7% of the respondents suggest that e-book reading devices should be able to directly access to databases. Tablet PC and e-book readers can support this function with the built-in networking components. Besides, more than a half of the respondents expect fine picture display in e-book reading devices to make reading more vivid and luxuriant.

In conclusion, we present a survey on e-books utilization in UESTC, and analyzed the results. It is found that, despite its rapid development, e-books are not as extensively and frequently utilized as print books among library patrons. However, some excellent e-books databases are becoming well known and utilized by users. Moreover, the using condition of e-book reading devices as well as users' expectation for such devices are investigated, results show that nice performance, inexpensive and easy-operating devices are key elements for e-books spreading among reader.

Reference

1. Dillon, D.: E-books: The University of Texas Experience, Part 2. Library Hi Tech. 19(4), 350–362 (2001); Gibbons, S.: Growing Competition for Libraries. Library Hi Tech. 19(4), 363–367 (2001)

Evaluation of Link System between Repository and Researcher Database

Kensuke Baba, Toshie Tanaka, Emi Ishita, Masao Mori, Eisuke Ito, and Sachio Hirokawa

Kyushu University, Fukuoka, Japan
baba@lib.kyushu-u.ac.jp

Abstract. This paper evaluates the effect of a Web system which activates institutional repositories. Institutional repository is an important service of libraries in academic institutions. The authors developed a link system between the institutional repository and the researcher database of their university. The system reduces the efforts of researchers by reusing the metadata in the researcher database for registrations of their papers to the repository. The authors observed the access log of the repository before and after the start up of the link system. The result shows that the system increased the number of access, however there was no significant change on the number of registration of papers.

Keywords: institutional repository, Web database, access log, library.

1 Introduction

"Open access" to scholarly information provides free availability of research outputs. One of the methods to realize the idea of open access is "self archiving" [2]. An *institutional repository* (*IR*) is a system to archive and open research outputs in an institution. By improving the IR in each institution, open access to scholarly information will be realized.

A problem of IR is the fact that the number of the archived papers is extremely small compared to the papers practically produced by researchers. We considered that one of the obstacles should be the efforts of researchers to register their papers to IR. To reduce the efforts of researchers, we developed a link system between the IR (QIR)[1] and the researcher database (DHJS)[2] in Kyushu University. The system is working from April 2009 and was improved in April 2010 [1].

In this paper, we analyze the number of access and the number of paper registration to QIR. As the result, the system increased the number of access, however there was no significant change on the number of paper registration. The result suggests that the number of paper registration to IR cannot be improved by only reducing the efforts of researchers.

[1] https://qir.kyushu-u.ac.jp/dspace/

[2] http://hyoka.ofc.kyushu-u.ac.jp/search/index_e.html

C. Xing, F. Crestani, and A. Rauber (Eds.): ICADL 2011, LNCS 7008, pp. 381–382, 2011.

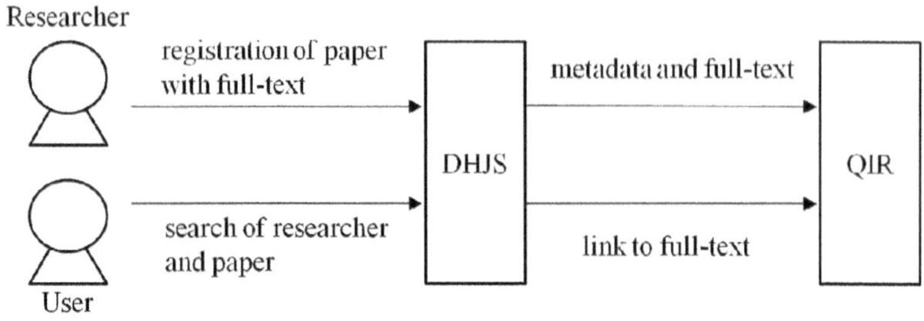

Fig. 1. The outline of the link system between QIR and DHJS

2 The Developed System

Fig. 1 is the outline of the developed system. The concept was implemented by two kinds of icons in the lists of published papers of each researcher in DHJS, one icon means that there exists the full-text of the paper in QIR and the other that there is no full-text. By the first one, a user of DHJS can obtain the full-text corresponds to the metadata. By the second one, the link system returns the result of a search by the author name in QIR. Additionally, if the user is an author of the paper, the link leads the user to the registration form of a paper to QIR. At the time, the metadata is automatically used to fill the registration form, therefore this function reduces the efforts for paper registration to QIR.

3 Evaluation

We analyzed the access log to QIR for each month from June 2008 to October 2010. For the number of access from DHJS, there existed two points of increases, one is at March 2009 (the link system started) and the other is at April 2010 (the system was improved). By the result, we can conclude that the system is effective to the number of access. As for the number of paper registration, we could not find any effect by the system.

We consider that, in addition to reducing the efforts of paper registration to IR, some returns for researchers are necessary to increase the number of paper registration. For example, showing researchers the results of access log analyses of their papers in IR can be an incentive for paper registration.

References

1. Baba, K., Mori, M., Ito, E.: A synergistic system of institutional repository and researcher database. In: The Second International Conferences on Advanced Service Computing, pp. 184–188. IARIA (2010)
2. Harnad, S., Brody, T., Vallieres, F., Carr, L., Hitchcock, S., Gingras, Y., Oppenheim, C., Stamerjohanns, H., Hilf, E.: The access/impact problem and the green and gold roads to open access. Serials Review 30(4), 310–314 (2004)

Characteristic Practice in the Construction of the Chinese Medical Digital Library -- Wanfang MED ONLINE as the Example of the Characteristic Resources Organization and Presentation as Well as Data Mining of the Medical Literature

Xiumei Zhang, Gongliang Yang, Xiaolei Li, and Jing Li

1 Introduction

This paper uses Wanfang Data MED Online (WFMO) [1] as an example to looks in depth into characteristic practice in the resources organization and presentation as well as data mining of the medical literature in CMDL.

2 Investigation Research

The research targets at scientific researchers, teaching researchers, clinical doctors and college students. The research methodology includes societal survey, expert interview, web metrics and evaluation. And the research finds all the users of WFMO not only need the medical literature in their research area, but also care about the up to date scientific and technical projects, outputs and journal literatures in related areas. They demand all the information to be up to day, comprehensive and in depth. In a word, all of the users want to find out the dynamic, updating developments, local and foreign, in their professional areas or in the related areas.

3 Characteristic Practice

3.1 Providing the Resources on a Platform

According to the investigation research, WFMO commit itself to collect as more journals as it can firstly to provide more resources. So far, WFMO collected around 1100 kinds of journals to satisfy the need of users in quantity of resources. Wanfang Data builds the exclusive strategic cooperation with Chinese Medical Association, Chinese Medical Doctor Association and other associations.

WFMO developed the navigation function based on the external characteristics, and developed journal evaluation system. What's more, the knowledge points of the papers are connected together to disclose the relationship of the literatures, in this way, the users can have a whole knowledge map in one specific. For instance, reference links, related articles links. And because of Digital Object Identifier (DOI), the searching system can provide the cross-platform literature links.

3.2 Information Organization and Presentation of the Platform

According to research above, the users of WFMO tends to search for author, organization, and funds in the simple searching window on the website which is like

C. Xing, F. Crestani, and A. Rauber (Eds.): ICADL 2011, LNCS 7008, pp. 383–384, 2011.

Google, Baidu and other searching engines. To improve the user experience, WFMO carries out further resource processing, standardization, organization according to the five elements of fetching information which are subject, author, organization, funding and theme aiming at satisfying the multiple-dimensional user requirements as you can see from Figure 1. [2] The five elements of the literature can treated as five objects individually. Any one of the paper can be related to the other four objects of the paper, meanwhile, the clustering of the papers which are related by the same properties of the same object.

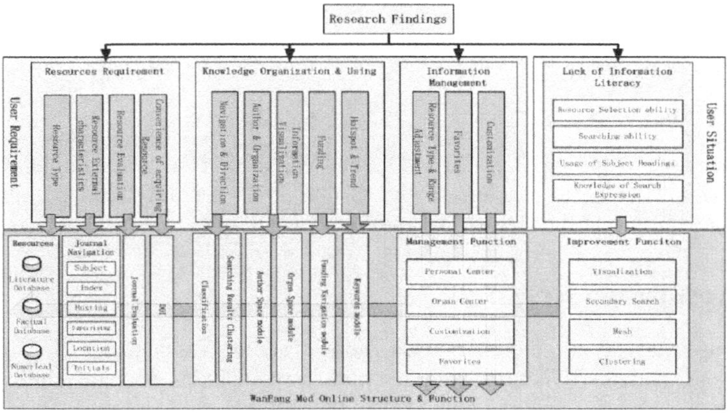

Fig. 1. WFMO Research findings & Structure mapping

3.3 Exploratory Efforts to Offer Customized Services

WFMO not only builds the CMDL in China, but also never stops trying to explore more services to improve the medical digital library in China.

About the customized push services, like personal customer center for each registered user, when the users order the area which they care about, and WFMO will push the newest information to them. And there are also organization center which is used to manage the users signed up in the organization.

4 Summary

This paper introduced how WFMO is designed based on the theory of five elements of fetching information, and the exploratory practice WFMO took in the construction of CMDL. This is the first time in history that WFMO applies the theory into practice in such a large-scale CMDL, and all the above indicates that WFMO plays a vital role in the development of the Chinese medical digital library.

References

1 Wanfang Med Online, http://med.wanfangdata.com.cn/
2 Wu, G.Y.: On Recall/Precision Ratio. J. Digital Library Forum (7, 2010)

Use of Information Technology in Library Service: A Study on Some Selected Libraries in Rajshahi District of Bangladesh

Md. Jamal Uddin

32, S.M. Murshed Sarani,
Agargaon, Dhaka-1207, Bangladesh
jamal_nlbd@yahoo.com

Abstract. Bangladesh is one of the least developed countries, which have been facing a lot of obstacles in introducing Information Technology (IT) in all information-related sectors, especially in its libraries and information centers. This current work is centered on the Rajshahi District of Bangladesh, perceived to be worse hit in terms of inadequacy of the application of IT in library and information centers, when compared with the capital city of Bangladesh. Therefore, this research work, made with a full study plan and procedures seeks to investigate the environment, identify barriers and explore possibilities of improving IT application in libraries. The study took into account the various systems, services, problems and prospects of these selected libraries. Besides this, the study also showed how to improve the entire condition and services of these libraries by adopting information technology (computer, fax, internet) including provision of phone, photocopier, television, microfiche, microfilm etc. This investigation will reveal the entire condition of these libraries including the problems encountered by them in applying IT and put forward necessary recommendation, which may be helpful in improving their services in information dissemination process.

Keywords: Information Technology (IT), Library Service, Library and Information Science, Information Centers.

1 Introduction

Information Technology (IT) is the modern buzz word. It has provided facilities for the free flow of information. The world has been come a global village with Information Superhighways created through networks like Internet. This has facilitated electronic librarian-ship with the diversification of library facilities and services to its user. The concept of virtual libraries OPAC, Hypertext, and teleconferences for the purpose of library and information services have become common. New technological developments have already profoundly affected libraries. Almost every function carried out in a library has been altered to some extent by advances in electronics, computerization, and telecommunications. The technological evolution in libraries has been called a "quiet revolution". The changes brought about by advances in technology have been so extensive that it is difficult to

C. Xing, F. Crestani, and A. Rauber (Eds.): ICADL 2011, LNCS 7008, pp. 385–389, 2011.

assess their total effect, but is clear that libraries are in the state of fundamental transformation. Information Technology is a generic term used to denote all activities connected with computer based processing, storage and transfer of information. Information Technology means not only a single unit of technology, but the integration and convergence of a large number of technologies which have come together to serve and meet the user's needs in the age of information revolution. It involves computers, electronic media, satellites, telecommunications and storage devices.

Ahmed (1998) carried out a study on 224 different libraries in Bangladesh and found out that only 58 libraries were using computers for their operations and services. Mannan (1998) also conducted a comprehensive study on 25 libraries found out that there was poor technological status in the country, Alam (1998) had made a thorough investigation on 13 large special libraries and expressed that there was an acute shortage of IT resources. IT trained manpower, and favorable IT related infrastructural facilities in the libraries of Bangladesh. Nasiruddin (1999) conducted a research project on 44 libraries in northern part of Bangladesh where he observed a very unsatisfactory technological status prevailed at district level. Most of the observation and study was made primarily on the libraries of the capital city than on the libraries in divisional areas. This study is carried out mainly to identify the present condition and technological position as well as the willingness and complexities toward the IT application.

2 Objectives of the Study

The main objective of this study was to examine the existing conditions of the libraries and information services in Rajshahi District of Bangladesh. The specific objectives of the study are to:

- find out the level of IT application and their extent of use in the library.
- investigate the benefit of information technologies to the librarian and library users.
- examine the available technologies in the selected libraries.
- identify the specific problems that are being faced that by the library professionals in the application of IT for information service.

3 Methodology

Eleven libraries were selected for this study on the basis of having minimum library equipment and facilities that are needed to provide basic library service. All of the surveyed libraries were located at greater Rajshahi of Bangladesh. A structured questionnaire was designed to interview the head of the library or the librarian or the person responsible for library sector. Besides personal observation, informal discussion and thorough visit of the site have helped the research to co-relate the necessary data needed for the study.

Study Area: The collection of data was made from all types of library including academic, special and public that are selected such as Rajshahi University Library, Rajshahi University of Engineering and Technology Library, Rajshahi Medical College Library, Rajshahi College Library, Institute of Bangladesh Studies Library, Bangladesh Council of Scientific and Industrial Research Library, Bangladesh Rice Research Institute Library, Rural Development Academy Library, Bangladesh Rural Advancement Committee Library, CARITAS Library, Divisional Public Library.

4 Findings from the Study

Most libraries and information centers provide such service with varying degrees of effectiveness and efficiency. An integrated information infrastructure links library systems, computer based data centers with academic, special and public libraries. Some of the important libraries and information centers of Rajshahi, Bangladesh have been using information technologies which are in embryonic stage, for their services. Users are not fully satisfied with their use and services they provided. The necessity of the application of IT in these libraries is strongly required for satisfying the demands of their clientele. It can be deduced that all the participated libraries in the survey desire to introduce IT in their services. But in the context of Bangladesh, it is not a very easy job. Bangladesh has been suffering from many acute problems such as political instability, poor economic condition and others. Inadequacy of trained and skilled personnel with mastery of application of IT in these libraries is also a great problem. All said and done, some of the respondent mentioned the futility of efforts in venturing to introduce technologies in the libraries and information centers in Rajshahi district of Bangladesh without considering their associated problems.

5 Problems

There are a lot of impediments to the application and use of IT in the library and information centers at Rajshahi District of Bangladesh. They include- Absence of Proper Planning, Lack of appropriate Initiative, Political Instability, Indifference of Administrative Body, Lack of Trained and Skill Personnel, Scary Approach to IT Environment, Insufficient Fund Support, Insufficient Initiative for Creating Public Awareness, Attitude of User Groups toward Information and IT, Lack of Professional and Social Awareness etc.

6 Recommendations

Information Technology has important implications for rural, industrial, cultural and research development. It is virtually impossible to carry out any meaningful educational activities without the help of libraries and information services. It is now encouraging to note that the present Government in Bangladesh has started to increasingly realize the role of libraries and information services in the various sectors of the country's research development, education and planning process. Considering that efforts involved in developing the application and use of information technologies for the library and information center calls for substantial supports in terms of time, equipment, money and expertise.

- The Government of Bangladesh should take initiatives, efforts and plans on the part of the national developments, which help in the modernization of library and information services. In fact, the present scenario of library automation in our country especially in northern part is still in a primary and experimental stage. In this regard, first of all the Government has to set a work plan to proceed with the various components of the planned activities and programs.
- Professional library associations in Bangladesh like LAB (Library Association of Bangladesh), BALID (Bangladesh Association of Librarians, Information Scientists and Documentalists) should organize short-term training program for library professionals in the field of IT application to library and information services.
- The library personnel should have learned how to use the new technological devices effectively. They should acquire the skill of exploiting information technologies to the best of their advantage for increased productivity.
- The policy makers should give an emphasis to allocate sufficient funds to support the purchasing of information technologies by libraries.
- The various political parties should have an approach to reach consensus regarding the development issues of library and information sector and push the movement for automation in a more dynamic stage.
- It is one of the important task of higher authority to remove the scary situation among the old professionals, who have no familiarity with IT application, by ensuring their job security. In this regard, the authority should rather take initiative to organize some effective program to familiar those professionals with IT application in libraries.

7 Conclusion

In most of the developing countries the mode of information generation, collection and organization are various. In this study and attempt has been made to determine the use of information technologies in library service. It is necessary to mention that the Information Technology (IT) has been tremendously influencing all spheres of our life.

References

1. Mannan, S.M.: The availability of Automation Facilities for Networking and Resource sharing in Bangladesh: a study on some selected libraries. In: Bangladesh Journal of Library and Information Science, pp. 101–118. University of Dhaka, Dhaka (1998)

2. Nasiruddin, M.: Assessing of Information Technology and its impact on libraries and information centers in Bangladesh. In: Bangladesh Journal of Library and Information Science, pp. 77–78. University of Dhaka, Dhaka (1998)
3. Nasiruddin, M.: Users Opinion on library and Information Services: a study of the Rajshahi University library. In: The Journal of Institute of Bangladesh Studies, pp. 123–139. Rajshahi University Press, Rajshshi (2000)
4. Zabed Ahmed, S.M.: Library Automation in Bangladesh: problems and prospects. In: Bangladesh Journal of Library and Information Science, pp. 39–46. University of Dhaka, Dhaka (1998)

Effect of the Number of Comments Inserted by Students during Each Lecture on Their Grades in the Course

Akihiro Motoki, Tomoko Harada, and Takashi Nagatsuka

Tsurumi University, Dept. of Library,
Archival and Information Studies,
Tsurumi 2-1-3, Tsurumi-ku, Yokohama, 230-8501, Japan
{motoki-a,harada-t,nagatsuka-t}@tsurumi-u.ac.jp

Abstract. When students read textbooks in the classroom, they usually apply active reading. Our previous study was to show the relation between the total number of comments inserted by students into their digital textbooks and their grades at the end of course. In this study, the number of comments inserted by students into their digital textbooks during each lecture is highlighted as one of main factors influencing their grades at the end of the course. Students who wrote a lot of comments during each lecture tend to receive a higher grade. Their grades at the end of the course are related to the number of comments inserted during each lecture in the early stages of the course. The finding suggests that if teacher can access information about comments that students wrote into digital textbooks just after each lecture, he may use it for improving students performance at the next lecture.

Keywords: lecture, note-taking, grade, digital textbook, comments.

Many studies or projects involving in-class and online educational technology include note-taking or annotation applications[1-2].

The authors chose the digital textbooks(DTs) using Microsoft Word 2003 because of its annotation and note-taking features. DTs are uploaded to the server and then students download the DTs to their laptops for use during each lecture. Students then write comments on DTs in their classrooms. Our previous study was to show the relation between the total number of comments inserted by students into their DTs and their corresponding grade at the end of course [3].

DTs in Japanese were prepared by students' teacher taking charge of the course "Introduction to Networks" with Microsoft Word 2003 in 2008 and Microsoft Word 2007 in 2010, respectively. The numbers of students who registered for the courses were seventy five in 2008 and eighty in 2010, respectively. At the end of each lecture in the course, students uploaded their DT with their comments to the server.

The cumulative numbers of comments in each grade in the course of "Introduction to Networks" are shown in Fig. 1. The cumulative number of comments at the end of the course "Introduction to Networks" shows a clear decline from grade S to grade C.

The average number of comments per student within a group of the same grade has a positive relation in most lectures of the course with their grades in the course.

Table 1 shows the correlation coefficient between the number of comments inserted by each student during each lecture and their corresponding grade point

C. Xing, F. Crestani, and A. Rauber (Eds.): ICADL 2011, LNCS 7008, pp. 390–391, 2011.

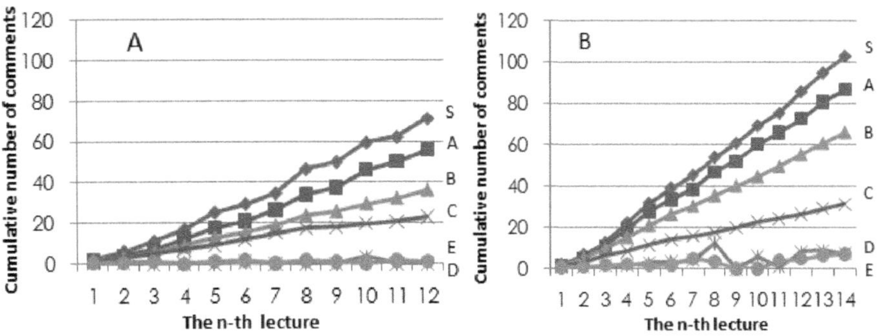

Fig. 1. The cumulative number of comments inserted by each student during each lecture and their grades in the course of "Introduction to Networks" in 2008(A) and 2010(B)

Table 1. The correlation coefficient between the number of comments inserted by each student during each lecture and their grade point values in the course of "Introduction to Networks" in 2008 and 2010

Year	The n-th lecture in the course of "Introduction to Networks"													
	1	2	3	4	5	6	7	8	9	10	11	12	13	14
2008	0.50	0.43	0.43	0.50	0.49	0.47	0.52	0.53	0.54	0.56	0.56	0.57	-	-
2010	0.41	0.65	0.68	0.69	0.72	0.75	0.75	0.76	0.77	0.76	0.74	0.73	0.72	0.71

values in the course of "Introduction to Networks" in 2008 and 2010. In both cases, the number of comments inserted by each student during each lecture is positively correlated with their grade point values in the course of "Introduction to Networks".

In this study, we found that the average number of comments per student within a group achieving the same grade during each lecture had a positive relation in most lectures of the course with their grades in the course. We also found that the values of the correlation coefficient during the early stages of the course were lower than those during mid to late stages of the course. The finding suggests that if teacher can access information about comments that students wrote into digital textbooks just after each lecture, he may use it for improving students performance at the next lecture. We should study student behavior and practices and be prepared for whatever changes are likely to occur through the increasingly widespread use of digital textbooks as opposed to the more traditional approach in the learning environment of the paper-and-pen on the classroom.

References

1. Hoff, C., Wehling, U., Rothkugel, S.: From paper-and-pen annotations to artifact-based mobile learning. Journal of Computer Assisted Learning 25(3), 219–237 (2009)
2. Hartley, J., Ivor, K.D.: Note-Taking: A Critical Review. Programmed Learning and Educational Technology 15(3), 207–224 (1978)
3. Motoki, A., Harada, T., Nagatsuka, T.: The relation between comments inserted into digital textbooks by students and grades earned in the course. In: Chowdhury, G., Koo, C., Hunter, J. (eds.) ICADL 2010. LNCS, vol. 6102, pp. 40–49. Springer, Heidelberg (2010)

Coordinating Concepts and Discourse in Model-Oriented Research Reports

Robert B. Allen

iSchool, Drexel University, Philadelphia, PA 19107, USA
rba@boballen.info

Abstract. Model-oriented research reports has been proposed as a highly structured approach that weaves together models for research methods and analyses, conceptual process models for the phenomena under investigation, and discourse structures for presenting the models.

1 Introduction

Allen [1, 2] has proposed the development of "model-oriented scientific research reports". Some parts of research reports are already structured but other parts have traditionally been presented just as text-based narrative. Much work on scientific communication has focused on specifying the discourse elements (which show the author's purpose). With model-oriented scientific research reports, the entire report would be structured. The model-oriented approach includes the development of conceptual process models to describe the phenomena under investigation. Discourse models and conceptual models are both essential for scientific research reports and any attempt to fully capture the content of the reports must have representations of both. There would be many advantages to model-oriented research reports. For instance, constraints are implicit in all research reports; they would be made explicit in the model-oriented approach. Moreover, recent work in digital preservation has emphasized the need to capture the details associated with scientific data sets. Some approaches explore associating the report with data as separate files. A model-oriented full research report which included the data would provide a clear basis for understanding that data. In this paper, we consider examples and issues for presenting tests of conceptual process models in model-oriented research reports.

2 Issues for Coordinating Research Report Component Models

We randomly selected a paper from *PLoS Biology* by Zhai et al. [7] as a test case to convert to a model-oriented research report. This paper presents a study exploring the molecular basis of protective factors associated with Wallerian Degeneration -- a disintegration of axons after they have been injured. The study compares two conceptual models for how the protective factors function. The two models are tested as indicated by the pseudo code below. NAD and Degeneration refer to the level of two factors in the models. The NAD level is manipulated as the independent variable and the level of Degeneration is the dependent variable. The use of conditionals for

C. Xing, F. Crestani, and A. Rauber (Eds.): ICADL 2011, LNCS 7008, pp. 392–394, 2011.

selection between model instances is related to meta-modeling, with the goal in this case to coordinate and contrast models rather than to identify a general model.

```
if ((NAD = =LOW) && (Degeneration = = LOW)) then {prefer Model1;}
else if ((NAD = =LOW) && (Degeneration = = NORMAL)) then {prefer Model2;}
```

To capture the reasoning behind the models being tested, discourse structures need to be linked to conceptual process models. The Introduction of a research report is structured as defining the problem to be addressed with CARS (Creating a Research Space) [7]. In the Zhai et al. article, the introduction starts with an applied problem and then presents possible mechanisms of that problem to be tested. We developed a notation for describing the alternative models being tested, with a visual presentation that is highly constrained.

After the Introduction, the procedures for testing the candidate conceptual process models are described in the Methods section. The description of the testing procedures involves the coordination of several report component models. The Research Design is distinguished from the Research Procedure as a distinct component model. One of the test procedures described in the Zhai et al. paper is denoted as $f_{AdoptAModelSystem}(\ )$ – testing *Drosophila* as a model system for the mechanisms controlling Wallerian Degeneration in mammals. Development, validation, and application of that model system requires substantial testing, and several pages of the research report are devoted to describing the Research Procedure to do so. Showing that those pages present the Research Procedure is helpful for readers to understand the overall structure of the paper.

Citations are another area in which discourse models mix with conceptual process models. There is often ambiguity in exactly what claim is being substantiated with a citation. Consider the following example from [7]:

```
In an attempt to dissect the mechanisms underlying the
neurodegeneration of mutant photoreceptors, we first tested
whether the degeneration is light dependent, because a common
cause for retinal degeneration is light stimulation [41, 42].
```

A reader who followed the citations would benefit from a quick, perhaps graphical, overview of the claim, the conceptual model about which the claim is being made, and how the evidence was collected to support that claim and model. For a reader who is familiar with the topic, this can sometimes be accomplished simply by skimming the cited source but it would often be more effective, especially for non-experts, to have a highly structured overview. Indeed, interactive research reports might be developed that are personalized to the background of the readers. Moreover, the conceptual models provide a tight interweaving of the content that is not the case for most adaptive hypertexts systems.

Although deduction in science (cf., [4]) can be seen as a type of argumentation (e.g., [6]), some feel that argumentation is antithetical to deduction. Moreover, graphical argumentation systems (e.g., [3]) have not included causal models. Generally, systematic argumentation, as well as other types of discourse modeling, needs to be coupled with conceptual process models.

References

1. Allen, R.B.: Highly Structured Scientific Publications. ACM/IEEE JCDL 472 (2007)
2. Allen, R.B.: Model-Oriented Scientific Research Reports. D-Lib. (May-June 2011)
3. Conklin, J., Begeman, M.L.: gIBIS: A Hypertext Tool for Exploratory Policy Discussion. ACM Transactions on Office Information Systems 4, 303–331 (1988)
4. Popper, K.R.: The Logic of Scientific Discovery. Harper & Row, New York (1968)
5. Swales, J.M.: Genre Analysis: English in Academic and Research Settings. Cambridge University Press, Cambridge (1990)
6. Toulmin, S.: Uses of Argument. Cambridge University Press, Cambridge (1958)
7. Zhai, G., Cao, Y., Hiesinger, P.R., Mehta, S.Q., Schulze, K.L., Verstreken, P., Zhou, Y., Bellen, H.J.: Drosophila NMNAT Maintains Neural Integrity Independent of its NAD Synthesis Activity. PloS Biology 4(12) (2006)

Author Index

GPSR Compliance

The European Union's (EU) General Product Safety Regulation (GPSR) is a set of rules that requires consumer products to be safe and our obligations to ensure this.

If you have any concerns about our products, you can contact us on ProductSafety@springernature.com

In case Publisher is established outside the EU, the EU authorized representative is:

Springer Nature Customer Service Center GmbH
Europaplatz 3
69115 Heidelberg, Germany

Batch number: 10844619

Printed by Printforce, the Netherlands